Genetics: Analysis and Principles

Genetics: Analysis and Principles

Edited by **Rosanna Mann**

◻SYRAWOOD
PUBLISHING HOUSE

New York

Published by Syrawood Publishing House,
750 Third Avenue, 9th Floor,
New York, NY 10017, USA
www.syrawoodpublishinghouse.com

Genetics: Analysis and Principles
Edited by Rosanna Mann

International Standard Book Number: 978-1-68286-178-3 (Hardback)

Printed in the United States of America.

Contents

Preface

Every book is a source of knowledge and this one is no exception. The idea that led to the conceptualization of this book was the fact that the world is advancing rapidly; which makes it crucial to document the progress in every field. I am aware that a lot of data is already available, yet, there is a lot more to learn. Hence, I accepted the responsibility of editing this book and contributing my knowledge to the community.

This book covers in detail some existent theories and innovative concepts revolving around the wide field of genetics. It is a compilation of chapters that discuss the most vital concepts, principles and emerging trends in this field of study. Different approaches, evaluations, methodologies and advanced studies on DNA sequencing, transcription pathways and gene interaction have been included in this book. There has been rapid progress in genetics in last few decades and its applications are finding their way across multiple industries. Researches and case-studies by internationally acclaimed researchers are included in this text to highlight the current developments in the field of genetics. This book, with its detailed analyses and data, will prove immensely beneficial to professionals and students involved in this area at various levels.

While editing this book, I had multiple visions for it. Then I finally narrowed down to make every chapter a sole standing text explaining a particular topic, so that they can be used independently. However, the umbrella subject sinews them into a common theme. This makes the book a unique platform of knowledge.

I would like to give the major credit of this book to the experts from every corner of the world, who took the time to share their expertise with us. Also, I owe the completion of this book to the never-ending support of my family, who supported me throughout the project.

Editor

Gene diversity and identification of putative hybridizing parents for root rot resistance in cassava using simple sequence repeats

Oluwasayo Kehinde Moyib[1,2,3]*, Jonathan Mkumbira[2], Oyeronke Adunni Odunola[3] and Alfred Godwin Dixon[2,4]

[1]Department of Petroleum and Chemical Sciences, Tai Solarin University of Education, Lagos-Benin Express Road, P. M. B. 2118, Ijebu-Ode, Ogun State, Nigeria.
[2]Cassava Breeding Unit, International Institute of Tropical Agriculture (IITA) P. M. B. 5320, Oyo Road, Ibadan, Oyo State, Nigeria.
[3]Department of Biochemistry, University of Ibadan, Ibadan, Oyo State, Nigeria.
[4]Sierra Leone Agricultural Research Institute, Tower Hill, P. M. B. 1313 Freetown, Sierra Leone.

The incidence of root rot diseases partly contribute to the currently observed low percentage increase in the yield of cassava. We estimated gene diversities and identified putative hybridizing parents for root rot resistance using 18 simple sequence repeats loci in 43 improved genotypes of cassava. Root rot was measured over 2 years as the percentage proportion of rotten roots to the total number of roots harvested at 12 month after planting. Estimated rot ranged from 1.2 to 21.2% with a mean of 5.7±0.5. Rank-sum analysis generated 8 rot classes and identified TMS 96/1089A as best genotype resistant to root rot. Gene diversity analysis revealed expected heterozygosity that ranged from 0.701 for very highly susceptible genotypes to 0.781 for moderately resistant and susceptible. Genetic differentiation ranged from -0.0178 (resistant and susceptible) to 0.0523 (very highly resistant and highly resistant genotypes). A total heterozygosity of 0.764 was estimated and was largely due to within class diversity (0.755). DNA analysis representatives for window (DARwin) identified 10 hybridizing groups with a dissimilarity coefficient that ranged from 0.18 to 0.81 on a mean of 0.60. The results obtained from the present study are useful for the genetic improvement of cassava against root rot disease.

Key words: Cassava genotypes, gene diversity, heterozygosity, resistant genotypes, root rot disease.

INTRODUCTION

Cassava, *Manihot esculenta* Crantz, is the main starchy staple of the lowland tropics that has recently gained global popularity because of its combined food, feed, fiber, and bio-fuel traits. It is therefore targeted for the reduction of food insecurity and poverty in Africa. Cassava production in Africa increased tremendously from 81.2 million tonnes in 1995 to 117.449 million tonnes in 2006 (FAO, 1995, 2006). This increased production could be attributed to the adoption of improved cultivars developed in International Institute of Tropical Agriculture (IITA) Ibadan combined with the introduction of good farming practices to the farmers. However, cassava production in Africa has little increased in recent years over what it was in 2006 to an estimate of 126.627 million tonnes in 2010 (FAO, 2006-2010). The currently observed stagnation in production might be attributed partly to stresses such as postharvest deterioration (physiologic or pathogenic) and disease (mosaic disease, bacterial blight, anthracnose, and more

*Corresponding author. E-mail: kmoyib@hotmail.com or okmoyib@gmail.com.

recently root rot) prevalent in the producing regions.

Cassava root rot is becoming a very important disease of cassava and was first discovered in West Africa by Mskita et al. (1998). It was reported to cause about 20 to 80% yield loss (Mskita et al., 2005). Aigbe and Remison (2010) reported that the starch content and quality of garri decreased with increasing incidence or severity of root rot among assessed varieties of cassava. Reported causes of cassava root rot diseases are water logged or flooded soils and microorganisms (fungi and bacteria) that are hydrophilic.

Lately, a large parasitic mushroom (*Polyporus sulphureus*) has been found to be causing severe root rot of cassava in some parts of Ghana and is capable of causing 100% yield loss in susceptible cultivars (Moses et al., 2005). Commonly reported symptoms associated with cassava root rot are brown and wilting plants defoliation, shoot or stem dieback. The major observation is at the harvest: the roots are swollen, soft, give out an offensive odour and often discoloured when cut. The use of resistant or tolerant cultivars has been listed among the measures to control cassava root rot (Moses et al., 2005). Therefore, efforts are geared towards evaluating and screening cassava germplasm for root rot to identify resistant or tolerant varieties (Onyeka et al., 2006; Okechukwu et al., 2009) that can be used as parents for the crop's genetic improvement.

Molecular marker biotechnology has proven useful and substantial in complementing conventional breeding over the years, especially for disease resistance and high yield in cassava (Fregene and Puonti-Kaerlas, 2005). Genomic tools such as genetic diversity studies either in revealing the genetic relatedness for broadening gene pools, genotype identification, or the elimination of redundancy/duplicates are crucial for any breeding objective in cassava genetic improvement. Molecular markers that are polymerase chain reaction-based are routinely used for genetic diversity studies in crops (Tautz, 1989; Williams et al., 1990). Simple sequence repeat markers (SSR) are preferred in cassava genomics because of their co-dominancy, reproducibility, and unambiguous data, and also, they are cheap and easy to use. SSR markers have been used over the years for genomics analyses in cassava which ranged from dissecting genetic relatedness, and molecular mapping, to marker-assisted selection (Mba et al., 2001; Fregene at al., 2003; Okogbenin et al., 2006; Moyib et al., 2007; Siqueira et al., 2009).

We, therefore, sought to employ SSR markers to evaluate the gene diversity parameters such as heterozygosity and genetic differentiation and also to identify putative parents with which hybridization and introgression of useful genes for root rot resistance is possible among the improved genotypes of cassava that are bred for disease resistances. This study is expected to enhance the molecular breeding of a large number of new cassava varieties that are resistant or tolerant to root rot diseases. Therefore, the study offers measurable benefits in improving the livelihood of cassava farmers in rural areas and also increasing cassava productivity in Africa.

MATERIALS AND METHODS

Plant materials

This study was conducted in 2004/2005, 2005/2006 and 2006/2007 at the research farm of IITA-Ibadan, Nigeria with an average annual rainfall of 1305 mm; an altitude of 243 m; mean annual temperature of 20 to 34°C; coordinates 7°31' N; 3°54' E; and ferric luvisol soil. The 43 improved genotypes of cassava developed at IITA for disease resistance and high yielding were used. The experimental design was a randomized complete block in four replications under rain-fed conditions. Mature stem cuttings of 0.25 m long of 43 genotypes were planted on plots of six ridges at a spacing of 1 m between ridges and 1 m within ridges. A ridge was 6 m long and 0.3 m wide, so each plot contained 36 plants. Pre-emergence herbicide as recommended (1% gramozone at 4 L/ha) and hand weeding were used to control weeds.

Evaluation of root rot disease

Data on root rot diseases were collected from 4 inner ridges on 36 plants at 12 month after planting (MAP). Root rot was measured as the percentage proportion of rotten roots to the total number of roots harvested that is, $Rootrot = (NR_tRs / TNRs) * 100$ (where Root rot = Rootrot, NR_tRs = Number of rotten roots, TNRs = Total number of roots harvested). The estimated data for root rot diseases were analyzed for precision measures using Statistical system of analysis software (SAS, 2002).

Grouping of genotypes into root rot resistant classes

Rank-sum procedure in SAS was used to group the 43 genotypes into different phenotypic classes for root rot resistance. Rank-sum first assigned rank $(rank = Xn)$ to the genotypes in descending order based on their estimated mean rot. Grand mean $(Gn =,$ the deviation $(D = Gn - Xn)$, standard deviation $(SD = \sqrt{Xn/N})$, and standardized mean $(stdzdmean = D/SD)$ were calculated for the ranking. The estimated stzdmean is then used to generate the classes among the genotypes using the following synthax as used in the rank-sum procedure: If the standardized mean is < = -3, then the class is 'very highly resistant' (VHR); otherwise, if it is < -3 but <= -2, the class is 'highly resistant' (HR); if it is <-2 but <=-1, the class is 'resistant' (R); if it is < -1 but < 0' it is 'moderately resistant' (MR); if it is <= 0 but < 1, then it is 'moderately susceptible' (MS); if it is <= 1 but <2, then the class is 'susceptible' (S); if it is <=2 but <3, it is 'highly susceptible' (HS). If the standardized mean is >=3, the class is 'very highly susceptible' (VHS).

SSR analysis

Genomic DNA was isolated using modified Dellaporta et al. (1983) for DNA mini-preparation. Eighteen SSR markers were used to amplify genomic DNA from the 43 genotypes of cassava evaluated for rot resistance. The amplified products were separated on 6% polyacrylamide gel electrophoresis using silver staining for visualization. The DNA bands were scored as 1 for the presence of

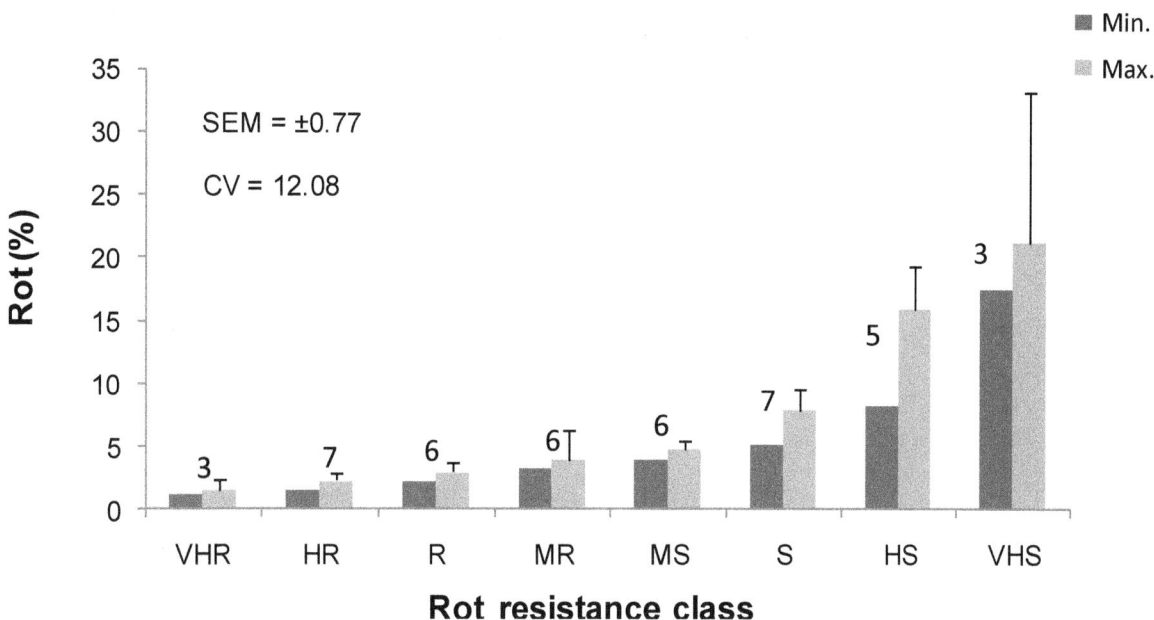

Figure 1. Variation of estimated minimum and maximum values of root rot and number of genotypes within each of the eight root rot resistance classes among the 43 CMD-resistant genotypes of cassava assessed. VHR, very highly resistant; HR, highly resistant; R, resistant; MR, moderately resistant; MS, moderately susceptible; S, susceptible; HS, highly susceptible; VHS, very highly susceptible. The genotypes were distributed into eight different classes of rot resistance based on the rank sum analysis in SAS. Min. and max. values within each class is shown at the center of data points. The number of genotypes in each class is given directly outside the data point of each class.

a DNA band and 0 for the absence of a DNA band. The raw single data was transformed into bi-allelic data and analyzed using F-statistics (FSTAT) version 2.9.3.2 software package (Goudet, 2002) for F-statistics such as mean number of alleles ($\bar{A}$), mean allelic richness ($\bar{A}R$), and percentage of polymorphic loci (P%) and also for gene diversity estimators, heterozygosity (He) such as expected heterosygosity ($\hat{H}_e$), observed heterozygosity (H_O), within class heterozygosity (H_S), among class heterozygosity (D_{ST}), total heterozygosity (H_T), and proportion of among class heterozygosity (G_{ST}) according to Nei (1978) and genetic differentiation based on fixation index (F_{ST}, theta) estimation over allele, locus and population as described by Weir and Cockerham (1984). Pairwise values of F_{ST} between rot classes were also estimated and the matrix was analyzed by cluster analysis using unweighted pair-group method of arithmetic in numeric taxonomy system of statistics (NTSYS) software package (Roulph, 2000). The single molecular data of SSR markers and the rank-sum of the genotypes were subjected to cluster analysis for the identification of genotypic groups with which easy hybridization is possible so as to generate large number of new varieties with possible root rot resistance among the genotypes using DARwin software package (Perrier and Jacquemoud-Collet, 2006).

RESULTS

Variation and ranking of rot among cassava genotypes

The harvested TNRs ranged from 5.00 - 139.00 with a mean of 63.36±1.33 and NR$_t$Rs ranged from 0 - 16 with a mean of 2.72±0.18. The mean TNRs and NR$_t$Rs at 12

MAP varied very highly significantly among the 43 genotypes of cassava studied (CV = 135.4). The estimated rot ranged from 0.0 to 90.9% with a mean of 5.7±0.5%. The estimated mean rot among genotypes ranged from 1.2±1.0% for TMS 96/1089A to 21.2±12.0% for TMS 92/0067 with an average of 5.71±0.48%. Based on the rank-sum procedure used, 8 classes of rot resistance were generated among the 43 genotypes with 3 genotypes, each, in VHR and VHS class and there were 7 HR, 6 H, 6 MR, 6 MS, 7S, and 5 HS, among the 43 genotypes (Figure 1). The 3 genotypes in the VHR were TMS 96/1089A (mean rot of 1.2±1.0%, 97/4779 (1.4±0.7%), and 94/0561 (1.5±0.8%) and were identified as the best three genotypes for rot resistance; and VHS genotypes were TMS 99/6012 (17.5 ±12.4), 94/0026 (19.1±6.5), and 92/0067 (21.2 ±12.0%). The descriptive statistics and rank-sum analysis for the classes of rot among the 43 genotypes are given in Table 1.

Gene diversity and identification of putative hybridizing parents for root rot resistance

A total number of 102 alleles were amplified from the genome of the 43 improved genotypes of cassava using 18 SSR loci, the profile pattern among the 43 cassava clones by SSRY101 is shown in Figure 2. F-statistics analysis results for rot classes showed an estimated $\bar{A}$

Table 1. Mean rot, standard deviation, standard error, coefficient of variation, range, minimum, maximum, ranking, standardized mean, and class for root rot resistance in 43 CMD-resistant genotypes of assessed rot at IITA-Ibadan in 2005 and 2006.

S/N	Genotype	Mean rot %	Descriptive measures						Rank-sum Analysis†			
			SD	SE±	CV	Range	Min.	Max.	Xn	D	stdmean	Class
1	96/1089A	1.2	2.4	1.0	199.2	5.9	0.0	5.9	43	-21	-3.4	VHR
2	97/4779	1.4	2.0	0.7	143.9	6.0	0.0	6.0	42	-20	-3.2	VHR
3	94/0561	1.5	2.1	0.8	137.7	5.0	0.0	5.0	41	-19	-3	VHR
4	TME419	1.6	2.6	1.1	160.8	5.9	0.0	5.9	40	-18	-2.8	HR
5	95/0289	1.7	1.0	0.4	59.6	2.8	0.0	2.8	39	-17	-2.8	HR
6	99/3073	1.7	2.6	0.9	153.8	7.0	0.0	7.0	38	-16	-2.6	HR
7	92/0057	2.0	1.9	0.7	95.8	5.4	0.0	5.4	37	-15	-2.4	HR
8	94/0039	2.1	3.2	1.6	147.8	6.7	0.0	6.7	36	-14	-2.2	HR
9	97/4769	2.2	1.2	0.5	57.4	3.5	0.0	3.5	35	-13	-2	HR
10	91/02324	2.3	1.2	0.5	54.1	3.4	0.0	3.4	34	-12	-2	HR
11	97/2205	2.3	3.7	1.5	161.8	9.4	0.0	9.4	33	-11	-1.8	R
12	96/1565	2.5	1.7	0.9	68.7	3.7	0.0	3.7	32	-10	-1.6	R
13	97/3200	2.5	3.9	1.6	153.9	10.0	0.0	10.0	31	-9	-1.4	R
14	95/0379	2.8	2.9	1.2	101.1	7.0	0.0	7.0	30	-8	-1.2	R
15	97/4763	2.8	0.8	0.3	27.2	2.1	2.2	4.3	29	-7	-1.2	R
16	4(2)1425	3.0	2.5	0.7	84.4	7.8	0.0	7.8	28	-6	-1	R
17	98/2226	3.2	3.0	1.0	91.3	7.5	0.0	7.5	27	-5	-0.8	MR
18	96/1569	3.3	3.1	1.1	94.1	9.1	0.0	9.1	26	-4	-0.6	MR
19	30572	3.4	6.9	1.0	201.4	46.2	0.0	46.2	25	-3	-0.4	MR
20	96/0603	3.4	3.4	1.5	99.3	8.7	0.0	8.7	24	-2	-0.4	MR
21	M98/0068	3.8	3.3	1.2	86.4	10.3	0.0	10.3	23	-1	-0.2	MR
22	M98/0028	3.9	6.4	2.4	165.7	17.5	0.0	17.5	22	0	0	MR
23	96/1632	4.0	3.6	1.3	91.0	8.9	0.0	8.9	21	1	0.2	MS
24	97/0162	4.1	4.0	1.6	96.5	10.0	0.0	10.0	20	2	0.4	MS
25	98/0581	4.3	6.2	2.2	145.5	18.2	0.0	18.2	19	3	0.4	MS
26	98/2101	4.3	5.4	1.9	126.7	14.9	0.0	14.9	18	4	0.6	MS
27	99/2123	4.7	4.9	1.7	103.7	13.8	0.0	13.8	17	5	0.8	MS
28	82/00058	4.8	3.0	0.7	62.7	10.5	0.0	10.5	16	6	1	MS
29	98/0002	5.2	4.0	1.4	75.4	13.2	0.0	13.2	15	7	1.2	S
30	97/0211	6.3	2.6	1.0	40.7	7.5	1.9	9.4	14	8	1.2	S
31	98/0510	6.4	11.5	4.1	179.3	34.4	0.0	34.4	13	9	1.4	S
32	95/0166	6.5	4.9	2.0	76.1	13.6	0.0	13.6	12	10	1.6	S
33	96/1642	7.3	2.8	1.4	38.7	6.5	4.0	10.5	11	11	1.8	S
34	M98/0040	7.5	5.9	2.2	79.3	15.4	3.1	18.5	10	12	2	S
35	92/0326	7.9	4.3	1.7	54.3	11.1	3.6	14.7	9	13	2	S
36	96/0523	8.3	7.9	2.8	95.1	20.0	0.0	20.0	8	14	2.2	HS
37	92B/00061	9.3	10.1	4.1	109.1	28.6	0.0	28.6	7	15	2.4	HS
38	92B/00068	13.4	13.9	5.3	103.8	37.2	0.0	37.2	6	16	2.6	HS
39	98/0505	14.3	9.0	3.4	63.1	24.4	5.6	30.0	5	17	2.8	HS
40	92/0325	15.9	9.9	3.7	62.4	23.8	7.4	31.3	4	18	2.8	HS
41	99/6012	17.5	32.8	12.4	187.8	89.4	1.5	90.9	3	19	3	VHS
42	94/0026	19.1	15.8	6.5	82.9	43.0	4.9	47.8	2	20	3.2	VHS
43	92/0067	21.2	29.5	12.0	139.4	80.0	0.0	80.0	1	21	3.4	VHS
	Overall	5.4	9.0	0.5	168.4	90.9	0.0	90.9	Gn	SD		
	Grand mean	5.7	5.1	0.8	88.4	20	1.2	21.2	22	12.6		

† Xn, rank, is based on mean rot in descending order; Gn, grand mean of rank; D, deviation=Xn-Gn; SD, standard deviation of rank; stdmean, standardized mean=D/SD; $_x$, not applicable; class: VHR, very highly resistant; HR, highly resistant; R, resistant; MR, moderately resistant; MS, moderately susceptible; S, susceptible; HS, highly susceptible; VHS, very highly susceptible.

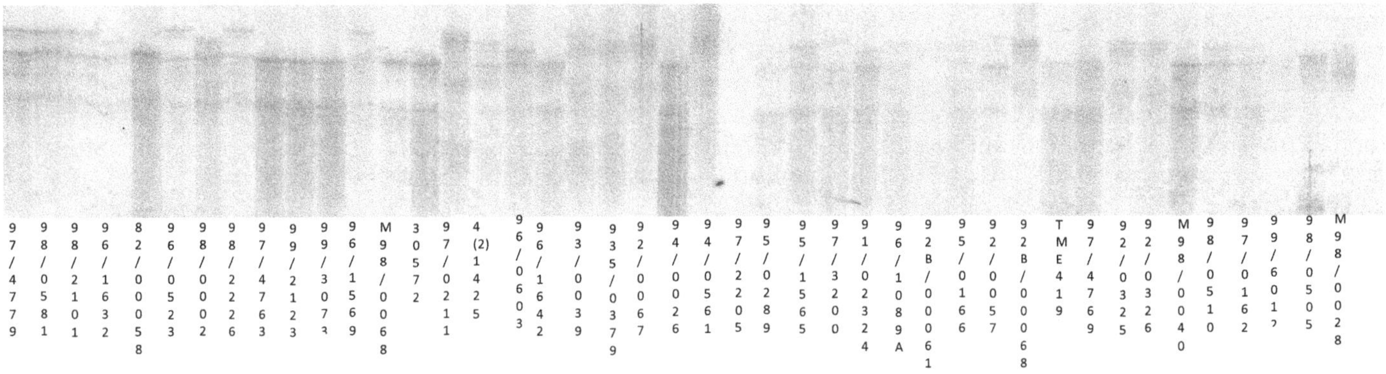

Figure 2. The genotyping profile pattern of 43 CMD-resistant genotypes of cassava by SSRY101 using 6% polyacrylamide gel electrophoresis and silver staining method.

Table 2. Estimates of genetic F-statistics, heterozygosity over population and genetic differentiation among the eight classes of root rot resistance generated within 43 CMD-resistant genotypes of cassava

Rot class	VHR	HR	R	MR	MS	S	HS	VHS
F-Statistics								
N	3	7	6	6	6	7	5	3
Ă	3.556	4.611	4.389	4.222	4.611	4.778	4.5	3.111
P%	100	100	100	100	100	100	100	94.4
AR	1.787	1.758	1.789	1.795	1.776	1.793	1.739	1.763
$\hat{H}_e$	0.759	0.745	0.771	0.781	0.764	0.780	0.724	0.701
F_{ST}	-0.171	-0.243	-0.196	-0.167	-0.186	-0.163	-0.271	-0.287

	H_O	H_T	H_S	D_{ST}	G_{ST}	F_{ST}
Heterozygosity over population						
Mean	0.911	0.764	0.755	0.009	0.011	0.011
SD	0.116	0.043	0.035	0.027	0.034	0.033
SE±	0.027	0.010	0.008	0.006	0.008	0.008

Genetic differentiation pairwise matrix based on F_{ST}, theta, estimator

	VHR	HR	R	MR	MS	S	HS	VHS
VHR	0	0.0331	-0.0103	0.0232	0.0424	0.0031	0.0047	0.0316
HR		0	-0.0034	0.0004	0.0112	-0.0174	0.0187	0.0523
R			0	0.0012	0.0214	-0.0178	0.0158	0.0256
MR				0	-0.0134	-0.0132	0.0145	0.0445
MS					0	-0.0009	0.0424	0.0377
S						0	-0.0058	0.0147
HS							0	0.0235
VHS								0

Mean	Min.	Max	SD	SE±
0.1186	-0.0187	0.0523	0.0194	0.0024

N, sample size; Ă. mean number of alleles; P%, percentage of polymorphic loci; AR, mean allelic richness; $\hat{H}_e$, mean gene diversity; H_O, observed heterozygosity; H_S, within class heterozygosity; H_T ($H_S + D_{ST}$), total heterozygosity; D_{ST}, between class heterozygosity; G_{ST} (D_{ST}/H_T), proportion of between class heterozygosity; and F_{ST}, (theta) fixation index for genetic differentiation; VHR, very highly resistant; HR, highly resistant; R, resistant; MR, moderately resistant; MS, moderately susceptible; S, susceptible; HS, highly susceptible; and VHS, very highly susceptible.

that ranged from 3.111 for VHS to 4.778 for S with a mean of 4.222; ĂR, 1.739 for HS to 1.795 for MR with an average of 1.775; P% was 100% for each class except for VHS that had 94.4%, with a mean of 99.3%. Nei's estimation of genetic diversities for class revealed Ĥe of 0.701 for VHS to 0.781 for MR and S on a mean of 0.754, and heterozygosity estimators over population were mean H_S of 0.755; between class D_{ST} of 0.009, which contributed a G_{ST} of 0.012; a H_T of 0.764; and H_O of 0.911 (Table 2). A total of 3 private alleles were detected from 3 loci within 3 classes, allele 3, from SSRY12 for VHR class; 3, SSRY101 for HS; and also 3, SSRY177 for R. Genetic differentiation (F_{ST}) distance by Weir and Cockerham (1984) estimation between pairs of rot class ranged from -0.0178 between R and S to 0.0523 for VHR and HR with a mean of 0.0119. NTSYS generated 7 clusters among the 8 classes of rot with HR and VHS in the same group (Figure 3).

The F-statistics for 18 SSR loci had Å that ranged from 4 for SSRY182 to 7 for SSRY69 and 164 with an average of 5.63; AR ranged from 1.701 for SSRY175 to 1.817 for SSRY108 with a mean of 1.764; and $\hat{H}e$ ranged from 0.681 for SSRY49 to 0.835 for SSRY164. For genetic diversity estimators, H_O ranged from 0.699 for SSRY4 and 175 to 1.000 for SSRY5, 45, 52, 61, and NS158 with a mean of 0.910; H_T, 0.700 for SSRY175 to 0.818 for SSRY164 with a mean of 0.764; H_S within rot class ranged from 0.680 for SSRY49 to 0.830 for SSRY164, an average of 0.755; and D_{ST} between class ranged from -0.008 for SSRY79 to 0.083 for SSRY64 with a mean of 0.009; and contributed a proportion of G_{ST} that ranged from -0.001 for SSRY69 to 0.104 for SSRY64 on an average of 0.012. Genetic differentiation estimator, F_{ST} (theta) over allele ranged from -0.103 for allele 5 SSRY12 to 0.255 for allele 3 SSRY49; over locus ranged from -0.034 for SSRY12 to 0.092 for SSRY49 with an overall mean of 0.011; and using jackknifing method, over locus was (-0.033 ±0.013) for SSRY12 to 0.098±0.064 for SSRY49 with an overall mean of 0.011±0.008 (Table 3).

DARwin analysis based on Jaccard-weighted-neighbour-joining gave a dissimilarity matrix with a coefficient that ranged from 0.18 to 0.81 with a mean of 0.60 and generated a tree that revealed 4 main clusters with sub-clusters that made up a total of 10 clusters of HZG. The first HZG (HZG1) consisted of 4 genotypes (2 VHR and 2 R); HZG2 also had 4 (1 HR, 2 R and 1 S); HZG3, 4, with one genotype each in HR, MR, MS, and S classes; HZG4 had 2 that belong to MS class, only; HZG5, 3, 2 MR and 1 S; HZG6, 5, 2 HR, 1 MR, and 2 MS; HZG7, 3, 1 HR, 1 R, and 1 HS; HZG8, 6, 2 HR, 1 MR, 1 S and 2 HS; HZG9, 5, 1 VHR, 1 R, 1 MR, and 2 S; and HZG10, 7, 1 MS, 1 S, 2 HS, and 3 VHS. Some groupings were unique for a few classes such as HZG4 for MS class, HZG1 and HZG2 for genotypes with a good level of resistance (but an outlier S in HZG 2), and HZG10 for those with a high level of susceptibility with an odd MS (Figure 4).

DISCUSSION

Variations of root rot within the cassava collection

Large variations of root rot disease have been previously detected in cassava collections, and there has been lots of documented work on the pathogens and their characterization. Onyeka et al. (2006) observed a large

variation of rot response and severity among improved varieties and landraces, and so also was Okechukwu et al. (2009) among improved varieties of cassava. The present study was also able to detect a large variation of root rot severity using the percentage estimate of rotten roots to total roots harvested at 12 MAP. Rank-sum grouping identified TMS 96/1089A (mean rot of 1.2%) as the best genotype for root rot resistance among the 43 genotypes of cassava assessed. The min. mean rot value of 1.2% obtained in this study is much lower than that by Okechukwu et al. (2009) of 7.58% for TMS 97/2205 but higher than that of Aigbe and Remison (2010) of 0.00% for TMS 4(2)1425. The best genotype in the present study was not assessed by Aigbe and Remison (2010) but by Okechukwu et al. (2009) and it took 4th position with mean rot of 11.42% and belongs to R class of the six classes. This genotype recorded a 0.00% rot in 8 out of 25 zones as reported by Okechukwu et al. (2009) across Nigeria. The best genotype for rot resistances by Okechukwu et al. (2009) and Aigbe and Remison (2010) had mean rot of 2.29 (HR) and 2.90% (R), respectively, in our present study. The differences observed in these values is mainly the result of differences in the number of location(s), which are highly influenced by G X E interactions, as explained by Egesi et al. (2007) for yield parameters and diseases in cassava breeding. Therefore, TMS 96/1089A is identified as the best for root rot resistance followed by 97/4779 and 94/0561 in agro-ecozones similar to the environment of IITA-Ibadan assessed in the present study and this finding is subject to further exploration. The first ten best genotypes in Table 1 are recommended for farmers around the agro-ecozones of IITA, Ibadan for increased productivity.

Gene diversity and identification of putative hybridizing parents using SSR markers

SSR markers have been used extensively for genetic diversity studies in cassava, either for genotypic identification or establishing genetic relationships and differentiation as a prerequisite to its molecular breeding improvement (Mba et al., 2001; Moyib et al., 2007; Siqueira et al., 2009). Each time, SSR markers were able to provide useful information that is employable for breeding objectives. In the present study, SSR markers were used for genetic diversity and the identification of putative hybridizing parents for root rot resistance. Each of the 18 SSR markers used showed P% of within each class except SSRY182 for the VHS class, which was due to lack of amplified alleles by TMS 94/0026, one of the 3 genotypes in the class. SSR markers detected high H_S within all the classes of rot with MR and S class having the highest (0.781) and the lowest diversity was observed for VHS (0.701) with a mean of 0.753. The estimated H_S was observed to depend more on the nature of Å, P%, ÅR, and $\hat{H}e$ of loci rather than the sample size within

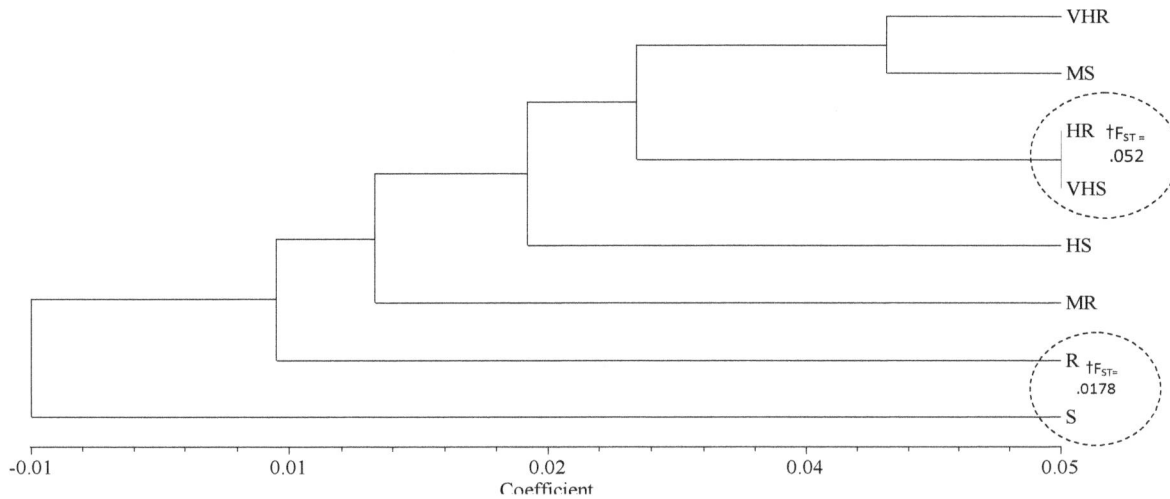

Figure 3. A dendrogram showing genetic differentiation (F_{ST}) between pairs of eight classes for root rot resistance among 43 CMD-resistant genotypes of cassava based on 18 SSR loci using unweighted pair-grouping method of arithmetic in NTSYS software analysis package. VHR, very highly resistant class; HS, highly susceptible; MR, moderately resistant; MS, moderately susceptible; R, resistant; S, susceptible class; VHR, very highly resistant; and VHS, very highly susceptible. Highest estimated genetic differentiation was between VHS and HR with a value of 0.052, while the least differentiation was between R and S, -0.0178.

Table 3. The locus name, locus map location in cassava genome and Nei's estimation of heterozygosity of the 18 SSR markers used for gene diversity analysis of root rot resistance among 43 CMD-resistant genotypes of cassava.

Locus	Map location†	$\bar{A}$	AR	Nei's estimation of heterozygosity						
				H_e	H_O	H_S	H_T	D_{ST}	G_{ST}	F_{ST}
SSRY12	H	6	1.729	0.745	0.957	0.748	0.725	-0.023	-0.031	-0.033
SSRY61	nd	5	1.756	0.752	1	0.753	0.75	-0.003	-0.004	0.002
SSRY52	H	5	1.736	0.732	1	0.732	0.741	0.009	0.012	0.007
SSRY182	UMA	4	1.754	0.719	0.958	0.723	0.746	0.023	0.03	-0.021
SSRY175	K	5	1.701	0.712	0.699	0.707	0.7	-0.007	-0.009	-0.014
SSRY79	nd	5	1.747	0.756	0.858	0.754	0.746	-0.008	-0.01	-0.011
SSRY5	J	5	1.798	0.823	1	0.82	0.8	-0.02	-0.024	-0.02
SSRY101	J	6	1.741	0.724	0.961	0.735	0.733	-0.002	-0.002	-0.002
SSRY4	nd	6	1.725	0.696	0.699	0.697	0.728	0.031	0.043	0.033
SSRY49	K	5	1.743	0.681	0.979	0.68	0.73	0.05	0.068	0.098
SSRY164	H	7	1.807	0.835	0.719	0.83	0.818	-0.011	-0.014	-0.014
SSRY69	D	7	1.782	0.788	0.914	0.787	0.786	-0.001	-0.001	0.004
SSRY45	nd	6	1.787	0.771	1	0.772	0.783	0.012	0.015	0.016
SSRY171	C	6	1.809	0.825	0.0723	0.821	0.802	-0.02	-0.024	-0.004
SSRY108	D	6	1.817	0.778	0.982	0.784	0.812	0.028	0.035	0.018
SSRY177	U	6	1.773	0.75	0.975	0.754	0.766	0.013	0.017	0.02
NS-158	nd	6	1.782	0.773	1	0.775	0.782	0.007	0.009	-0.001
SSRY64	nd	6	1.776	0.699	0.958	0.711	0.794	0.083	0.104	0.073
	Mean	5.7	1.764	0.753	0.91	0.755	0.763	0.009	0.012	0.011
	Min.	4	1.701	0.681	0.699	0.68	0.7	-0.023	-0.031	-0.033
	Max.	7	1.817	0.835	1	0.83	0.818	0.083	0.104	0.098
	SD	0.767	0.033	0.045	0.116	0.043	0.035	0.027	0.034	0.033
	SE±	0.181	0.008	0.011	0.027	0.01	0.008	0.006	0.008	0.008

†Source: Mba et al., 2001. $\bar{A}$, the mean number of alleles; AR, allelic richness; H_O = observed heterozygosity; $\hat{H}_e$, gene diversity; H_S = within class heterozygosity; H_T, total hetrozygosity; D_{ST} = among class heterozygosity; G_{ST}, proportion of among class diversity ($G_{ST} = D_{ST}/_{HT}$); FST, fixation index estimator for genetic differentiation; nd, no linkage data.

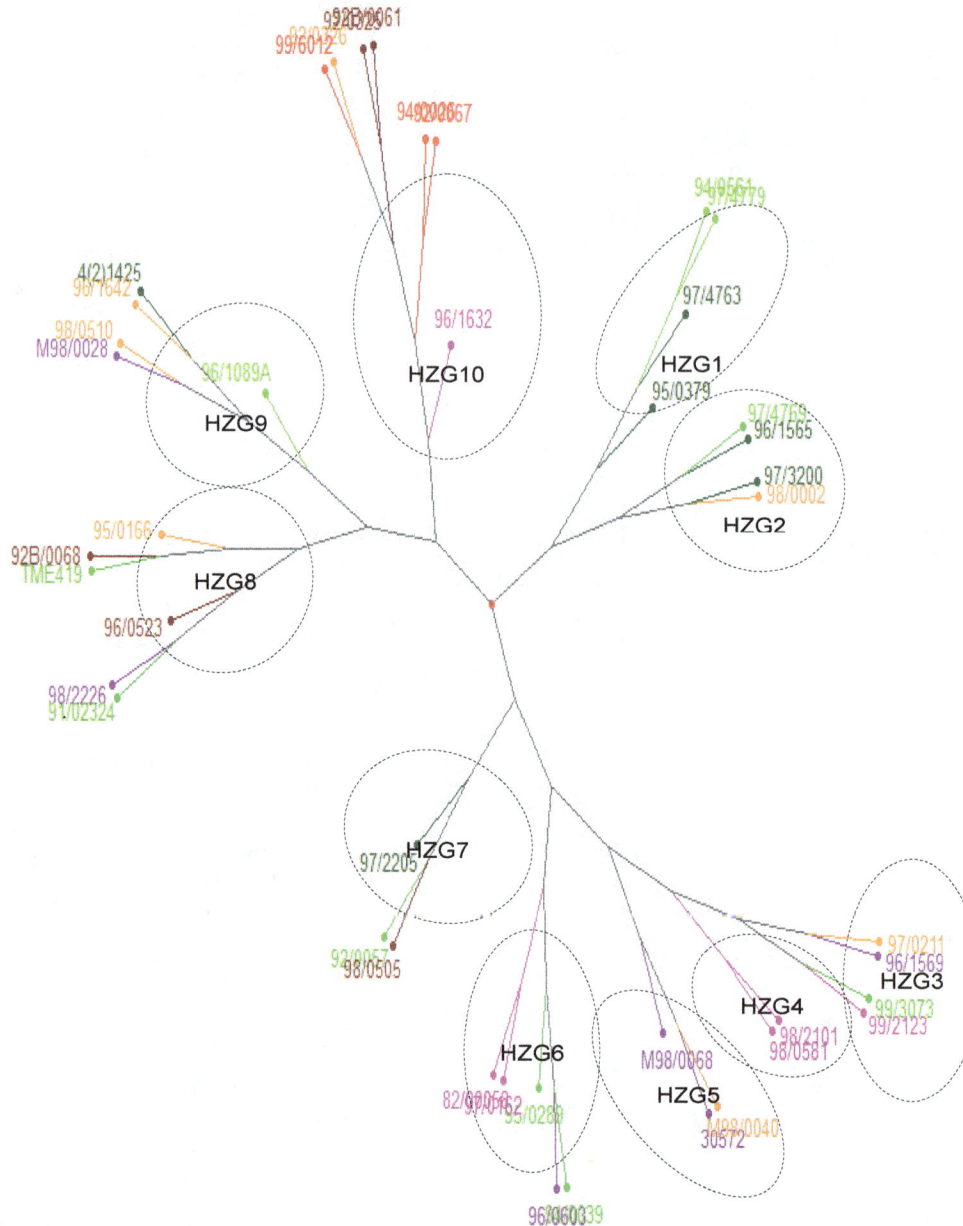

Figure 4. A phylogenetic tree showing putative hybridizing groups for root rot resistance among 43 CMD-resistant genotypes of cassava using 18 SSR markers based on weighted neighbor-joining Jaccard dissimilarity index in DARwin software package. Red, VHS genotypes; deep brown, HS; brown, S; lilac, MS; purple, MR; deep green, R; green, HR and lemon green, VHR; HZG, hybridizing group; VHR, very highly resistant; HR, highly resistant; R, Resistant; MR, moderately resistant; S, susceptible; HS, highly susceptible; VHS, very highly susceptible. Ten putative HZGs were identified among the 43 genotypes of cassava. There is a mixed of more than two rot classes in a HZG except HZG1, 4, and 5.

classes. The mean value of Ĥe, 0.754, obtained in the present study is higher than the mean value of 0.57 obtained for collections in regions of Brazil (Siqueira et al., 2009); 0.477 for African landraces (Lokko et al., 2006); and 0.535 for African and Neotropical countries (Fregene et al., 2003). This higher diversity obtained in the present study may be attributed to the diverse exotic sources of the 43 improved genotypes. Other studies assessed landraces that may be representative of local farmers' accessions. Furthermore, the study revealed that cassava still maintains high genetic diversity which favors production of large number of new varieties that will be resistant to prevalent biotic and abiotic stresses in targeted areas.

According to Nei's estimation, the SSR markers revealed a high observed heterozygosity (H_O = 0.911) and total heterozygosity (H_T = 0.764) among the genotypes used for the study. The high estimated H_T was largely contributed by the diversity within the rot class (H_S = 0.755), while a small proportion was due to diversity between classes (mean G_{ST}, D_{ST}/H_T = 0.012). The percentage proportion of G_{ST} (12%) obtained in the present study is higher than the 10% obtained by Fregene et al. (2003) but lower than 13% by Siqueira et al. (2009). All these values are lower than expected for predominantly out-crossing crops such as cassava, according to Hamrick and Godt (1997). The average genetic differentiation, F_{ST} (0.011) based on Jackknifing was low and similar to the estimated G_{ST}, 0.012, between class, which took into account variation in sample size. The lowest pairwise F_{ST} differentiation (-0.0178) was found between S and R, while the highest (0.0523) was between VHS and HR. These values were lower than the min. (0.0076) and max. (0.2696) average F_{ST} among regions (mean = 0.079) by Fregene et al. (2003). The authors indicated that their maximum value was odd but expected to be higher than the present study because of the larger number of genotypes and regions they assessed. Furthermore, the lower differentiation obtained in the present study is expected because of exchange of cassava across regions and more importantly, the particular collection assessed for this study was bred for common goal (resistance to CMD).

The 18 SSR markers used were able to identify 10 HZGs among the 43 genotypes of cassava in the present study. The observed clusters were not representatives of the rot classes; therefore, each HZG is composed of genotypes that are closely related genetically and not morphologically, which would facilitate hybridization and transfer of genes among the genotypes within a cluster.

This further explains why many HZGs were having genotypes from very different classes of root rot but a few were distinct for some related classes, such as HZG1 for VHR, HR, and R; HZG10 for VHS, HS, and MS. This observed distinction might be partly caused by some of the private alleles that were unique such as allele 3 from SSRY 12 for VHR, 177 for HR and 101 for HS. However, no private allele was distinct for VHS despite the fact that all the 3 genotypes were found clustered together in HZG10. Also, none was found for MS in HZG4, though they were also found in other HZGs. The clustering of the 3 VHS genotypes in the same HZG is fully supported by the lowest heterozygosity of 0.701 observed for the class. The results indicated that there are much more contributing factors than the identified private alleles for the observed clustering in the HZG1 and HZG10, which may have nothing to do with the root rot trait. Moreover, SSR are random markers and therefore, they do not possess a strong phenotypic variability relationship with agronomic traits (Dwivedi et al., 2007), which necessitate the need for new and large number of functional tools for

genetic analysis in cassava.

More impotantly, SSR markers generated groupings among the genotypes for root rot resistance, which would facilitate generation of large numbers of new varieties that could be resistant to diseases and pests prevalent in targeted regions among the improved genotypes assessed. The estimated low G_{ST} and F_{ST} between the classes would also facilitate crosses among the genotypes in the different classes of rot resistance and it is also useful for easy introgression of useful genes among them. Therefore, SSRs still remain important markers for routine evaluation of genetic diversity such as estimation of genetic variation, genotypic groupings and identification, development of core germplasm, and elimination of duplicates or domants within cassava collections and population. The findings of the present study are therefore useful for improved breeding programs for root rot resistance in cassava.

Abbreviations: A, Number of alleles; **Â,** mean number of allele; **AR,** allelic richness; **CMD,** cassava mosaic disease; **DARwin,** dis-similarity analysis representatives for windows; **D_{ST},** between class heterozygosity; **Freq.,** frequency; **FSTAT,** F-statistics; **F_{ST},** (theta) fixation index, estimator for genetic differentiation; **G_{ST},** proportion of heterozygosity between class; **He,** heterozygosity; **Ĥe,** expected heterozydosity; **H_O,** observed heterozygosity; **H_S,** within class heterozygosity; **H_T,** total heterozygosity; **HS,** highly susceptible; **HZG,** hybridizing group; **IITA,** international institute of tropical agriculture; **MR,** moderately resistant; **MS,** moderately susceptible; **NR_tR,** number of rotten roots; **NTSYS,** numeric taxanomy system of statistics; **P,** polymorphic loci; **R,** resistant; **Rootrot,** root rot; **S,** susceptible; **SAS,** statistical analysis system; **SSR,** simple sequence repeat; **Stdzdmean,** standardized mean; **TNR,** total number of roots harvested; **VHR,** very highly resistant; **VHS,** very highly susceptible.

REFERENCES

Aigbe SO, Remison SU (2010). The influence of root rot on starch content of cassava planted in different ecological environment of Nigeria. Nig. Annal. Nat. Sci. 10(1):60-70.

Dellaporta SL, Wood J, Hicks JB (1983). A plant DNA mini-preparation: Version II. Plant Mol. Biol. Rep. 1:19-21.

Dwivedi SL, Crouch JH, Mackill DJ, Xu Y, Blair MW, Ragot M, Upadhyaya HD, Ortiz R (2007). The Molecularization of public sector crop breeding: Progress, problems, and prospects. Adv. Agron. 95:163-318.

Egesi CN, Ilona P, Aigbe FO, Akoroda MO, Dixon AGO (2007). Genetic variation and GXE interaction for yield and other agronomic traits in cassava in Nigeria. Agron. J. 99:1137-1142.

FAO (1995). Food and Agriculture Organization.Production and trade yearbook, 1995. FAO, Rome,

FAO (2006). Food and Agriculture Organization. Food Outlook: Global market Analysis. http//www.fao.org//.

Fregene MA, Suarez M, Mkumbira J, Kulembeka H, Ndedya E, Kulaya A, Mitchel S, Gullberg U, Rosling H, Dixon AGO, Dean R, Kresovich S (2003). Simple sequence repeat marker diversity in cassava andraces: genetic diversity and differentiation in an asexually propagated crop. Theor. Appl. Genet. 107:1083-1093.

Fregene M, Puonti-Kearlas J (2005). Cassava biotechnology. In Hillocks RJ, Thresh JM, Bellotti AC (Eds) Cassava: Biology, production and utilization. CABI International, UK.

Goudet J (2002). FSTAT (version. 2.9.3.2.) A computer software to calculate F-statistics in genetics studies J. Hered 86:485-486.

Hamrick H, Godt MJW (1997). Allozyme diversity in cultivated crop. Crop Sci. 37:26-30.

Lokko Y, Dixon A, Offei S, Danquah E, Fregene M (2006). Assessment of genetic diversity among African cassava (*Manihot esculenta* Crantz) accession resistant to the cassava mosaic disease using SSR markers. Genet. Resour. Crop Evol. 53:1441-1453.

Mba REC, Stephenson P, Edwards K, Melzer S, Nkumbira J, Gullberg U, Apel K, Gale M, Tohme J, Fregene M (2001). Simple sequence repeat (SSR) markers survey of the cassava (*Manihot esculenta* Crantz) genome: Towards an SSR-based molecular genetic map of cassava. Theor. Appl. Genet. 102:21-31.

Moses E, Asafu-Agyei JN, Ayueboteng F (2005). Identification and control of root rot diseases of cassava: Disease guide. First year report of International Society for Plant Pathology (ISSP) Congress Challenge on the development of appropriate strategies to control cassava diseases in Ghana. http://www.isppweb.org/foodsecurity_cassavaghana.asp. accessed 13 Jan. 2011.

Moyib OK, Odunola OA, Dixon AGO (2007). SSR markers revealed genetic variation between improved cassava cultivars and commonly grown Nigerian landraces within a collection of Nigerian cassava germplasm. Afr. J. Biotechnol. 6(23):2666-2674.

Mskita W, James B, Ahounou M, Baimey H, Facho BG, Fagbemisi R (1998). Discoveries of new diseases of cassava in West Africa. Trop. Agric. 75:58-63.

Mskita W, Bissang B, James BD, Baimey H, Wilkinson HT, Ahounou M, Fagbemisi R (2005). Prevalence and severity of *Nattrassia mangiferae* root and stem rot pathogen of cassava in Benin. Plant Dis. 89:12-16.

Nei M (1978). Estimation of average heterozygosity and genetic distances from a small number of individuals. Genetics 89:583-590.

Okechukwu RU, Dixon AGO, Akoroda MO, Mwangi M, Bandropadhyay R (2009). Root rot resistance in new cassava varieties introduced to farmers in Nigeria. Exp. Agric. 45:15-24.

Okogbenin E, Marin J, Fregene M (2006). An SSR-Based molecular genetic map of cassava. Euphytica 147(3):433-440.

Onyeka TJ, Dixon AGO, Ekpo EJA (2006). Identification of levels of resistance to cassava root rot disease (*Botryodiplodia theobromae*) in African landraces and improved germplasm using *in vitro* inoculation method. Euphytica 145(3):281-288.

Perrier X, Jacquemoud-Collet JP (2006). DARwin software for windows Version 5.0 CIRAD-Agropolis International, Montpellier, France. http://darwin.cirad.fr/darwin/.

Roulph FJ (2000). Numeric Taxonomy and Multivariate Analysis System (NTSYS) for PC Version 2.0j. University of London, UK. Exeter Publishing Ltd. New York, USA.

SAS (2002). Statistical Analysis System for windows Version 9.2. SAS Institute Inc. Cary. NC, USA.

Siqueira MVBM, Queiroz-Silva JR, Bressan EA, Borges A, Perreira KJC, Pinto JG, Veasey EA (2009). Genetic characterization of cassava (*Manihot esculenta*) landraces in Brazil assessed with simple sequence repeats. Genet. Mol. Biol. 32.1. http//www.gmb,scielopub.org//. doi: 10.1590/S1415-47572009005000010. Accessed 24 Aug. 2010.

Tautz D (1989). Hypervariability of simple sequences as a general source for polymorphic DNA markers. Nucleic Acids Res. 17:6463-6471.

Weir BS, Cockerham CC (1984). Estimating F-statistics for the analysis of population structure. Evolution 38:1358-1370.

Williams SGK, Kuselik AR, Livak KJ, Rafalski JA, Tiney SV (1990). DNA polymorphisms amplified by arbitrary primers are useful as genetic markers. Nucleic Acids Res. 18:6531-6535.

2

Micropropagation of the Indian Birthwort *Arsitolochia indica* L.

Syed Naseer Shah[1], Amjad M. Husaini[2] and Fatima Shirin[1]

[1]Genetics and Plant Propagation Division, Tropical Forest Research Institute, Mandla Road, Jabalpur 482 021, India.
[2]Centre for Plant Biotechnology, Division of Biotechnology, SKUAST-K, Shalimar, Srinagar-191121, India.

Aristolochia indica L. is a medicinal woody perennial climber plant of immense pharmaceutical value. The species is endangered with possible extinction due to its indiscriminate harvesting as raw material for pharmaceutical industry, to manufacture drugs against cholera, inflammation, biliousness, dry cough and snake bite. A rigorous attempt has been made for development of *in vitro* propagation procedure for this species, involving four steps, namely: culture establishment, shoot multiplication, rooting and hardening. Aseptic cultures were established by growing nodal segments (1 to 1.5 cm) as explants on Murashige and Skoog (MS) medium containing 5.0 μM N6-Benzyladenine (BA). Five nutrient media, MS, Woody Plant Medium (WPM), Gamborg Medium (B5), Nitsch and Nitsch Medium (NN), and Schenk and Hildebrandt Medium (SH) supplemented with different cytokinins and auxins at a concentration of 10.0 μM were used in this study. Ads at 10.0 μM proved optimum for *in vitro* shoot multiplication. The treatment resulted in 100% shoot number per explant at 15 days and 61.9% at 30 days on MS medium, 65.2% node number per shoot at 15 days and 196.2% at 30 days on WPM medium and 147.5 and 366.6% node number per explant at 30 days after inoculation on MS medium. The *in vitro* multiplied shoots were used for rooting experiment. Five nutrient media (MS, WPM, B$_5$, NN and SH) and three auxin sources 10.0 μM each (IBA, IAA and NAA). SH medium with 10.0 μM NAA induced 327.8% rooting at 21 days and 654.8% at 28 days and root number per explant 4300% at 21 and 394% at 28 day after inoculation. The *in vitro* propagated hardened plants exhibited excellent growth on transfer to natural condition.

Key words: *Aristolochia indica* L, *in vitro* propagation, N^6-Benzyladenine.

INTRODUCTION

Aristolochia indica L. (family- Asclepiadaceae.) is a perennial climber with greenish whitish woody stem growing throughout India especially in the tropical and sub-tropical regions. The active constituent "Aristolic acid" is potent drug used in Ayurvedic, Sidda and Homeopathy systems of medicines. Roots are widely used in joint pains and seeds in inflammation, biliousness, dry cough and dyspepsia. The juice of leaves or roots is said to be a specific antidote for cobra poisoning (Kirtikar and Basu, 1987). The species is rare and endangered with extinction due to its indiscriminate collection and over exploitation from natural resources for commercial purpose by pharmaceutical industries (Rahman, 2001). The conventional propagation is hampered due to low seed viability and poor rooting of vegetative cuttings and emphasizes need for the alternative *in vitro* propagation method for large scale multiplication, improvement and conservation of the

Figure 1. Explant collection, culture establishment and shoot multiplication in *Aristolochia indica* L.; (a) mother plant, (b) a twig, (c) nodal explants, (d) the *in vitro* culture establishment and (e-f) the *in vitro* shoot multiplication.

species. The objective of the study was to develop an efficient protocol for its micropropagation.

MATERIALS AND METHODS

The selected (mother) plant from Jabalpur area of Madhya Pradesh, India (Figure 1a) was used to collect twig (s) (Figure 1b), which were washed thoroughly for 15 min under running water for removing the surface debris. The washed twigs were defoliated and cut into nodal explants (approximately 1 to 1.5 cm long and 0.5 to 0.6 cm diameter) (Figure 1c). These explants were washed with 2% Cetrimide® and kept for 10 min with constant vigorous (shaking 150 rpm) on an orbital shaker incubator followed by rewashing 4 to 5 times with distilled water to remove traces of Cetrimide®. The washed explants were sterilized for 5 min with $HgCl_2$ (0.1%) and

Bavistin® (1.0%) in the laminar flow cabinet. Finally, the surface sterilized nodal explants were rinsed 4 to 5 times with sterile distilled water and inoculated on MS medium (Murashige and Skoog, 1962) supplemented with 5.0 µM BA for culture establishment (Figure 1d).

The *in vitro* shoot multiplication (Figure 1e-f) was standardized through a factorial randomized experiment, using single nodal segments from established cultures. In this experiment we screened five nutrient media [MS (Murashige and Skoog, 1962), WPM (Lloyd and McCown, 1980), B_5 (Gamborg et al., 1968), NN (Nitsch and Nitsch, 1969) and SH (Schenk and Hildebrandt, 1972)] along with 10.0 µM each of three cytokinins (BA, TDZ, Ads), and their combinations on shoot number per explant, node number per shoot and node number per explant at 15 and 30 days after inoculation. In second experiment, five nutrient media (MS, WPM, B_5, NN and SH) and three auxins (IBA, IAA and NAA) at

Figure 2. The *in vitro* induction of adventitious root in *Aristolochia indica* L. roots formation on semi-solid medium at (a) 21 and (b) 28 days after inoculation.

concentrations of 10.0 µM and their effect on rooting and root number was recorded at 21 and 28 days after inoculation (Figure 2).

Culture conditions

The inorganic salts used for preparation of culture medium were obtained from Qualigens Pvt. Ltd., India and phytohormones and B vitamins from Sigma Chemicals Pvt. Ltd., India. The medium contained 3% (w/v) sucrose, 0.8% (w/v) agar (Hi-Media chemical Ltd., India). The pH of the medium was adjusted to 6.0 before autoclaving for 15 min at 1.06 kg cm^{-2} (121°C). Explants were cultured in a 150 ml conical Borosil® flasks containing 40 ml semi-solid medium. For *in vitro* shoot multiplication and rooting experiment, the cultures were incubated at 25 ± 2°C under 16 h illuminations with fluorescent light (50 µmol Em^{-2}s^{-1}).

Hardening and transplantation

The *in vitro* raised plantlets were removed from rooting medium washed with distilled water and the plantlets were subsequently transferred to root trainers containing autoclaved soilrite (Figure 3a) and covered with perforated polythene to maintain humidity which were kept under culture room conditions for about 10 days. Subsequently, they were transferred to perforated polythene bags and kept initially in washing room for 5 days and finally transferred to natural condition (Figure 3b-d).

Statisticaly analysis

Each experiment had three replicates for *in vitro* shoot multiplication and rooting. Each replicate had 10 propagules. The data were subjected to two way (factor) analysis of variance for both the experiments with "F" test for ascertaining level of significance. If the data were found significant at p ≤ 0.05, LSD$_{0.05}$ was computed for comparison of treatment means.

RESULTS

In vitro shoot multiplication

The effect of cultured media, cytokinin sources and their all possible combinations on shoot number per explant, node number per shoot and node number per explant at both the stages of sampling was recorded.

Shoot number per explant

SH medium produced maximum shoot number per explant at 15 days and MS medium at 30 days (Table 1).The enhancement of shoot number per explant in SH medium was 100% in comparison to B$_5$ medium at 15 days and 61.90% in MS medium at 30 days after inoculation in comparison to that in B$_5$ medium. Further BA had significantly maximum shoot number explant^{-1} at both stages and was statistically equalled by Ads at 15 days. Shoot number per explant in BA was 1333%, 76% more than that of TDZ at 15 and 30 days, respectively. MS medium with 10.0 µM BA produced maximum shoot numbers per explant which was 189% at 30 days after inoculation.

Node number per shoot

SH medium produced maximum node number per shoot at 15 days and WPM medium at 30 days (Table 2). The enhancement of node number per shoot in SH medium was 65.2% more than that obtained in NN medium at 15 days and 139% at 30 days in comparison to that in B$_5$ medium which produced the lowest value for the parameter at both stages of sampling . BA induced maximum node number per shoot, which was enhanced by 61% at 15 days and 239% at 30 days in comparison with TDZ. Maximum node number per shoot was observed on SH medium with Ads, at 15 days and WPM 10.0 µM Ads at 30 days. NN medium produced minimum effect on node number per shoot at both the stages of sampling at 15 days and B$_5$ medium at 30 days.

Figure 3. Hardening and acclimatization of the *in vitro* raised plantlets of *Aristolochia indica* L. Plantlets transferred to root trainers (a) and covered with polythene (b) placed in the culture room, (c) hardened plantlets transferred to polythene bags and (d) growth of the plantlets in the open environment.

Table 1. Effect of culture media and different cytokinins on shoot number per explant in *Aristolochia indica* L. at two stages of sampling.

Cytokinin source (C)	Culture media(M)											
	Inoculation days											
	15 Days						30 Days					
	MS	WPM	B5	NN	SH	Mean	MS	WPM	B5	NN	SH	Mean
BA	1.00	1.00	0.39	0.56	1.11	**0.81**	**2.89**	1.73	1.06	1.67	1.45	**1.76**
TDZ	0.33	0.00	0.00	0.00	0.00	0.06	1.00	1.00	1.00	1.00	1.00	1.00
Ads	0.78	1.00	0.72	0.72	1.11	**0.86**	1.17	1.17	1.11	1.78	1.45	1.33
Mean	**0.70**	**0.66**	0.37	0.42	**0.74**		**1.70**	1.30	1.05	1.48	1.30	

Variable	LSD $_{(0.05)}$	
	15Days	30Days
C	0.16	0.13
M	0.20	0.17
C *M	NS	0.29

Node number per explant

SH and MS medium induced maximum node number per explant at 15 days and at 30 days respectively (Table 3). The enhancement of node number per explant in SH medium was 147.6% at 15 days and 366.6% in MS medium at 30 days as compared to B$_5$ medium. BA and Ads had significantly maximum node number per explants at 15 and 30 days, respectively. BA enhanced node number per explants by 48% at 15 days and Ads by

Table 2. Effect of culture media and different cytokinins on node number per shoot in *Aristolochia indica* L. at two stages of sampling.

Cytokinin sources (C)	Culture media(M)											
	Inoculation days											
	15 Days						30 Days					
	MS	WPM	B5	NN	SH	Mean	MS	WPM	B5	NN	SH	Mean
BA	1.28	1.55	1.17	1.28	2.78	**1.61**	4.04	2.67	1.11	2.28	2.72	2.57
TDZ	1.00	1.00	1.00	1.00	1.00	1.00	1.78	1.06	1.00	1.00	1.06	1.18
Ads	1.56	**2.22**	1.06	1.17	1.83	1.56	3.50	**5.78**	1.11	4.28	5.32	**4.00**
Mean	1.28	1.60	1.80	1.15	**1.90**		3.10	**3.17**	1.07	2.52	3.03	

Variable	LSD (0.05)	
	15 Days	30 Days
C	0.19	0.47
M	0.25	0.61
C *M	0.43	1.06

Table 3. Effect of culture media and different cytokinins on node number per explant in *Aristolochia indica* L. at two stages of sampling.

Cytokinin sources (C)	Culture media(M)											
	Inoculation days											
	15 Days						30 Days					
	MS	WPM	B5	NN	SH	Mean	MS	WPM	B5	NN	SH	Mean
BA	1.28	1.56	0.81	0.67	**3.07**	**1.48**	11.72	4.33	1.36	3.97	4.27	5.13
TDZ	1.00	1.00	1.00	1.00	1.00	1.00	1.06	1.00	1.00	1.00	1.06	1.02
Ads	1.22	2.22	0.76	0.81	2.03	1.41	4.00	6.80	1.27	7.55	**7.73**	**5.47**
Mean	1.16	1.60	0.86	0.82	**2.03**		**5.60**	4.04	1.20	4.18	4.35	

Variable	LSD (0.05)	
	15 Days	30 Days
C	0.21	0.85
M	0.28	1.10
C *M	0.48	1.91

436% at 30 days in comparison to TDZ, which produced the lowest value for the parameter. As for interaction, SH medium with 10.0 µM Ads registered the highest value for the parameter at 30 days after sampling.

In vitro adventitious rooting

Auxin sources and their combinations with different media induced significant rooting and root number per explant at both the stages of sampling.

Percent rooting

SH medium produced significantly high percent of rooting. The enhancement of rooting in SH medium was 327.8% at 21 days and 655% at 28 days in comparison to MS medium. MS, B_5, NN and WPM produced minimum effect on rooting. NAA produced significantly maximum rooting (%), which was 228 at 21 days and 443.7% at 28 days after inoculation in compared to IAA producing minimum value for rooting. SH medium with 10.0 µM NAA maximum rooting at both stages of sampling (Table 4).

Root number per explants

SH medium produced maximum root number per explant at both the stages of sampling. The enhancement of root number per explant was 4300% at 21 days and 394% at 28 days after inoculation in comparison with WPM, MS and NN medium. NAA was found to have significant effect on root number per explant at both stages of sampling and resulted in 800% at 21 days and 2900% at 28 days more than that obtained in IAA. SH medium

Table 4. Effect of culture media and different auxins on percent of rooting in *Aristolochia indica* L. at two stages of sampling.

Auxin sources (A)	Culture media (M)											
	Inoculation days											
	Rooting at % 21 Days						Rooting at % 28 Days					
	MS	WPM	B5	NN	SH	Mean	MS	WPM	B5	NN	SH	Mean
IBA	0 (4.16)	0 (4.16)	0 (4.16)	16.66 (24.06)	0 (4.16)	3.33 (8.14)	0 (4.16)	0 (4.16)	5.55 (10.79)	16.66 (24.06)	0 (4.16)	4.44 (9.47)
IAA	0 (4.16)	0 (4.16)	0 (4.16)	0 (4.16)	0 (4.16)	0 (4.16)	0 (4.16)	0 (4.16)	0 (4.16)	5.55 (10.79)	0 (4.16)	1.11 (5.49)
NAA	5.55 (10.79)	0 (4.16)	0 (4.16)	0 (4.16)	50 (45)	11.11 (13.65)	22.22 (24.4)	0 (4.16)	16.66 (24.06)	5.55 (10.79)	100 (85.84)	28.89 (29.85)
Mean	1.9 (6.4)	0 (4.16)	0 (4.16)	5.6 (10.8)	16.7 (17.8)		7.4 (10.9)	0 (4.16)	7.4 (13.00)	9.3 (15.2)	33.3 (31.4)	

Variable	LSD (0.05)	
	21 Days	28 Days
A	2.21	5.49
M	2.85	7.08
A*M	4.94	12.27

Table 5. Effect of culture media and different auxins on root number per explant in *Aristolochia indica* L. at two stages of sampling.

Auxin sources (A)	Culture media(M)											
	Inoculation days											
	Root number 21 Days						Root number 28 Days					
	MS	WPM	B5	NN	SH	Mean	MS	WPM	B5	NN	SH	Mean
IBA	0.00	0.00	0.00	0.16	0.00	0.03	0.00	0.00	0.28	0.39	0.00	0.13
IAA	0.00	0.00	0.00	0.00	0.00	0.00	0.00	0.00	0.00	0.17	0.00	0.03
NAA	0.05	0.00	0.00	0.00	1.33	0.27	0.66	0.00	0.44	0.11	3.28	0.90
Mean	0.01	0.00	0.00	0.05	0.44		0.22	0.00	0.24	0.22	1.01	

Variable	LSD (0.05)	
	21 Days	28 Days
A	0.16	0.27
M	0.20	0.35
A*M	0.36	0.61

along with 10.0 µM NAA was found to have significant effect on root number per explant at 21 and 28 days after inoculation (Table 5).

DISCUSSION

The micro-propagation of *A. indica* comprises four steps, namely: establishment of culture from nodal explants, shoot multiplication, root induction and hardening and acclimatization. The present investigation was intended for the standardization of culture medium and plant growth regulators at second and third steps followed by hardening procedure. For shoot multiplications, the best *in vitro* combination was SH medium supplemented with 10.0 µM Ads. There is no published report on the *in vitro* shoot multiplication of *A. indica* using SH medium. The suitability of SH medium in the present study contrasts with earlier reports of micropropagation for this species wherein MS medium was found to be the most effective

(Siddique et al., 2006a; 2006b; Pattar and Jayraj, 2012). The results indicate that the species requires low amount of nitrogen for growth and differentiation of new shoots. Adenine sulphate was found as the most suitable cytokinin for shoot multiplication. Similar results have been reported in the medicinal plant *Cichorium intybus* also, where multiple shoots proliferation was observed on medium supplemented with BA, IAA and adenine sulphate (Nadagopal and Ranjitha Kumari, 2006).

For *in vitro* rooting also, better performance was obtained on SH medium. High concentration of thiamine (Vitamin B_1) included in SH medium seems to be synergistic with auxins for facilitation of rhizogenesis as reported in teak by Ansari et al. (2002). Of the various auxin treatments, NAA was found to be the best auxin for *A. indica*. Superiority of NAA for *in vitro* rooting may be attributed to its synthetic nature and stability. Further, NAA also eludes the auxin oxidizing/ degrading enzyme systems of the plants (Jacobs, 1972). IAA was found to be inferior to both NAA and IBA. In literature also there are reports of IBA and NAA being more effective than IAA, because of the instability of the latter (Gaspar and Coumans, 1987).

Conclusion

The study demonstrates successful development of *in vitro* propagation procedure for *A. indica*. The procedure offers a potential system for conservation and mass propagation using explants derived from mature plants. SH (medium supplemented with 10.0 µM Ads has been found the best for efficient and rapid multiplication of *in vitro* shoots, while SH medium supplemented with 10.0 µM NAA for optimum induction of *in vitro* adventitious roots. Further, the hardening procedure reported here ensures 70 to 80% field survival of micropropagated plants of *A. indica*.

REFERENCES

Ansari SA, Sharma S, Pant NC, Mandal AK (2002). Synergism between IBA and thiamines for induction and growth of adventitious roots in *Tectona grandis"*. J Sustain. For. 15:99-112.

Gamborg OL, Miller RA, Ojima K (1968). Nutrient requirements of suspension cultures of soya bean root cells, Exp. Cell Res. 50:151-58.

Gaspar T, Coumans M (1987). Root formation In: J.M. Bonga and D.J. Durzan (eds). Cell and Tissue Culture in Forestry, Specific Principles and Methods: Growth and Developments, Martinus Nihoff Publishers, Dordrecht, 2:202-217.

Jacobs WP (1972). The movement of plant hormones: auxins, gibberellins and cytokinins". In D.J. Carr (ed.), Plant Growth Substances, Springer, New York. pp. 701-709.

Kirtikar KR, Basu LM (1987). Indian Med. Plants. pp. 2117-2118.

Lloyd G, McCown B (1980) Commercially feasible micropropagation of mountain laurel *Kalmia latifolia* by use of shoot tip culture, Comb. Proc. Int. Plant prop. Soc. 30:421-27.

Murashige T, Skoog F (1962) A revised medium for rapid growth and bioassays with tobacco tissue cultures, Physiol. Plant. 15:473-97.

Nadagopal S, Ranjitha Kumari BD (2006). Adenine sulphate induced high frequency shoot organogenesis in callus and *in vitro* flowering of *Cichoriumintybus* L. cv. Focus - a potent medicinal plant, Acta Agric. Slov. 87(2):415-425.

Nitsch JP, Nitsch C (1969). Haploid plants from pollen grains, Sci. 163:85-87.

Pattar PV, Jayraj M (2012). *In vitro* Regeneration of Plantlets from Leaf and Nodal explants of *Aristolochia indica* L. - An important threatened medicinal plant. Asian Pac. J. Trop. Biomed. 2(2):488-493.

Rahman M (2001). Red data Book of Vascular plants. Bangladesh National Herbarium. Dhaka, Bangladesh.

Schenk RU, Hildebrandt AC (1972). Medium and technique for induction and growth of monocotyledonous and dicotyledonous plant cell culture, Can. J. Bot. 50:199-204.

Siddique NA, Bari MA, Pervin MM, Nahar N, Banu LA, Paul KK, Kabir MH, Huda AKMN, Ferdaus KMKB, Hossin MJ (2006a). Plant Regeneration from Axillary Shoots Derived Callus in *Aristolochia indica* Linn. an Endangered Medicinal Plant in Bangladesh, Pak. J Biol. Sci. 9:1320-323.

Siddique NA, Kabir MH, Bari MA (2006b). Comparative *in vitro* study of plant regeneration from nodal segments derived callus in *Aristolochia indica* Linn. and *Hemidesmus indicus* (L.) R. Br. endangered medicinal plants in Bangladesh, J. Plant Sci. 1(2):106-118.

3

Patterns of colonization and immune response elicited from interactions between enteropathogenic bacteria, epithelial cells and probiotic fractions

Mariana Carmen Chifiriuc[1]*, Coralia Bleotu[2], Diana-Roxana Pelinescu[1], Veronica Lazar[1], Lia-Mara Ditu[1], Tatiana Vassu[1], Ileana Stoica[1], Olguta Dracea[3], Ionela Avram[1] and Elena Sasarman[1]

[1]Department of Microbiology, MICROGEN (Center for Research in Genetics, Microbiology and Biotechnology), Faculty of Biology, University of Bucharest, Ale. Portocalilor 1-3, Sector 5, 77206-Bucharest, Romania.
[2]Institute of Virology Stefan S. Nicolau, 285 Mihai Bravu Ave. 030304, Bucharest, Romania.
[3]Cantacuzino Institute, Sp. Independentei 103, Bucharest, Romania.

The purpose of this study was to investigate by *in vitro* studies the antimicrobial activity of eight lactic acid bacterial (LAB) strains belonging to *Lactobacillus paracasei* spp. *paracasei*, *L. plantarum* and *L. rhamnosus* species against *Salmonella enteritidis*, *Shigella flexneri* and EPEC pathogenic strains isolated from pediatric diarrhoea cases, simultaneously with the assessment of the cytotoxicity and immunomodulatory potential of the respective strains. The study of the adherence capacity to the cellular substrate represented by HeLa cells was performed by Cravioto's adapted method. The cytotoxicity was determined on HeLa cells and the level of soluble pro- and anti-inflammatory cytokines was assessed by ELISA. Our *in vitro* studies are demonstrating that the selected probiotic strains are inhibiting the adherence and colonization of HeLa cells by the enteropathogenic strains isolated from pediatric diarrhoea mainly by direct competition for adherence sites, demonstrating their potential use in the treatment of pediatric gastro-intestinal disorders, as an alternative to/in association with antibiotics. A great advantage of the selected probiotic strains is their low cytotoxicity and ability to trigger a beneficial cytokine response in the epithelial cells, which potentiates their antimicrobial activity by stimulating the occurrence of a rapid immune response following the intestinal injury.

Key words: Antimicrobial activity, immunomodulatory, cytokines, invasive bacteria, *Lactobacillus*, enteropathogenic bacteria, antimicrobial effect, cytotoxicity, probiotics.

INTRODUCTION

In recent years, several articles have reviewed the efficacy, mechanism of action, and safety of probiotics in the treatment of infectious disease (Vanderhoof and Young, 2002; Alvarez-Olmos and Oberhelman, 2001; Salminen and Arvilommi, 2001; Borriello et al., 2003; Ishibashi and Shoji, 2001).

The term probiotic refers to a product or preparation containing viable, defined microorganisms in numbers thought to be sufficient to alter the host's microbiota (by implantation or colonization) and thereby exert beneficial effects (Havenaar and Huis, 1992). The microorganisms most frequently used as probiotic agents are lactic acid bacteria (species of *Lactobacillus*, *Enterococcus* and *Bifidobacterium*) and nonpathogenic, antibiotic-resistant, ascosporic yeasts, especially *Saccharomyces boulardii* (FAO, 2001). *Lactobacillus rhamnosus* strain GG (ATCC 53103), which was originally isolated from human intestinal flora, is the most widely studied probiotic agent for adults and children (Gorbach, 2000). *L. rhamnosus* strain GG can prevent diarrhoea and atopic diseases among children (Szajewska and Mrukowicz, 2001; Kalliomaki et al., 2001; Majamaa and Isolauri, 1997). A

---

*Corresponding author: E-mail: carmen_balotescu@yahoo.com.

Abbreviations: LAB, Lactic acid bacterial; **IL,** interleukins; **ATCC**, american type culture collection; **FCS**, foetal calf serum; **EPEC**, *Enteropathogenic E. coli*.

recent meta-analysis suggested that *Lactobacillus* is a safe effective treatment for children with acute infectious diarrhoea, serious infections attributable to probiotic lactobacilli being extremely rare (Van Neil et al., 2002; Rautio et al., 1999; Mackay et al., 1999; Bradley , 2006).

For the best use of the probiotic microorganisms, the mechanisms by which they work should be better understood. The selection of an appropriate probiotic strain for its inclusion in a probiotic preparation should be made on the basis of its capacity to induce an improved gut immune response without modification of the intestinal homeostasis. To achieve this task, probiotic strains should have the following properties: (i) high cell viability, thus they must be resistant to low pH and bile acids; (ii) ability to persist in the intestine even if the probiotic strain cannot colonize the gut (continuous or prolonged administration may be necessary); (iii) adherence to the gut epithelium to avoid the flushing effects of peristalsis; In this last aspect, there are many relevant literature reports of the adhesive property of the probiotic bacteria to epithelial cells proved by *in vitro* studies (iv) also, they should be able to interact or send signals to the immune cells associated with the gut (Perdigon et al., 1995). There are reports from *in vitro* assays that show the activation of immune cells after stimulation by probiotics (Spanhaak et al., 1998).

The mechanisms of probiotics action are most probably multi-factorial, involving a variety of effector signals, cell types and receptors (Jack et al., 1995; Link-Amster et al., 1994; Schiffrin et al., 1995). But, strains may differ in their respective ability to trigger these signals, which is depending on both immunocompetence and the intestinal epithelial cells (Schiffrin et al., 1997; De Simone et al., 1992).

It is commonly suggested that probiotics must "persist and multiply" in the target ecosystem to be efficient.

However, the interaction of orally ingested probiotics with the intestinal epithelium or other immunologically active intestinal cells has just begun to be rigorously studied. However, for requirement of probiotics persistence to be efficacious, it depends on the specific immunological activities (prophylactic or therapeutic) and mechanisms by which ingested probiotics prevent or cure enteric disorders (Malin et al., 1996; Majamaa and Isolauri, 1997; Link-Amster et al., 1994).

It has been proposed that some probiotics are able to prevent reduction of or restore intestinal homeostasis after a prolonged antibiotherapy in connection with immunological disorders, improving mucosal barrier functions as well as down-regulation of the inflammatory responses (Malin et al., 1996). It was observed that strong strain-specific variations of the *in vitro* cytokine induction profiles after stimulation of immuno-competent host cells (Schiffrin et al., 1995; Schiffrin et al., 1997).

Our objective in this study was to investigate by *in vitro* studies the antimicrobial activity of some lactic acid bacterial (LAB) strains against pathogenic bacterial strains isolated from pediatric diarrhoea, simultaneously assessing the cytotoxicity and immunomodulatory potential of the respective strains.

MATERIALS AND METHODS

LAB strains

Eight (8) *LAB* strains, that is, 6 *Lactobacillus paracasei subsp. paracasei* strains encoded CMGB 18, CMGB 19, CMGB 20, CMGB 21, CMGB 22 and CMGB 23 isolated from infant child feces, one *L plantarum* strain CMGB 24 isolated from fermented vegetal debris and one *b. rhamnosus* CMGB 29 strain isolated from milk products were studied. The strains were isolated and grown in Man Rogosa Sharp (MRS) medium. All strains were identified by standard morphological, cultural and biochemical (API 50 CHL) features and stored in the collection of the Center for Research, Education and Consulting in Microbiology, Genetics and Biotehnology - MICROGEN (acronym CMGB) at -70°C in appropriate medium represented by MRS and supplemented with 20% glycerol.

Fresh cultures were obtained and thereafter cultivated in MRS liquid medium in order to obtain mid-logarithmic phase cultures that were further used in our experiments (Smarandache et al., 2004; Lazar et al., 2004).

Enteropathogenic strains

The pathogenic strains used in our study were recently isolated from acute diarrhoeal cases in children under 2 years of age and identified following stool sample enrichment and seeding on specific selective media (EMB Levine, SS). The biochemical identification was performed comparatively by classical and API 20E microtests (BioMérieux) and the species were serologically confirmed by agglutination with polyvalent and monovalent sera as: two *Shigella flexneri*, one *Salmonella enteritidis* and one enteropathogenic *Escherichia coli* (EPEC) strain.

Study of the adherence and invasion capacity to the cellular substrate represented by HeLa cells (Cravioto's adapted method) (Smarandache et al., 2004; Lazar et al., 2004)

In this purpose, HeLa cells were routinely grown in Eagle's minimal essential medium (Eagle MEM) supplemented with 10% heat-inactivated (30 min at 56°C) foetal bovine serum (Gibco BRL), 0.1 mM nonessential amino acids (Gibco BRL), and 0.5 ml of gentamycin (50 μg/ml) (Gibco BRL) and incubated in a 5% CO_2 humidified atmosphere, at 37°C for 24 h (Kalliomaki et al., 2001). HeLa cell monolayers grown in 6 multi-well plastic plates were used at 80 - 100% confluence. Bacterial strains from an overnight culture on 2% nutrient agar were diluted at 10^7 CFU/ml in Eagle MEM with no antibiotics. The HeLa cell monolayers were washed 3 times with Phosphate Buffered Saline (PBS) and 2 ml from the bacterial suspension were inoculated in each well. The inoculated plates were incubated for 3 h at 37°C. After incubation, the monolayers were washed 3 times with PBS, briefly fixed in cold ethanol (3 min), stained with Giemsa stain solution (1:20) (Merck, Darmstadt, Germany) and incubated for 30 min. The plates were washed, dried at room temperature overnight, examined microscopically (magnification, ×2500) with the immersion objective (IO) and photographed with a Contax camera (Company, City, Country) adapted for Zeiss (Axiolab 459306) microscope (Zeiss, City, Country).

For the quantitative assay of adhesion and invasion capacity, the infection step was performed in duplicates for each strain, and after

Table 1. The results of testing the adherence ability to HeLa cells of the enteropathogenic strains.

Enteropathogenic microbial strains	Adherence pattern	Adherence index (%)
EPEC	Aggregative (AggA)	40
S. enteritidis 361	Localized (LA)	5
S. flexneri 29833	Diffuse-aggregative	10
S. flexneri 29834	Diffuse (DA)	5

3 h incubation of the HeLa monolayer in the presence of microbial strains, the first well plates were washed four times in PBS, the cells were permeabilized by Triton X 1% (Sigma) and incubated for 5 min at 37°C for the release of intracellular invasive bacteria. Thereafter, serial ten-fold dilutions in saline solution were performed and 20 µl from each dilution was spotted in triplicates on solid media; in the second plate, after 2 h of incubation the monolayer was washed 4 times in PBS and 1 ml of 100 mg/ml gentamycin solution was added; the plates were further incubated for 1 h, in order to kill all adherent extra-cellular bacteria. Thereafter, the second plate was treated as the first one.

It is to be mentioned that in the case of lactic acid bacteria, the adherence capacity was investigated in three variants: 1) integral mid-logarithmic phase cultures; 2) microbial suspensions obtained from the washed sediment; and 3) heat inactivated microbial suspensions (30 min at 100°C).

In order to investigate how the probiotic strains influence the adherence and invasion of the HeLa cells by the pathogenic strains, the adherence and invasion assays were performed following the steps mentioned above, the monolayer infection being done in the presence of equal volumes of different LAB culture fractions (that is, integral cultures, washed sediments and heat-inactivated washed sediments).

Qualitative screening of the antimicrobial properties of the LAB culture supernatants

The qualitative screening was performed by an adapted disk diffusion method. Petri dishes with Mueller Hinton medium were seeded with bacterial inoculums as for the classical antibiotic susceptibility testing disk diffusion method (Kirby-Bauer); a 5 µL drop of the LAB culture supernatants were placed on the seeded medium, at 30 mm distance. The plates were left at room temperature for 20 - 30 min and then incubated at 37°C for 24 h. The positive results were read as the occurrence of an inhibition zone of microbial growth around the dried liquid spot (CLSI, 2008; Lazar et al., 2005).

Cytotoxicity of LAB strains

An approximate quantity of 5×10^3 HeLa cells were seeded in each well of a 96- well tissue culture plate, in DMEM supplemented with 10% foetal calf serum (FCS). After 24 h the pH adjusted culture supernatants and respectively heat-inactivated microbial suspensions obtained from washed sediments were added. The cytotoxicity effect was read using the Cell Titer 96 Aqueous One Solution Cell Proliferation Assay kit (Promega), according to manufacturer's indications (Balotescu et al., 2005).

Annexin-V immunostaining and flow cytometric analysis

HeLa cells were treated for 24 h with pH adjusted LAB supernatants in DMEM with 10% foetal bovine serum (1:1). Cells from the supernatant and monolayer were harvested and 5×10^5 cells were stained with annexin V and propidium iodide using the Immunotech Annexin V-FITC Kit and following the manufacturer's instructions (Beckman Coulter Company, France). Cells were analyzed by flow cytometry using a Coulter EPICS XL flow cytometer (Beckman Coulter). Green fluorescence (525 nm; FITC annexin V) and red fluorescence (613 nm; propidium iodide) were measured (Balotescu et al., 2005).

Immunomodulatory activity of LAB strains

In order to assess the potential immunomodulatory activity of the studied LAB strains, as well as correlate the obtained results with a certain fraction (that is, soluble fraction located in supernatants or cell-associated fraction located in the bacterial sediment) the levels of the main soluble pro- and anti-inflammatory cytokines, that is, IL-1, IL-2, IL-6, IL-8, IL-10, TNF-alpha and INF-gamma induced by different fractions of probiotic cultures were assessed by ELISA (Pierce Endogen kit) according to manufacturer's indications.

RESULTS

Adherence to the cellular substrate represented by HeLa cells and competition studies

Our results showed that the enteropathogenic tested strains exhibited different adherence abilities for colonizing the HeLa cells, as demonstrated by different adherence patterns and rates (Table 1).

It is to be mentioned that the adherence rates observed for these pathogenic strains are only apparently low and they are not reflecting a reduced ability to colonize the cellular substratum, but on the contrary, their ability to be rapidly internalized in the host cell following the initial adherence step. This hypothesis was confirmed by our quantitative adhesion and invasion assay studies, showing high invasion indexes for the selected enteropathogenic strains (data not shown).

Concerning the LAB strains, the ability to adhere to HeLa cells was different, depending on the analyzed fraction, that is, integral mid-logarithmic phase cultures, microbial suspensions obtained from the washed sediment as well as heat inactivated microbial suspensions (Table 2).

The live cells from the integral cultures generally exhibited higher adherence rates than the washed sediment and heat-inactivated cells.

However, all strains exhibited adherence abilities, with higher dherence rates for *L. paracasei* ssp. *paracasei*

Table 2. The results of testing the adherence ability to HeLa cells of different fractions of the LAB cultures.

Microbial strain	Adherence pattern			Adherence		
	Integral fraction	Washed sediment	Heat inactivated washed sediment	Integral fraction (%)	Washed sediment (%)	Heat inactivated washed sediment (%)
L. paracasei ssp. *paracasei CMGB 18*	Diffuse - aggregative	Diffuse-localized	Aggregative	100	100	100
L. paracasei ssp. *paracasei CMGB 19*	Aggregative	Localized - aggregative	Aggregative	50	70	80
L. paracasei ssp. *paracasei CMGB 20*	Diffuse - aggregative	Localized - aggregative	Aggregative	100	10	25
L. paracasei ssp. *paracasei CMGB 21*	Aggregative	Localized - aggregative	Localized	80	10	90
L. paracasei ssp. *paracasei CMGB 22*	Aggregative	Diffuse-aggregative	Diffuse-localized	50	50	50
L. paracasei ssp. *paracasei CMGB 23*	Aggregative	Diffuse	Diffuse	80	100	90
L. plantarum CMGB 24	Diffuse - localized	Diffuse	Diffuse	10	10	10
L. rhamnosus CMGB 29	Diffuse	Non-adherent	Non-adherent	5	0	0

(50 to 100%) than for *L. rhamnosus* and *L. plantarum* (5 to 10%) (Table 2).

Three distinct patterns of adherence have been investigated during this study: localized adherence (LA), in which bacteria attach to and form microcolonies in distinct regions of the surface; diffuse adherence (DA), in which bacteria adhere evenly to the whole cell surface, and aggregative adherence (AggA), in which aggregated bacteria attach to the cell in a stacked-brick arrangement.

The adherence index was expressed as the ratio between the numbers of the eukaryotic cells with adhered bacteria: 100 eukaryotic cells counted on the microscopic field. The adherence pattern of the *L. paracasei* whole cultures was predominantly aggregative, while for *L. plantarum* and *L. rhamnosus* a diffuse one (Figures 1-4). However, *L. paracasei* exhibited a significantly higher ability to colonize the cellular substrate compared with the other two tested species. *L. paracasei* exhibited the highest adherence for all tested fractions (100%) (Table 2). The adherence pattern was not significantly different for different fractions of the same strain. The slight changes in the adherence patterns consisted in the shift from aggregative (observed for the integral cultures) to a diffuse pattern (when other fractions were used) in case of *L. paracasei* strains and the loss of adherence ability in the case of *L. plantarum* strain (Table 2).

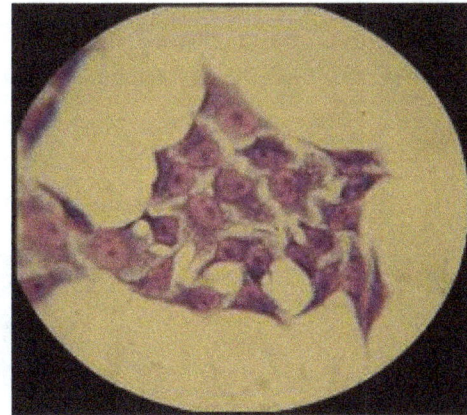

Figure 1. HeLa cells control (Giemsa staining, x2500).

Competition studies

All tested fractions of the selected probiotic strains, that is, whole bacterial culture, washed viable cells and heat-inactivated bacterial cell suspensions inhibited almost totally the adherence ability of the pathogenic strains to the cellular substrate (Figures 5 -10).

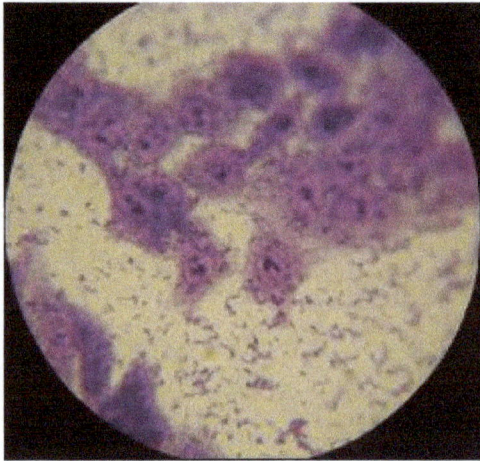

Figure 2. HeLa cells infected with *L. paracasei ssp paracasei* CMGB 18 (Giemsa staining, x2500).

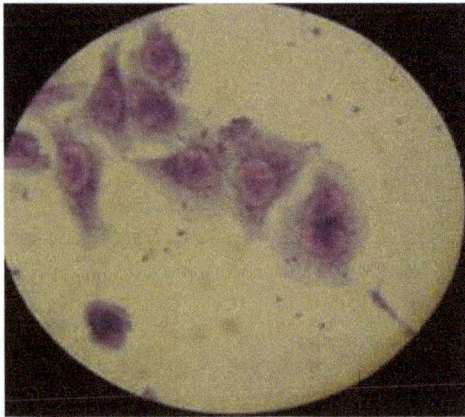

Figure 3. HeLa cells infected with *L. plantarum* CMGB 24 (Giemsa staining, x2500).

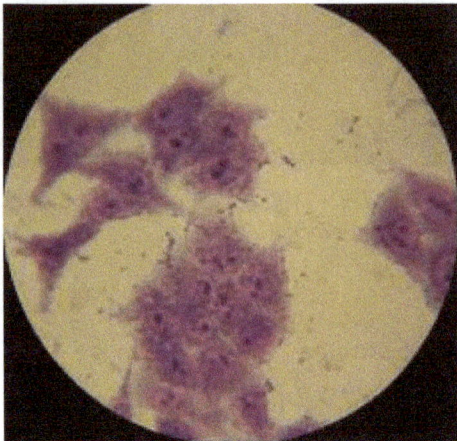

Figure 4. HeLa cells infected with *L. rhamnosus* CMGB 29 (Giemsa staining, x2500).

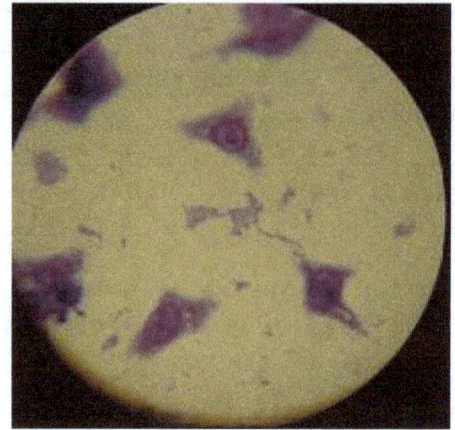

Figure 5. HeLa cells infected with *EPEC* (Giemsa staining, x2500).

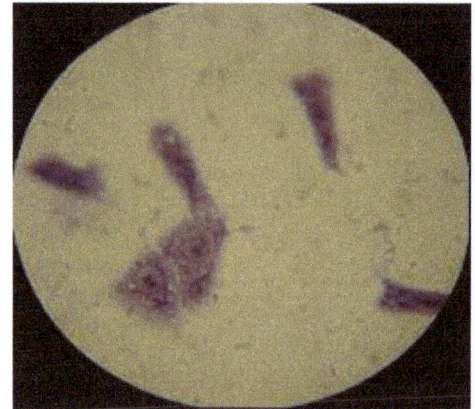

Figure 6. HeLa cells infected with *Salmonella enteritidis* (Giemsa staining, x2500).

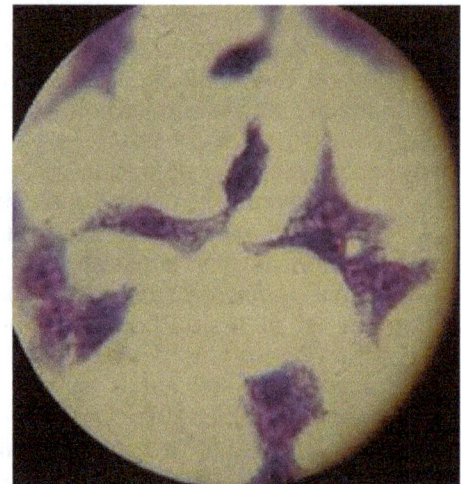

Figure 7. HeLa cells infected with *S. flexneri 29834* (Giemsa staining, x2500).

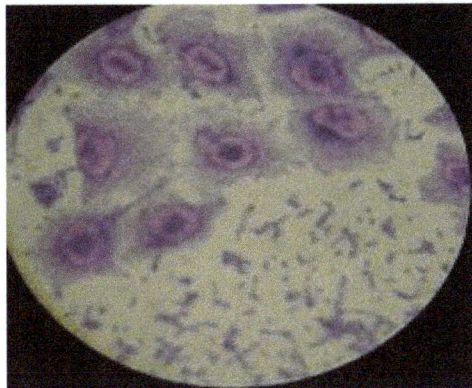

Figure 8. HeLa cells infected simultaneously with *EPEC* and whole culture of *L. paracasei ssp paracasei* CMGB 18 (Giemsa staining, x2500).

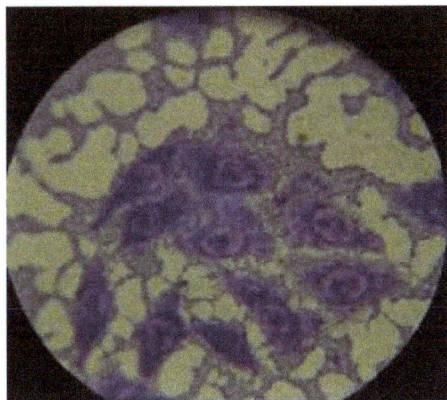

Figure 9. HeLa cells infected simultaneously with *Salmonella enteritidis* and washed viable cell suspension *L. paracasei* ssp *paracasei* CMGB 19 (Giemsa staining, x2500).

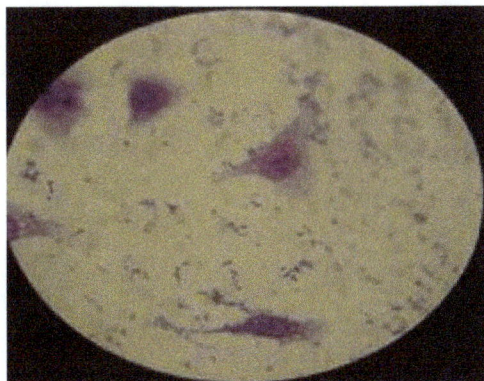

Figure 10. HeLa cells unfected simultaneously with *S. flexneri* 29834 and washed viable cell suspension of *L. paracasei* ssp *paracasei* CMGB 21 (Giemsa staining, x2500).

Antimicrobial activity of the LAB cultures supernatants

The tested strains did not exhibit any quantifiable antimicrobial activity, so they probably did not produce bacteriocins or other soluble factors with intrinsic antimicrobial activity.

Cytotoxicity

All tested supernatants with adjusted neutral pH and heat inactivated LAB washed cells suspension showed no cytotoxicity when tested by Cell Titer method. However, when assessed by Annexin V immunostaining, it was noticed that they are inducing the apoptosis of HeLa cells (Table 3).

Immunomodulatory activity of LAB supernatants and heat-inactivated LAB washed cells suspension

One of the purposes of this study was to investigate the influence of different probiotic fractions on the secretory pattern of the HeLa cells, by quantifying the most important pro - and anti-inflammatory cytokines.

Our results have demonstrated that the cellular fractions of tested lactic bacteria specifically modulate the expression of pro and anti-inflammatory cytokines in epithelial HeLa cells.

The level of the IL-1, the major pro-inflammatory cytokine and respectively IL-10 which have anti-inflammatory properties such as *in vitro* suppression of production of other pro-inflammatory cytokines could not be detected by our experimental model.

The general effect of bacterial cells and supernatants was the increase in TNF-alpha expression (Figure 11). The expression level of this cytokine was different, depending on the probiotic strain and fraction, the highest level being induced in the presence of two probiotic strains supernatants, that is, *L. plantarum* CMGB 24 and *L. paracasei paracasei* CMGB 19.

The increase of TNF-alpha level and the subsequent pro-inflammatory effect is balanced by the decreased levels of IL-6 (Figure 2) and IL-8 (Figure 13), also observed for all tested fractions, regardless of the analyzed strain.

Our results showed that the probiotic fractions are inhibiting the expression of IL-8 (interleukin with chemotactic properties) (Figure 12).

It should also be noted the positive correlation between increasing expression of IL-2, TNF-alpha and IFN gamma-induced by the cellular fractions, as well as by supernatants of three *L. paracasei* ssp. *paracasei* CMGB 18, 19, 23 strains, having in mind that IL-2 stimulates

Table 3. The effect of the probiotic strains supernatants on HeLa cells quantified by flow cytometry.

Crt. No.	Microbial strains	Viable cells (%)	Apoptosis	Advanced apoptosis	Necrosis
1.	HeLa cell control	99.20	0.37	0.34	0.10
2.	*L. paracasei* ssp. *paracasei CMGB 18*	83.36	12.93	3.47	0.23
3.	*L. paracasei* ssp. *paracasei CMGB 19*	84.34	10.02	2.89	2.76
4.	*L. paracasei* ssp. *paracasei CMGB 20*	83.46	11.02	4.72	0.20
5.	*L. paracasei* ssp. *paracasei CMGB 21*	81.34	13.52	4.76	0.23
6.	*L. paracasei* ssp. *paracasei CMGB 22*	89.19	7.29	0.76	2.76
7.	*L. rhamnosus CMGB 29*	84.10	12.80	3.06	0.07
8.	*L. paracasei* ssp. *paracasei CMGB 23*	84.19	11.52	3.96	0.33
9.	*L. plantarum CMGB 24*	84.55	9.10	5.85	0.50

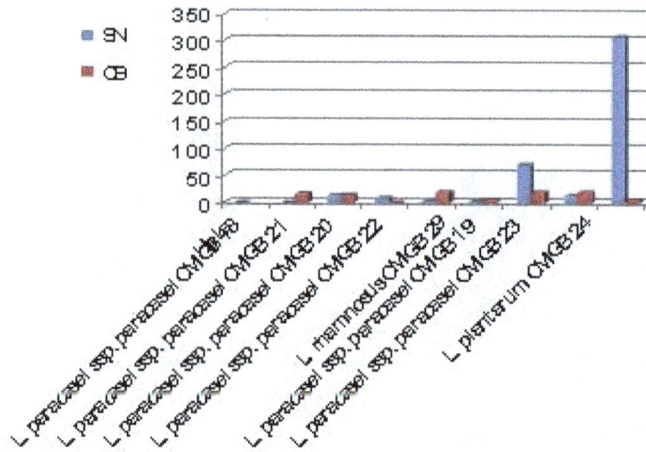

Figure 11. The comparative representation of TNF-alpha expression levels in HeLa in the presence of LAB supernatants (SN) and heat-inactivated LAB washed cells suspension (CB).

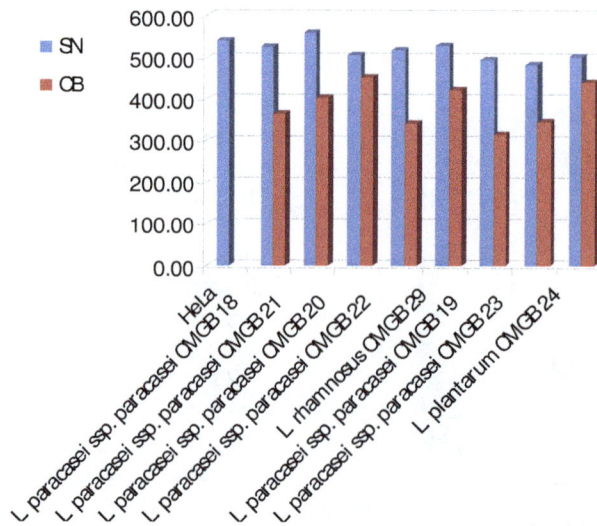

Figure 12. The comparative representation of IL-6 expression levels in HeLa in the presence of LAB supernatants (SN) and heat-inactivated LAB washed cells suspension (CB).

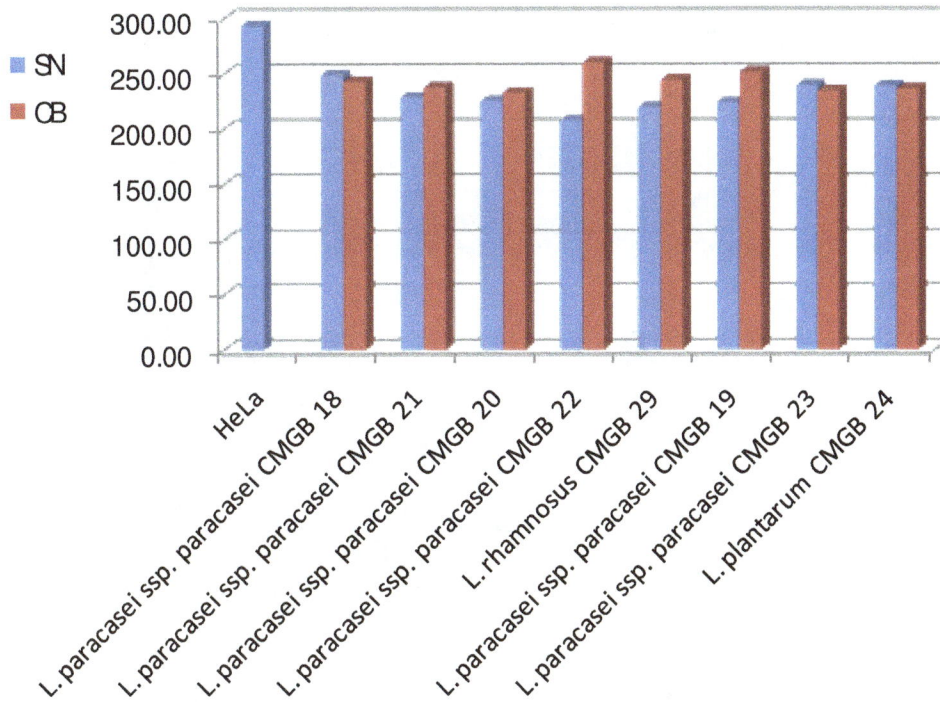

Figure 13. The comparative representation of IL-8 expression levels in HeLa in the presence of LAB supernatants (SN) and heat-inactivated LAB washed cells suspension (CB).

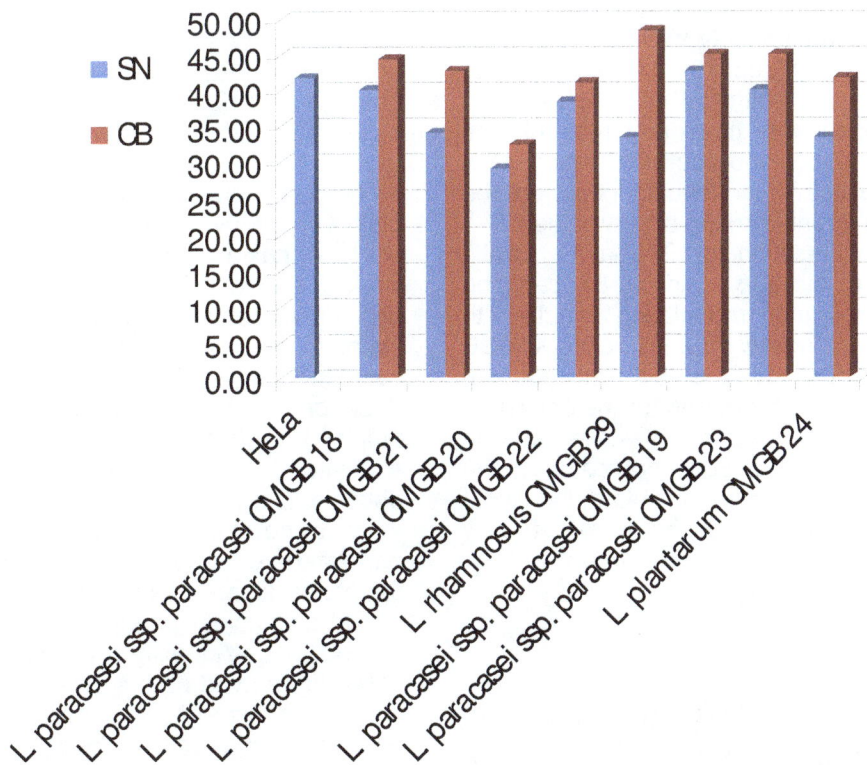

Figure 14. The comparative representation of IL-2 expression levels in HeLa in the presence of LAB supernatants (SN) and heat-inactivated LAB washed cells suspension (CB).

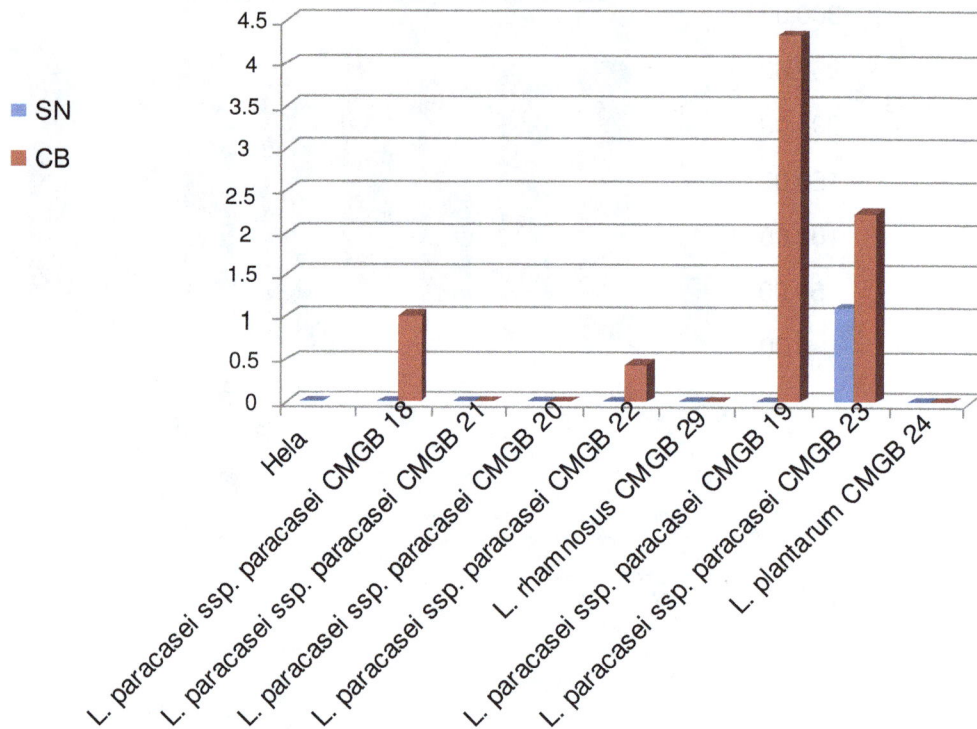

Figure 15. The comparative representation of IFN-gamma expression levels in HeLa in the presence of LAB supernatants (SN) and heat-inactivated LAB washed cells suspension (CB).

production of interferon gamma (IFN γ), IL-4, TNF by *T. lymphocytes* (Figures 11, 14 and 15). The induction of this secretory pattern in the epithelial cells by probiotic fractions might decrease the time required for the activation of *T. lymphocytes* and the occurrence of a specific immune response, which could speed up the elimination of infectious agent from the host organism.

In conclusion, the stimulation of TNF-alpha, while decreasing expression of IL-6 and IL-8, can be considered a beneficial immunomodulating effect. With special reference to *L. paracasei* ssp. *paracasei* CMGB 18, 19, 23 strains, the induction in the epithelial cells of a secretory pattern similar to that required for the activation of lymphocytes could be of significant advantage in probiotic administration in therapeutic purposes at the beginning of infection, in order to stimulate the rapid occurrence of a specific immune response mediated by *T. lymphocytes*.

DISCUSSION

The probiotic arsenal includes multiple mechanisms for preventing infection and enhancing the immune system, but each mechanism of action is strain-dependent (Majamaa and Isolauri, 1997; Link-Amster et al., 1994).

Our study is clearly demonstrating that one of the main mechanisms of probiotic antimicrobial activity is direct competition with the probiotic bacterial cells for the adhesion sites at the intestinal level triggering the pathogens exclusion.

The reduced cytotoxicity of the tested probiotic strains is representing an important advantage in the selection process of probiotic products to be administered in children. The cytotoxicity was assessed by flow cytometry using Annexin-V assay carried out in conjunction with vital dye Propidium Iodide (PI) staining in order to distinguish between apoptosis and necrosis, because PI staining can detect DNA that has leaked from the necrotic cell (Liu et al., 2003). Annexin V is a 35-to 36-kD calcium-dependent, phospholipid binding protein that has an affinity for phosphatidyl serine and bind to cells with exposed phosphatidyl serine. Because the membrane phospholipid phosphatidyl serine is translocated from the inner to the outer leaflet of the plasma membrane at the early stage of apoptosis, an Annexin-V assay can identify apoptosis at an earlier stage than the TUNNEL assays based on nuclear changes such as DNA fragmentation. In additions, Annexin-V is conjugated to fluorochrome FITC that served as a sensitive probe for flow cytometric analysis in this Annexin-V assay.

The first stage of infectious process is the adherence of pathogenic microorganisms to epithelial cells or extracellular matrix molecules, which will lead to the colonization at the entrance gate.

Host colonization involves complex cell to cell

communication systems, both between bacterial cells, and cross communication between bacterial and host cells. One of the aspects of this cell to cell communication is the modulation of the epithelial cells secretory pattern (Wilson et al., 1998).

It is to be noticed that the eukaryotic cells are never exposed to a single cytokine and different cytokines could act synergistically or be antagonist; their combined action could result in a certain behavior of the target cells, having as consequence the amplification of inhibition of different biological activities.

Interaction between bacterial pathogens and host cells invariably results in the release of one or more cytokines, the actual cytokines produced depending mainly on the nature of the bacterium and host cells involved. Several bacterial pathogens have the capacity to alter host cell cytokine synthesis, degrade pro-inflammatory cytokines, or use cytokine receptors as portals of entry for cellular invasion (Link-Amster et al., 1994). Manipulation of the cytokine networks to the advantage of the invading pathogen offers a further example of the importance of pro-inflammatory cytokines in the protection against microbial invasion. The resulting cytokine network, of course, constitutes an important part of the innate immune response and represents the host's attempt to deal with that particular organism (Eckmann et al., 1993). As such, therefore, the ability of probiotic bacterial components (supernatants or bacterial cells) to modulate the cytokine release from host cells can be regarded as an aspect of improving the host response to pathogenic bacteria aggression and to decrease the intensity and/or chronicity of infectious diseases pathology.

All tested strains increased the level of tumor necrosis factor (TNF) expression. It is well known that TNF acts synergistically and have similar biological activities with IL-1, which is the major mediator of the acute phase response, inducing the production of other cytokines during infection, the expression of adhesion molecules and chemokines secretion by endothelial cells (Jung et al., 1995).

Regarding IL-6, one of the main *in vivo* roles of this cytokine is to initiate the acute phase response by inducing the production of acute phase proteins by hepatocytes (Barton, 1997; Barton et al., 1996). IL-6 is detected in blood of patients with inflammation, thus the decreasing of IL-6 expression by probiotics, could prevent the occurrence of an intense inflammatory response with lesional effects on host.

A wide variety of cells produce IL-8 as a response to IL-1β, TNF, LPS, microbial adhesins, radiation, infection, other cytokines and retinoic acid. The decrease induced by probiotics in Il-8 level could be considered as a by probiotics in Il-8 level could be considered as a positive effect, having in view that increased levels of the IL-8 were observed in case of gastro-intestinal infections with invasive bacteria such as *Shigella* spp. or *Salmonella* spp. (Eckmann et al., 1993; Schulte et al., 1996).

Taken together, the immunomodulatory studies allowed us to state that the most appropriate immunomodulatory activity was exhibited by three strains belonging to the same species, that is, *L. paracasei* ssp. *paracasei* (Malin et al., 1996; Majamaa and Isolauri, 1997; De Simone et al., 1992), the strains modulating the expression of the most important cytokines in the development of the anti-infectious immunity against enteric pathogens. The immunomodulation findings are also supported by the good results exhibited by these strains in the competition studies and also by their low cytotoxicity on HeLa cells, this last aspect being of great importance having in view that we intend to develop a probiotic product for the infant use.

Conclusion

Our results showed that the interactions between epithelial cells and probiotic bacteria are complex and involve complex cross-talk communication mechanisms involving both the physical interaction between bacterial and epithelial cells that is, adhesions and complementary receptors and the modulation of pro-inflammatory and anti-inflammatory molecules. Our *in vitro* studies are demonstrating that the selected probiotic strains are antagonizing the adherence and colonization of HeLa cells by the enteropathogenic strains isolated from pediatric diarrhoea mainly by direct competition for the adherence sites, demonstrating their potential use in the treatment of pediatric gastro-intestinal disorders, as an alternative or in association with antibiotics. A great advantage of the selected probiotic strains is their low cytotoxicity and ability to induce a beneficial cytokine response in the epithelial cells, which potentiates their antimicrobial activity by stimulating the rapid immune response following the intestinal injury. Different probiotic fractions have shown specific features, demonstrating the utility of this approach for the selection of the most appropriate LAB culture fraction, in order to maximize their beneficial, antimicrobial and immunomodulatory activities and to reduce the side effects on the host.

REFERENCES

Alvarez-Olmos MI, Oberhelman RA (2001). Probiotic agents and infectious diseases: a modern perspective on a traditional therapy. Clin. Infect. Dis., 32: 1567-1576.

Barton BE (1997). IL-6: insights into novel biological activities. Clin. Immunol. Immunopathol., 85: 16-20.

Barton BE, Shortall J, Jackson JV (1996). Interleukins 6 and 11 protect mice from mortality in a staphylococcal enterotoxin-induced toxic shock model. Infect. Immun., 64: 714-718.

Balotescu C, Oprea E, Petrache LM, Bleotu C, Lazar V (2005). Antibacterial, antifungal and cytotoxic activity of Salvia officinalis essential oil and tinctures. RBL., 10: 2471-2479.

Borriello SP, Hammes WP, Holzapfel W (2003). Safety of probiotics that contain lactobacilli or bifidobacteria. Clin. Infect. Dis., 36: 775-780.

Bradley C, Johnston AL (2006). Supina and Sunita Vohra. Probiotics for pediatric antibiotic-associated diarrhoea: A meta-analysis of randomized placebo-controlled trials. CMAJ., 175: 4.

Clinical and Laboratory Standards Institute (CLSI) (2008). Performance Standards for Antimicrobial Susceptibility Tests; Eighteen informational supplement M100-S18.

De Simone C, Ciardi A, Grassi A (1992). Effect of *Bifidobacterium bifidum* and *Lactobacillus acidophilus* on gut mucosa and peripheral blood B lymphocytes. Immunopharmacol. Immunotoxicol., 14: 331–40.

Eckmann L, Kagnoff MF, Fierer J (1993). Epithelial cells secrete the chemokine interleukin-8 in response to bacterial entry. Infect. Immun., 61: 4569-4574.

Gorbach SL (2000). Probiotics and gastrointestinal health. Am. J. Gastroenterol., 95: S2-S4.

Havenaar R, Huis in 't Veld JHJ (1992). Probiotics: a general view. In: Wood, BJB, editor. Lactic acid bacteria in health and disease. of the series, The lactic acid bacteria. Amsterdam (Elsevier Applied Science Publishers).

Ishibashi N, Shoji Y (2001). Probiotics and safety. Am. J. Clin. Nutr., 73: 465S-470S.

Jack RW, Tagg JR, Ray B (1995). Bacteriocins of gram-positive bacteria. Microbiol. Rev., 59: 171–200.

Jung HC, Eckmann L, Yang SK, Panja A, Fierer J, Morzyckawroblewska E, Kagnoff MF (1995). A distinct array of pro-inflammatory cytokines is expressed in human colon epithelial cells in response to bacterial invasion. J. Clin. Invest., 95: 55-65.

Kalliomaki M, Salminen S, Arvilommi H, Kero P, Koskinen P, Isolauri E (2001). Probiotics in primary prevention of atopic disease: a randomized placebo-controlled trial. Lancet. 357: 1076-1079.

Lazar V, Balotescu MC, Smarandache D, Vassu T, Sasarman E, Petrache LM, Orasanu M, Cernat R (2004). *In vitro* study of the interference of some Enterococcus strains with the adhesion of human and poultry enteropathogens to HeLa cells. Roum. Biotech. Lett., 9(3): 1675-1681.

Lazar V, Balotescu MC, Moldovan L, Alexandru V, Ditu LM, Bulai D, Cernat R (2005). Comparative evaluation of qualitative and quantitative methods used in the study of antifungal and antibacterial activity of hydroalcoholic vegetal extracts. RBL., 9(6): 2225-2232.

Link-Amster H, Rochat F, Saudan KY, Mignot O, Aeschlimann JM (1994). Modulation of a specific humoral immune response and changes in intestinal flora mediated through fermented milk intake. FEMS Immunol. Med. Microbiol., 10: 55–64.

Liu J, Yagi T, Sadamori H, Matsukawa H, Sun DS, Mitsuoka N, Yamamura M, Matsuoka J, Jin Z, Yamamota I, Tanka N (2003). Annexin V assay-proven anti-apoptotic effect of ascorbic acid 2-glucoside after cold ischemia/reperfusion injury in rat liver transplantation. Acta Medica Okayama, 57: 5.

Mackay AD, Taylor MB, Kibbler CC, Hamilton-Miller JM (1999). *Lactobacillus* endocarditis caused by a probiotic organism. Clin. Microbiol. Infect., 5: 290–292.

Majamaa M, Isolauri E (1997). Probiotics: a novel approach in the management of food allergy. J. Allergy Clin. Immunol. 99: 179–185.

Malin M, Suomalainen H, Saxelin M, Isolauri E (1996). Promotion of IgA immune response in patients with Crohn's disease by oral bacteriotherapy with *Lactobacillus* GG. Ann. Nutr. Metab., 40: 137–145.

Perdigon G, Alvarez S, Rachid M, Agüero G, Gobbato N (1995). Immune stimulation by probiotics. J. Dairy Sci., 78: 1597–606.

Rautio M, Jousimies-Somer H, Kauma H (1999). Liver abscess due to a *Lactobacillus rhamnosus* strain indistinguishable from *L. rhamnosus* strain GG. Clin. Infect. Dis., 28: 1159 –1160.

Schiffrin EJ, Brassart D, Servin AL, Rochat F, Donnet-Hughes A (1997). Immune modulation of blood leukocytes in humans by lactic acid bacteria: criteria for strain selection. Am. J. Clin. Nutr., 66: 515S-520S.

Schiffrin EJ, Rochat F, Link-Amster H, Aeschlimann J., Donnet-Hughes A (1995). Immunomodulation of human blood cells following the ingestion of lactic acid bacteria. J. Dairy Sci., 78: 491–497.

Schulte R, Wattiau P, Hartland EL, Robins-Browne RM, Cornelis GR (1996). Differential secretion of interleukin-8 by human epithelial cell lines upon entry of virulent or nonvirulent *Yersinia enterocolitica*. Infect Immun., 64: 2106-2113.

Salminen S, Arvilommi H (2001). Probiotics demonstrated efficacy in clinical settings. Clin.Infect.Dis., 32: 1577-1578.

Smarandache D, Lazar V, Balotescu MC, Vassu T, Ghindea R, Sasarman E, Petrache LM, Orasanu M, Cernat R (2004). Characterization of adhesion properties to the cellular substratum of some *Enterococcus strains* selected for potential use in probiotic products or food products. RBL.; 9(3): 1669-1674.

Spanhaak S, Havenaar R, Schaafsma G (1998). The effect of consumption of milk fermented by *Lactobacillus casei* strain Shirota on the intestinal microflora and immune parameters in humans. Eur. J. Clin. Nutr., 52 :1–9.

Szajewska H, Mrukowicz J (2001). Probiotics in the treatment and prevention of acute infectious diarrhoea in infants and children: a systematic review of published randomized, double-blind, placebo-controlled trials. J. Pediatr. Gastroenterol.Nutr., 33: S17-S25.

Van Neil CW, Feudtner C, Garrison MM (2002). *Lactobacillus* therapy for acute infectious diarrhoea in children: a meta-analysis. Pediatrics, 109: 678-684.

Vanderhoof JA, Young RJ, (2002). Probiotics in pediatrics. Pediatrics., 109: 956-958.

Wilson M, Seymour R, Henderson B (1998). Bacterial perturbation of cytokine networks. Infect. Immun., 66: 2401-2409.

4

Use of SSR markers for genetic diversity studies in mulberry accessions grown in Kenya

Nderitu Peris Wangari, Kinyua Miriam Gacheri, Mutui Mwendwa Theophilus and Ngode Lucas

Moi University, P.O. Box 1125-30100, Eldoret, Kenya.

The knowledge and understanding of the extent of genetic variation of mulberry germplasm is important for conservation and improvement. The objective of this study was to analyze the genetic diversity between and within two mulberry species widely grown in Kenya which include *Morus alba and Morus indica.* Five individuals per species were genotyped with 13 simple sequence repeat (SSR) markers. The SSR markers presented a high level of polymorphism and detected a total of 35 polymorphic bands and 74.47% polymorphic loci. The mean observed heterozygosity per primer was 0.3670 suggesting a high degree of variation. The analysis of molecular variance (AMOVA) revealed that 95% of the variation was found within the species and only 5% between the species. Principal coordinates analyses (PC$_O$A) clearly distinguished three groups. It was evident from this study that the mulberry accessions did not cluster on the basis of their geographical origin, and neither according to the group of species they fall into. The study showed a close relationship between the two species and therefore mulberry improvement should target sampling more individuals within species rather than among species.

Key words: Mulberry, simple sequence repeat markers, genetic diversity.

INTRODUCTION

Genetic diversity assessment has potential uses in evolution, breeding and conservation of genetic resources (Wu et al., 1999). The extent of diversity in the accessions is paramount for improvement and utilization of genetic resources. Genetic diversity is therefore the backbone of conservation of plant genetic resources for both present and future use (Quedraogo, 2001). Genetic diversity of species also helps formulate appropriate sampling strategies for both *in situ* and *ex situ* conservations, with the aim to protect the variability of taxa so as to preserve ecological processes, rate of establishment, survival and fecundity (Millar et al., 2000).

Information on the levels and distribution of genetic variation in the natural populations of mulberry is of great importance. Mulberry bears different sex types, is monoecious or dioecious, with sex expression varying among species and varieties due to their cross fertilization (Das

et al., 1994; Dandin, 1998). Such a high degree of cross species reproductive success is not encountered often in nature, thereby creating variation within and among mulberry populations (Awasthi et al., 2004).

Advancement of molecular biology techniques in the last two decades has aided in the concise classification of individual plants and species. Molecular markers have been used in screening of germplasm, genetic diversity, identifying redundancies in collections, testing accession stability and integrity as well as resolving taxonomic relationships (Kameswara, 2004). SSRs are advantageous over many other markers as they are highly polymorphic, robust, can be automated; only very small DNA is required, highly abundant, analytically simple, readily transferable and have a co-dominant inheritance (Matsuoka et al., 2002).

The genetic structure of mulberry species has been

studied extensively (Tikader et al., 2009). These studies have revealed a high genetic divergence among mulberry accessions. For instance, Awasthi et al. (2004) and Vijayan et al. (2004, 2005, 2006) examined the genetic structure of indigenous mulberry accessions in India, and found a great degree of polymorphism amongst them. Similarly, Bhattacharya and Ranade (2001) and Sharma et al. (2000) also found a wide variation in genetic distance among different *Morus* genotypes in India and Japan, respectively.

Genetic diversity provides plant breeders with options to develop productive crops that are resistant to virulent pests and diseases and adapted to changing environments. However, a significant proportion of mulberry accessions in Kenya are from other countries such as India, China, Thailand, etc, that have entirely different geographical and ecological features, therefore, when these accessions are subjected to an environment totally different from their natural habitat, it is expectable that they may undergo genetic changes for adaptation (Tikader et al., 2009).

The ability to identify genetic variation is indispensable to effective management and use of genetic resources. This work, though very essential, has however not been undertaken in Kenya. It was therefore prudent to subject the mulberry accessions to genetic characterization using molecular markers to understand the extent of available and to design strategies to eliminate duplicates and nearly similar accessions in the national gene bank.

MATERIALS AND METHODS

Plant materials

Ten mulberry accessions which included Thailand, Thika, Embu, Ithanga, Limuru (*Morus alba* L.), S41, S13, S54, S36 and Kanva-2 (*Morus indica* L.) were obtained from The International Centre of Insect Physiology and Ecology (ICIPE) mulberry germplasm site at Nairobi headquarters' and Kenya Agricultural Research Institute (KARI) at Thika.

DNA extraction

Genomic DNA was extracted from the young fresh leaf samples using cetyl trimethyl ammonium bromide (CTAB) procedure described by FAO/IAEA (2007). The DNA quantity and quality was visually quantified using the agarose gel electrophoresis method as described by Manniatis et al. (1982). DNA samples were stored at -20°C.

PCR amplification reaction for the simple sequence repeat (SSR) primers

Fifteen microsatellite primers used to amplify the genomic DNAs extracted from the ten mulberry accessions were of two sets: SS and Mulstr primers obtained from Invitrogen, United Kingdom. SS primers amplification was performed in a final volume of 15 µl containing 10 mM Tris-HCl (Ph 8.4), 50 mM KCl, 1.5 mM MgCl$_2$, 0.6 µl each primer, 10 ng of genomic DNA, 1 U *Taq* polymerase and molecular biology water. This was run using a touchdown protocol. The amplification cycle included a 3 min denaturation at 95°C, fol-

followed by 16 cycles of denaturation at 94°C for 30 s, 30 s primer annealing at 63°C and decreasing the annealing temperature by 0.5°C at each succeeding cycle, and 1 min elongation at 72°C. Then, 24 cycles of 30 s at 94°C, 1 min at 56°C and 1 min at 72°C were conducted, followed by 5 min at 72°C. The Mulstr primer amplification was carried out in 15 µl reactions containing 10 ng template DNA, 2 µl each primer and 1U *Taq* polymerase on a thermocycler programmed as follows: initiatial denaturation at 94°C for 10 min, followed by 35 cycles of 94°C for 1 min denaturation, primer specific annealing temperature for 1 min and 72°C for 1.5 min extension and a final extension step at 72°C for 5 min and a cooling step at 4°C. The amplified SSR products were electrophoresed in 2% agarose gel in 0.5 XTBE buffer at 150 V for 2 h, stained in ethidium bromide for 30 min, and photographed on a UV transilluminator at 312 nm.

Data analysis

Scoring of bands was done visually. Fragments were scored as present (1) or absent (0) for each of the SSR markers. Nei's (1973) gene diversity, Shannon information index (Shannon and Weaver, 1949), number of alleles (Na), expected heterozygosity (H$_E$), observed heterozygosity (Ho), genetic similarity and genetic distance estimated by Nei's coefficient between pairs were analyzed using popgene software version 3.5 (Yeh et al., 1999). The data matrix was then subjected to analysis of molecular variance (AMOVA) to partition the genetic variation into within and among the populations' components using GenAlEx software (Peakall and Smouse, 2006). Principal coordinate analysis (PCoA) and scores for the first and second components were also plotted using GenALEx software.

RESULTS

Genetic diversity among accessions

From a total of fifteen SSR primers used to investigate ten mulberry accessions, only thirteen produced clear amplification products and polymorphisms. The thirteen SSR primers generated a total of 35 polymorphic bands. The highest number of bands was 15 which were generated by SS09 primer while the lowest number was 2 bands generated by SS19, SS20, M2, M6 and M5 primers. A total of 74.47% polymorphic loci were generated by the 13 primers studied. The bands produced by these primers varied in size from 110 to 420 bp. The observed number of alleles per locus in all mulberry accessions ranged from 2 to 15 with a mean of 6.3. The observed heterozygosity per primer ranged from 0.0500 to 0.7000 with an average of 0.3670 suggesting a high degree of variation. Expected heterozygosity per primer ranged from 0.0473 to 0.5053 with an average of 0.3610 (Table 1).

The mean observed heterozygosity per accession varied from 0.15 to 0.4595, hence, suggesting a high degree of variation among the accessions. The Shannon information index showed a high diversity across the accessions ranging from 0.1040 in S36 accession to 0.3185 in Thailand accession, with a mean of 0.4399. In all the accessions, Thailand accession was the most highly diverse according to the observed heterozygosity and Shannon information index.

Table 1. Band sizes, observed number of alleles (Na), observed (Ho) and expected (He) heterozygosity for thirteen microsatellite markers assayed across ten mulberry accessions.

Primer	Repeat motif	Primer sequence (Aggarwal et al., 2004; Zhao et al., 2005)	Band size (bp)	Na	Ho	He
SS05	$(CA)_5CC(CA)_{27}$	F:TCCAGCAAAGATGTGACAAAAGTT R:TTGCCTTCCCGATTATGCTG	320-400	5	0.05	0.0473
SS02	$(CA)_{49}$	F:GCTTCGATCAATCTAGCTTCCC R:GCAAACTACGCCACCCCG	310-420	12	0.2973	0.3318
SS04	$(TG)_{27}$	F:CGAGGGAGGGATGAGGAGC R:CACATTCATCCACCCTCCTATA	190-250	6	0.5444	0.445
SS17	$(CA)_{26}$	F:TACAGGGCTCGGGCAAATG R:TGATCCGAAGCTTGGGGTCT	220-300	9	0.2674	0.3643
SS06	$(TG)_{18}$	F:ACTCAAAATGAAGGAAAAGGAATTATAC R:TTTACTTAAATCCCAGCCACA	180-220	10	0.545	0.4845
SS19	$(TG)_{18}$	F:TTCTGTCGTGTCCTCCGTCAA R:TGAGAACATACACTAATAGGTGAAAAC	300-370	2	0.4000	0.3368
SS09	$(CA)_{56}$	F:AGAACCCTTCCGCCCTATG R:CCTTGGCGTAGGCAAAGTTG	200-290	15	0.2125	0.4992
SS20	$(CA)_{14}ACAA(CA)_{12}$	F:CCCTTTCATCGCCTCCTCC R:CTCTGCCATTCAGTAGCGG	300-350	2	0.500	0.3947
SS18	$(CA)_{27}$	F:TCTTCGCCCGTTGTTCGC R:AGCAATTTTCTTCAACTCACCTTCT	180-210	10	0.1805	0.2916
M2	$(GTT)_{11}$	F:CGTGGGCTTAGGCTGAGTAGAGG R:CACCACCACTACTTCTCTTCTTCCAG	190-210	2	0.4000	0.5053
M6	$(GT)_{15}$	F:TCCTTAGGTTTTTGGGGTCTGTTTACAT R:CCTCATTCTCCTTTCACTTATTGTTG	110-190	2	0.5000	0.3947
M4	$(GAA)_6$	F:GGTCAAGCGCTCCAGAGAAAAG R:GGTGCAGAGGATGAAAGATGAGGT	140-160	5	0.175	0.1197
M5	$(CCA)_8$	F:CCCCCTGCAATGCCCTCTTTC R:TGGGCGAGGCAGGGAAGATTC	160-180	2	0.7000	0.4789

Genetic diversity among species

The number of polymorphic loci generated by the SSR primers per single species varied from 29 in *M. alba* to 21 in *M. indica*. The percentage of polymorphic loci in *M. alba* was 61.7% while that of *M. indica* was 44.68%. The observed heterozygosity varied from 0.2415 to 0.3404 while Shannon information index ranged from 0.3051 to 0.3905 in *M. indica* and *M. alba*, respectively. Therefore these showed little variation between the two species studied.

Genetic similarity among and within species

The genetic similarity among and within species was estimated on the basis of analysis of molecular variance (AMOVA), genetic identity (I) and distance (D), UPGMA dendrogram and Principal coordinates analysis (PCoA).

Molecular variance among and within species (AMOVA)

Of the total species diversity, 5% was among the species, while 95% was within the species (Table 2).

Nei's (1978) genetic identity (I) and distance (D) among and within the two species of mulberry.

The genetic similarity coefficient among the accessions estimated on the basis of Nei's (1978) unbiased, varied from 0.8455 between Limuru and Ithanga accessions to 0.4077 between Limuru and Thailand accessions with an average genetic similarity of 0.5362. Where I = 0.0 shows no common alleles while I = 1.0 shows equal gene frequencies (Appendix 3c). Therefore, Limuru and Ithanga accessions shared most of the alleles while Limuru and Thailand were more diverse. The maximum genetic distance (0.8972) observed between any two accessions was that between the Limuru and Thailand accessions while the smallest distance (0.1678) was between Ithanga and Limuru accessions. The mean genetic distance was 0.5916. At species level, the genetic similarity coefficient was 0.7613 while their genetic distance was 0.2727 (Table 3).

Molecular analysis using the SSR data matrix to cluster the ten accessions and show genetic variation using Principal coordinate's analysis (PCoA) is shown in Figure 2. The PCoA coordinates one and two extracted 35.45

Table 2. Analysis of molecular variance (AMOVA) for the two species of mulberry (*Morus* sp.) based on SSR markers.

Source of variation	d.f.	SS	MS	Variance component	Total variation (%)	p-value*
Among species	1	43.200	43.200	1.750	5	<0.001
Within species	8	275.600	34.450	34.450	95	<0.001
Total	9	318.800	77.650	36.200	100	

*After 999 random permutations.

Table 3. Nei's (1978) unbiased measure of genetic identity (I) (above diagonal) and genetic distance (D) (below diagonal) for the two mulberry species studied.

	M. alba	*M. indica*
M. alba	****	0.7613
M. indica	0.2727	****

and 17.49% of the total variation, respectively.

DISCUSSION

The observed number of alleles (*Na*) per locus in the mulberry accessions studied ranged from 2 to 15 with an average of 6.3 (Figure 1). This mean was less than 18.6 reported by Aggarwal et al. (2004) but larger than 5.13 reported by Zhao et al. (2007) using SSR markers. The mean observed heterozygosity (Ho) of 0.3670 in this study was lower than 0.4296 reported in 27 mulberry genotypes by Zhao et al. (2005). Zhao et al. (2007) further reported a higher value of 0.4912 within cultivated species of mulberry for the same. Aggarwal et al. (2004) also reported a higher Ho value of 0.59 across 45 mulberry genotypes from species of diverse origin using SSR markers. The highest mean observed heterozygosity across the accessions was observed in Thailand accession with 0.4595 while the least was 0.1500 in S36 accession. The mean observed heterozygosity among the species varied from 0.3404 to 0.2415 in *M. alba* and *M. indica* species, respectively. These findings showed a high genetic variability among the accession than among species. This implies, there is a high genetic variability among mulberry accession grown in Kenya. In addition, SSR markers gave a mean Shannon information index of 0.4399 for these mulberry accessions, a value higher than 0.3200, 0.3000 and 0.2900 reported by Bhattacharya et al. (2005) using DAMN, ISSR and RAPD markers, respectively in both exotic and indigenous mulberry varieties. However, a higher value of 0.7399 was detected in cultivated species of mulberry using SSR markers (Zhao et al., 2007). Thailand accession with a Shannon index value of 0.3185 was the most varied while S36 with a value of 0.1040 was least varied.

The Shannon information index for both *M. alba* and *M. indica* species was 0.3905 and 0.3051, respectively.

Vijayan et al. (2004a) reported smaller values of 0.21 ± 0.27 and 0.25 ± 0.29 in *M. alba* and *M. indica* species using ISSR markers, respectively. In contrast, using RAPD markers, Vijayan (2004b) reported smaller values of 0.19 ± 0.27 and 0.17 ± 0.26 in *M. alba* and *M. indica* species; hence indicating a high diversity in the mulberry accessions and species.

The percent polymorphic loci revealed an average of 74.47% in all the mulberry accessions studied using SSR markers. This value was lower than 85 and 91% shown in RAPD and DAMD markers by Bhattacharya and Ranade (2001). Earlier, Sharma et al. (2000) had reported a higher average value of 81.2%. Additionally, a higher value of 86% was reported by Vijayan et al. (2004a) and Zhao et al. (2007). These values suggest a high level of genetic variability among the accessions studied.

Such high level of genetic diversity can be attributed to the mode of reproduction of these accessions. Vegetatively propagated species such as mulberry often have minute unobserved changes in passage of time and accumulation of these changes can lead to significant changes in plant population. Such variability could come from mutation (Vijayan, 2009) and/or long cultivation periods (Bhattacharya and Ranade, 2001). Mulberry is a highly heterozygous and outcrossing species and therefore it is expectable that its accessions exhibit a high level of polymorphism (Bhattacharya and Ranade, 2001; Awasthi et al., 2004).

Molecular variance between and within the mulberry species was high at 95% within the species (*M. alba* and *M. indica*) and low at 5% between the species. These findings are significant in outcrossing woody crops as mulberry. This is supported by Hamrick and Godt (1996) who reported a high genetic variation within populations than between populations of outcrossing woody plants. Similar results were reported by Alvarez et al. (2001) on tomato using microsatellite analysis. Steiger et al. (2002) also demonstrated a high genetic diversity within the cultivar catimor of *Coffea arabica* using AFLP markers. Earlier, Erstad (1996) studying vegetative and generative characters in ecotypes of woody spp. of *Ribes rubrum* found greater variation within populations than between populations and Heaton et al. (1999) too found comparable variabilities in a RAPD study of chicozapote tree (*Manilkara zapota*), thus confirming our results.

High level of genetic variability within populations such as those detected in this study, are attributed to mating

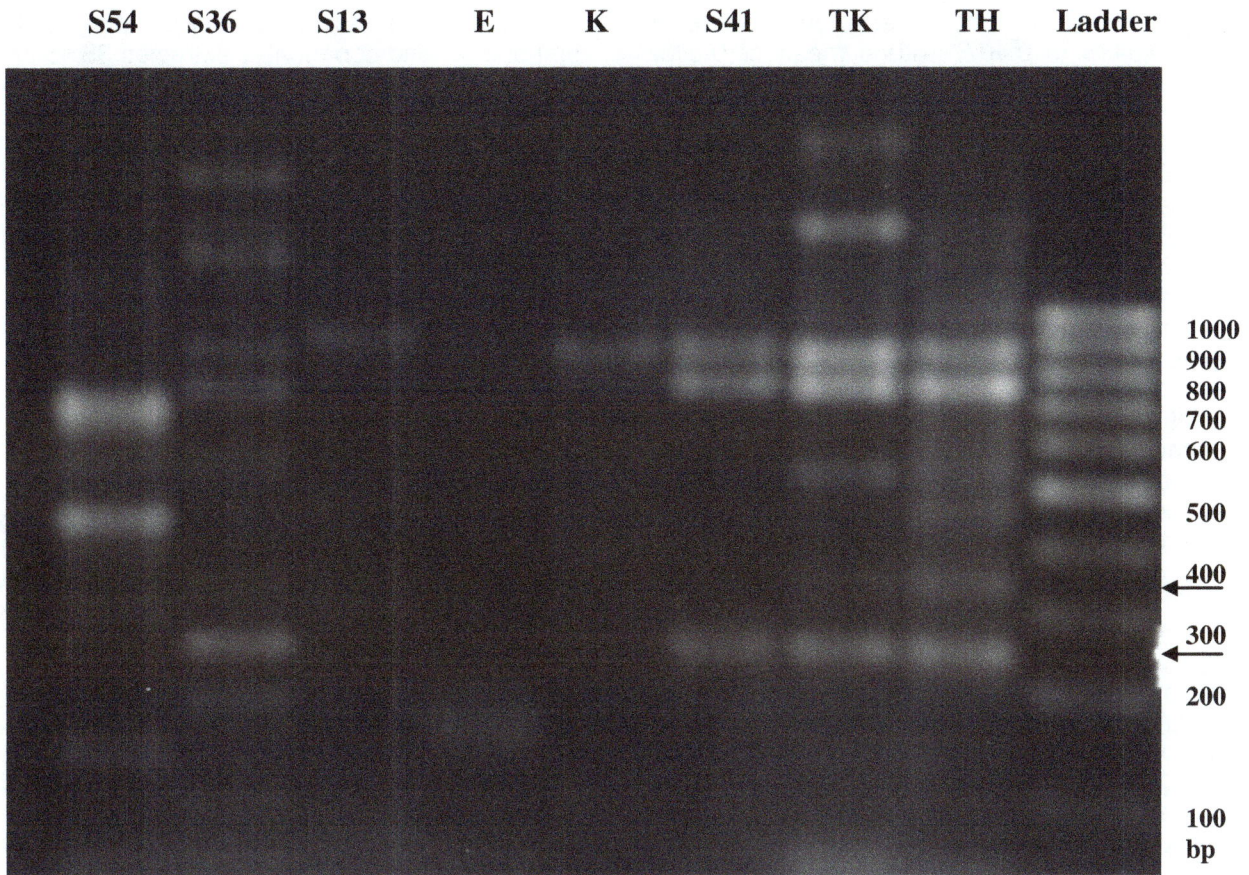

Figure 1. SSR electrophoresis profiles of 8 mulberry accession abbreviated at the top of the lanes, amplified using the primer SS02. The ladder is a DNA marker (100 bp). Arrows indicate some of the polymorphic bands.

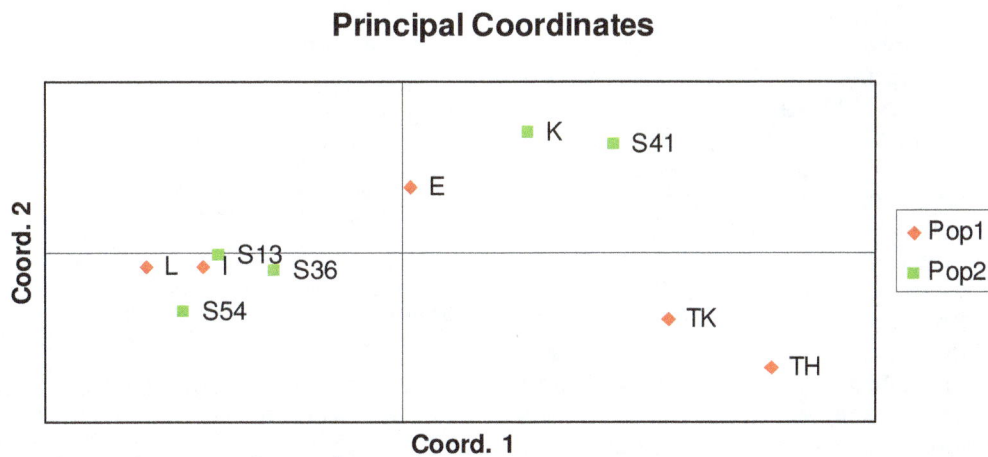

Figure 2. Scatter plot of 10 mulberry accessions based on first and second components of principal coordinate analysis using SSR data. Pop1 = *M. alba*, Pop 2 = *M. indica*.

systems. Cross pollinating species have a higher gene diversity as compared to self-pollinating species (Alvarez et al., 2001). This kind of species exhibit large gene pools, possibly as a result of widespread pollen and seed dispersal (Schierenbeck et al., 1997). Mulberry is believed to be pollinated mainly by wind (Datta, 2000). In addition, fruits are dispersed by birds and other large animals (Martin et al., 2002). This kind of long distance pollination and seed dispersal could explain why there is high genetic variability within species than among species.

The genetic similarity coefficients among accessions varied from 0.8455 to 0.4077 with a mean of 0.5362. Similar variation among mulberry genotypes with a genetic similarity ranging from 0.904 to 0.544 with a mean of 0.728 using ISSR markers has been observed (Vijayan et al., 2004a). Sharma et al. (2000) also reported a similar coefficient varying from 0.58 to 0.99 across 45 mulberry genotypes using AFLP markers. Using SSR markers, Zhao et al. (2007) reported a high mean genetic similarity of 0.6131 between wild and cultivated mulberry genotypes. This suggests that the genetic identity across the mulberry accessions studied represents a genetically diverse population.

The genetic distance across the mulberry accessions studied ranged from 0.1678 to 0.8972 with a mean of 0.5916. These results show a high extent of polymorphism across the ten mulberry accessions. Moreover, genetic divergence of 58 mulberry varieties was found to be high as determined by D^2 statistics of multivariate analyses. The genetic distance varied from 0.13 to 0.78 (Fotadar and Dandin, 1998). Other results on genetic distance of both exotic and indigenous mulberry varieties using DAMD, ISSR and RAPD markers showed a distance of 0.30 to 0.68, thereby indicating a wide genetic variability in mulberry. In addition, a minimum genetic distance of 0.049 and a maximum of 0.504 among mulberry genotypes has been reported (Prasanta et al., 2008). In general, the genetic distance reported in this study concurs with results of Prasanta et al. (2008) suggesting that mulberry accessions acquire a considerable amount of genetic diversity.

The genetic identity (I) between M. alba and M. indica species was 0.7613 suggesting close relationship between the two species. This is further supported by the low genetic distance between the species of 0.2727. Similar findings were reported by Tikader and Kamble (2008) and by Vijayan et al. (2004a).

The kind of genetic variation exhibited in this study can be attributed to the fact that mulberry accessions have been established and adapted in areas distant from their origin thereby considered "naturalized" (Sharma et al., 2000). The mulberry accessions in this study showed a close relationship between Thailand and Thika accessions while S41, Kanva-2 and Embu accessions clustered together. Further, S13, S36, S54, Ithanga and Limuru accession clustered together as evident from PCoA. These form of clustering showed a wide variability among the accessions regardless of their difference in species origin. Limuru, Ithanga, S13, S36 and S54 accessions grouped together revealing the fact that the two species M. alba and M. indica are closely related. These results are similar to those reported by Prasanta et al. (2008) using ISSR markers. Vijayan et al. (2006) also reported a genetic similarity between M. alba and M. indica species. Furthermore, Vijayan et al. (2004b) reported genetic closeness of M. indica and M. alba species using ISSR and RAPD markers.

According to principal coordinates analysis (PCoA), the first and second coordinates extracted 35.45 and 17.49% of the total variation, respectively. Similar results were reported by Zhao et al. (2007) with 30.3 and 25.5% of the total variation in the first and second coordinates using SSR markers in mulberry, respectively. A higher variation was detected by the first and second coordinates with 52.5 and 10.0%, respectively in mulberry using ISSR markers by Zhao et al. (2007). Thus, there is a high genetic variation among the mulberry accessions studied, however the two species, M. alba and M. indica are closely related.

Conclusion

A high polymorphism exists within than between mulberry species regardless of their geographical origins. The dendrogram and principal coordinates analysis clearly showed that these accessions are genetically diverse despite the two species (M. alba and M. indica) being genetically related.

ACKNOWLEDGEMENT

We express our hearty thanks to the VLIR AGIBIO Project for financial funding of this research. We also gratefully acknowledge the anonymous referees for corrections and suggestions to the manuscript.

REFERENCES

Aggarwal RK, Udayakumar D, Hendre PS, Sarkar A, Singh LI (2004). Isolation and Characterization of six novel microsatelite markers for mulberry (Morus indica), Mol. Ecol. 4: 477-479.

Alvarez AE, Van de wiel CCM, Smulders MJM, Vosman B (2001). Use of microsatellites to evaluate genetic diversity and species relationships in the genus Lycopersicon, Theor. Appl. Genet. 103:1283-1292.

Awasthi AK, Nagaraja GM, Naik GV, Kanginakudru S, Thangavelu K, Nagaruja J (2004). Genetic diversity in mulberry (Genus morus) as revealed by RAPD and ISSR marker assays, BMC Genetics. 5: 1 (http://www.biomedcentral.com/1471-2156/5/1)

Bhattacharya E, Dandin SB, Ranade SA (2005). Single primer amplification reaction methods reveal exotic and indigenous mulberry varieties similarly diverse, J. Biosci. 30:669-675.

Bhattacharyya E, Ranade SA (2001). Molecular distinction amongst varieties of mulberry using RAPD and DAMD profiles, BMC Plant Biol. 1-3:1471-2279.

Dandin SB (1998). Mulberry; a versatile biosource in the service of mankind, Acta sericologica sinica. 24:109-113.

Das BK, Das C, Mukherjee K (1994). Auxin:gibberellin balance – Its role in the determination of sex expression in mulberry (Morus spp.), Indian J. Seric. 33:188–190.

Datta RK (2000). Mulberry Cultivation and Utilization in India. FAO Electronic conference on mulberry for animal production (Morus-L). Available online http://www.fao.org/DOCREP/005/X9895e04.htm#TopOf Page.

Erstad JL (1996). Ecotype differentiation of Ribes rubrum in Norway, Euphytica. 88:201-206.

FAO/IAEA (2007). Mutant Germplasm Characterization using molecular markers. A manual prepared by the Joint FAO/IAEA Division of Nuclear Techniques in Food and Agriculture. International Atomic Energy Agency (IAEA), Vienna.

Fotadar RK, Dandin SB (1998). Genetic divergence in the mulberry Canye Kexue. 24:180-185.

Hamrick JL, Godt MJW (1996). Effects of life history traits on genetic diversity in plant species, Philosophical Transactions of the Royal Society of London Biol. Sci. 351:1291-1298.

Heaton HJ, Whitkus R, Gomez-pompa A (1999). Extreme ecological and phenotypic differences in the tropical tree chizopate (*Manilkara zapota* L.) are not matched by genetic divergence: A random amplified polymorphic DNA (RAPD) analysis, Mol. Ecol. 8:627-632.

Kameswara RN (2004). Plant genetic resources: Advancing conservation and use through biotechnology, Afr. J. Biotechnol. 3:136-145.

Manniatis T, Fritsch EF, Sambrook J (1982). Molecular cloning a laboratory manual. New York: Cold Spring Hobart Publications.

Martin G, Reyes F, Hernandez I, Milera M (2002). Agronomic studies with mulberry in Cuba. In: Sánchez, M. D. (ed.) 2002. Mulberry for Animal Production FAO Animal Production and Health Paper 147. Rome, p.103-113.

Matsuoka Y, Mitchell SE, Kresovich S, Goodman M, Doebley J (2002). Microsatellites in zea-variability, patterns of mutations, and use for evolutionary studies, Theor. Appl. Genet. 104:436-450.

Millar MA, Brine M, Coates DJ, Stuckely MJC, McComb JA (2000). Mating system studies in Jarrah, *Eucalyptus marginata* (Myrtaceae), Australian J. Botany. 48:475-479.

Peakall R, Smouse PE (2006). GENALEX 6: Genetic analysis in Excel. Population genetic software for teaching and research, Mol. Ecol. 6:288-295.

Prasanta KK, Srivastava PP, Awasthi AK, Raje SU (2008). Genetic variability and association of ISSR markers with some biochemical traits in mulberry (*Morus spp.*) genetic resources available in India, Tree Genet. Genomes. 4:75-83.

Quedraogo AS (2001). Conservation, management and use of forest genetic resources. Recent Research and development in Forest genetic Resources. Proceedings of training workshop on the conservation and sustainable use of forest genetic resources in Eastern and Southern Africa, December 1999, Nairobi, Kenya. pp.1-14

Schierenbeck KA, Skupski M, Lieberman D, Lieberman M (1997). Population structure and genetic diversity in four tropical tree species in Costa Rica, Mol. Ecol. 6:137-144.

Shannon CE, Weaver W (1949). The Mathematical Theory of Communication. University of Illinois Press, Urbana.

Sharma A, Sharma R, Machii H (2000). Assessment of genetic diversity in a Morus germplasm collection using fluorescence-based AFLP markers, Theor. Appl. Genet. 101:1049-1055.

Steiger D, Nagal C, Moore P, Morden C, Osgood R, Ming R (2002). ALFP analysis of genetic diversity within and among Coffea Arabica cultivars, Theor. Appl. Genet. 105:209-215.

Tikader A, Vijayan K, Kamble CK (2009). Conservation and management of mulberry germplasm through biomolecular approaches, Biotech. Mol. Biol. Rev. 3(4):92-104.

Tikader A, Roy BN (2001). Multivariate analysis in some mulberry (*Morus spp.*) germplasm accessions, Indian J. Seric. 40:71-74.

Tikader A, Kamble CK (2008). Mulberry wild species in India and their uses in crop improvement, Australian J. Crop Sci. 2(2):64-72.

Vijayan K (2009). Approaches for enhancing salt tolerance in mulberry (*Morus* L), Plant Omics J. 2(1):41-59.

Vijayan K, Kar PK, Tikader A, Srivastava PP, Awasthi AK, Thangavelu K, Saratchandra B (2004a). Molecular evaluation of genetic variability in wild populations of mulberry (*Morus serrata Roxb.*), Plant Breed. 123:568-572.

Vijayan K, Srivastava PP, Awasthi, AK (2004b). Analysis of phylogenetic relationship among five mulberry (*Morus*) species using molecular markers, Genome 47:439-448.

Vijayan K, Tikader A, Kar PK, Srivastava PP, Awasthi AK, Thamgavelu K, Saratchandra B (2006). Assessment of genetic relationships between wild and cultivated mulberry (*Morus*) species using PCR based markers, Genet. Res. Crop Evol. 53:873-882.

Vijayan KS, Chatterjee N, Nair CV (2005). Molecular characterization of mulberry genetic resources indigenous to India, Genet. Res. Crop Evol. (In press).

Wu J, Krutovskii KV, Strauss SH (1999). Nuclear DNA diversity, population differentiation, and phylogenetic relationships in the Califonia closed-clone pines based on RAPD and allozyme markers, Genome 42:893-908.

Yeh FC, Yang RC, Boyle TBJ, Ye ZH, Mao JX (1999). POPGENE 3.5, the user- Friendly Shareware for Population Genetic Analysis. Molecular Biology and Biotechnology Center, University of Alberta, Edmonton.

Zhao W, Miao X, Jia S, Pan Y, Huang Y (2005). Isolation and characterization of microsatellite loci from the mulberry, *Morus* L, Plant Sci. 16:519-525.

Zhao W, Zhou Z, Miao X, Zhang Y, Wang S, Huong J, Xiang H, Pan Y, Huang Y (2007). A comparison of genetic variation among wild and cultivated Morus species (Moraceae: *Morus*) as revealed by ISSR and SSR markers, Biodivers. Conserv. 16:275-290.

5

Proposal for access to the femur in rats

Clayton Miguel Costa[1], Geraldo Bernardes[1], Sandro Melim Sgrott[1], Jorge Bins Ely[1], Luismar Marques Porto[2] and Armando José d'Acampora[1]*

[1]Department of Surgery, Federal University of Santa Catarina, UFSC, Brazil.
[2]Graduate Program in Chemical Engineering, Federal University of Santa Catarina, UFSC, Brazil.

Reconstructions, implantation procedures and bone synthesis materials used to treat bone loss caused by tumors or fractures require *in vivo* testing. In this study, we created an access route to the femur in rats, allowing us to conduct *in vivo* bone experiments with less cost, more easily obtainable materials and simplified storage and care. Eighteen male Wistar rats, 180 days of age and weighting 350 to 370 g were surgically operated. Their femurs were accessed and exposed quickly and safely. This proved to be an effective method for experiments in which the bone needs to be approached and studied. The approach proposed in this study is suitable for full exposure of the rat femur.

Key words: Rats, femur, models, animal.

INTRODUCTION

The use of animals in scientific experiments has always been and continues to be controversial. However, to make progress in the fields of disease prevention, evaluation of diagnostic techniques and procedures for the treatment of disease, investigators cannot dispense with the use of laboratory animals (d'Acampora et al., 2008). When surgical interventions are performed, they must rigorously observe the technical norms and international standards of animal research (d'Acampora et al., 2008; Princípios éticos na experimentação animal, 1991; Festing and Wilkinson, 2007; National Centre for Replacement, 2010; National Research Council, Institute for Laboratory Animal Research, 1996).

It is not possible to experiment with new therapeutics, surgical techniques or materials without prior exhaustive testing of all possible forms *in vivo* (d'Acampora et al., 2008). Despite recent efforts by regulating authorities and the scientific community to develop *in vitro* and *in silico* models to replace animal models in pharmaceutical and biochemical research (National Centre for Replacement,

2010; Westmoreland and Holmes, 2009) the development of new biomaterials for subsequent use in prosthetics and/or stents still demands *in vivo* implant methodology. However, existing models for the study of immunologic/inflammatory reactions during complex host-implant interactions have limitations.

The use of *in vivo* tests implies the need for animal models. Some animal species, especially larger ones such as dogs, rabbits, sheep and pigs, can be used as experimental models. However, the larger the animal to be used in the experiment, the greater the total operational cost. The choice of the specific animal model, therefore, must match the type of experiment to be performed while still achieving the proposed objectives with the least cost and highest possible efficiency (Schossler, 1993). The space required for the maintenance of cages and diet, management of the waste produced, the cleaning material and the specialized human resources required to carry out hygiene care and support increase proportionally to the size of the animal model. Thus, there is a clear difficulty in maintaining larger animals such as dogs, pigs or rabbits, as all animals must be raised while maintaining the legal and ethical levels of adequate space and

---

*Corresponding author. E-mail: dacampora@gmail.com.

comfort.

Another important aspect in the cost of an experiment is the quantity and type of anesthesia, analgesics and post-surgical pharmaceutical support used for the different animal species. Each animal reacts in a particular way when subjected to stress, thus demanding different techniques and anesthesia associations. An animal surgical experiment becomes better when the following requirements are fulfilled:

(i) Scientifically accepted minimal number of animals;
(ii) Animals that are easily obtained;
(iii) Animals resistant to infections;
(iv) Animals that are easily housed;
(v) Animals that are easily handled and cared for;
(vi) Acceptable animal, diet and anesthesic costs.

Given the aforementioned considerations, the rat (*Rattus norvegicus*) is an excellent choice for scientific use because of its low maintenance costs compared to other animals. It occupies less laboratory space and requires the use of fewer anesthetics.

The rat is resistant to infections and has a short life cycle. Importantly, repair processes occur much more rapidly in the rat than in humans. Reactions that would take months to observe in humans take days to occur in this animal model. Having said this, the objective of this article is to demonstrate a surgical method of quickly, easily and safely accessing the femur of rats, allowing for ample exposure of the metaphyses and femoral diaphysis. A literature search did not identify a similar method.

Studies such as that published by Vialle et al. (2004) describe fractures of the femur with posterior access, while Giordano et al. (2006) performed halo grafting but did not describe how the rat femur was accessed. In the same way, many other studies did not provide a clear and systematic description of the femur access route. A standardized access route would be the quickest and most effective for the performance of *in vivo* biomaterial and medication reaction tests encompassing the bone compartment. The route proposed in this study can be of great added value for conducting experimental models of craniofacial, orthopedic and odontological implants. In particular, this method can be useful for testing biomaterials and/or scaffolds in *in vivo* simulations for the filling of faults or bone loss resulting from trauma or tumors, testing slow-release and progressive medication incorporated to the biomaterials or for evaluating the inflammatory reaction and/or biocompatibility of a new material.

METHODS

This study was previously submitted for approval to the Ethics Committee for Animal Use (Comissão de Ética para Utilização de Animais - CEUA) of the University of Southern Santa Catarina (Universidade do Sul de Santa Catarina - UNISUL) under the number 10.597.4.01.IV.

Sample

Eighteen male Wistar *Rattus norvegicus albinus* were used. All rats were 180 days of age and weighed between 350 and 370 g. All rats were healthy and housed in individual plastic cages under ambient temperature and noise conditions in the UNISUL Laboratory of Operating Procedure and Experimental Surgery (Laboratório de Técnica Operatória e Cirurgia Experimental - TOCE).

The animals were numbered 1 to 18 and individually maintained in cages with forage of sterile wood shavings, full access to water and a species-specific diet under ambient temperature and natural light. The animals were provided by the UNISUL Biotery and identified by means of numbering provided by the veterinarian in charge. Animals were weighed before surgery.

Technique

Upon identification and weighing, all animals underwent pre-anesthesia sedation through inhaled ethyl ether. Anesthesia was performed with an aqueous solution of 5% dextroketamine hydrochloride (Ketamin S+®) and an aqueous solution of 2% Xylazine hydrochloride (Rompum®) at doses of 30 and 8 mg/kg, respectively. Anesthesic was administered intramuscularly in the inner face of the animals' left hind leg (National Centre for Replacement, 2011). The animal was considered anesthetized after having lost the corneal-palpebral reflex and when it lacked any motor reaction. The test for verification of the ideal anesthesia plan was performed by pinching the animals' adipose plantar pad (De Luca et al., 1996). The animal was positioned on a wooden plank protected with a sterile surgical cloth in the left lateral decubitus position. An ample trichotomy was performed over the entire extension of the posterior limb, part of the back and the abdomen. The femorotibiopatellar joint was considered the distal reference point of the rat and the major trochanter was the proximal reference point (Figure 1).

Upon incision and opening of the skin, we identified a fine white line corresponding to the intermuscular septus, which separates the biceps femoral muscle from the superficial gluteal muscle (Hebel and Stromberg 1976; Hunt, 1924; Rowett, 1957) (Figure 2). An incision was made over the intermuscular septus with a cold blade scalpel, and the muscles were separated with *Metzembaum* spreading scissors while respecting the integrity of the tissue (Figure 3). Access to the anterior face of the femur throughout its length was created, identifying the major trochanter to the femoral distal metaphysic and allowing for visualization of the entire bone (Figure 4).

The parting and mobilization of the surgical site was performed using two small *Backhauss* tweezers placed around the bone, one located proximal and the other distal to the total area of the rat femur, totally exposing the bone. Synthesis was performed by anatomical planes with the muscular plane first closed with 3 to 0 gut string followed by the skin with 3 to 0 nylon, both with simple stitches. After the observation period of the study, the animals were subjected to the 'assisted painless death' procedure through induced anesthesia with sodium thiopental (Thiopentax®) at a dose of 50 mg/kg through intraperitoneal injection. Upon obtaining the ideal anesthesia plane, 19.1% potassium chloride was administered by intracardiac delivery.

RESULTS

All the animals were weighed immediately before surgery and at the time of sacrifice. The follow-up details are shown in Table 1. The animals subjected to painless death at 30 days had an average mass gain of 0.46%

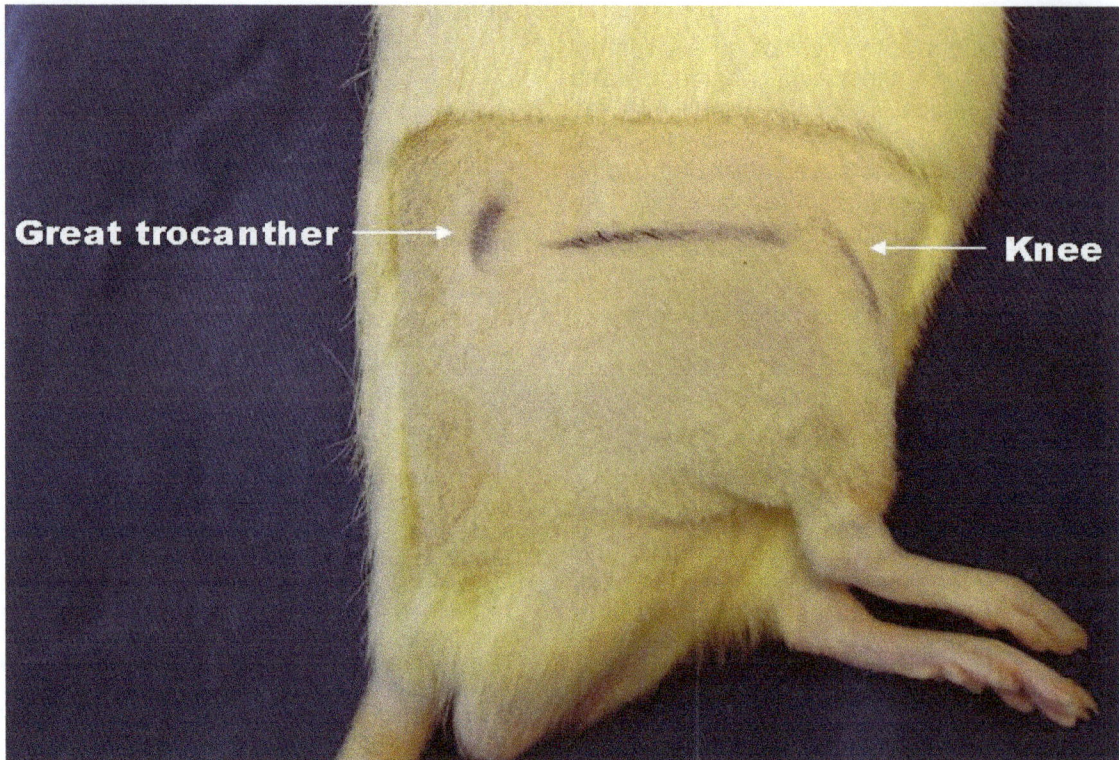

Figure 1. Animal positioned with the marked reference points.

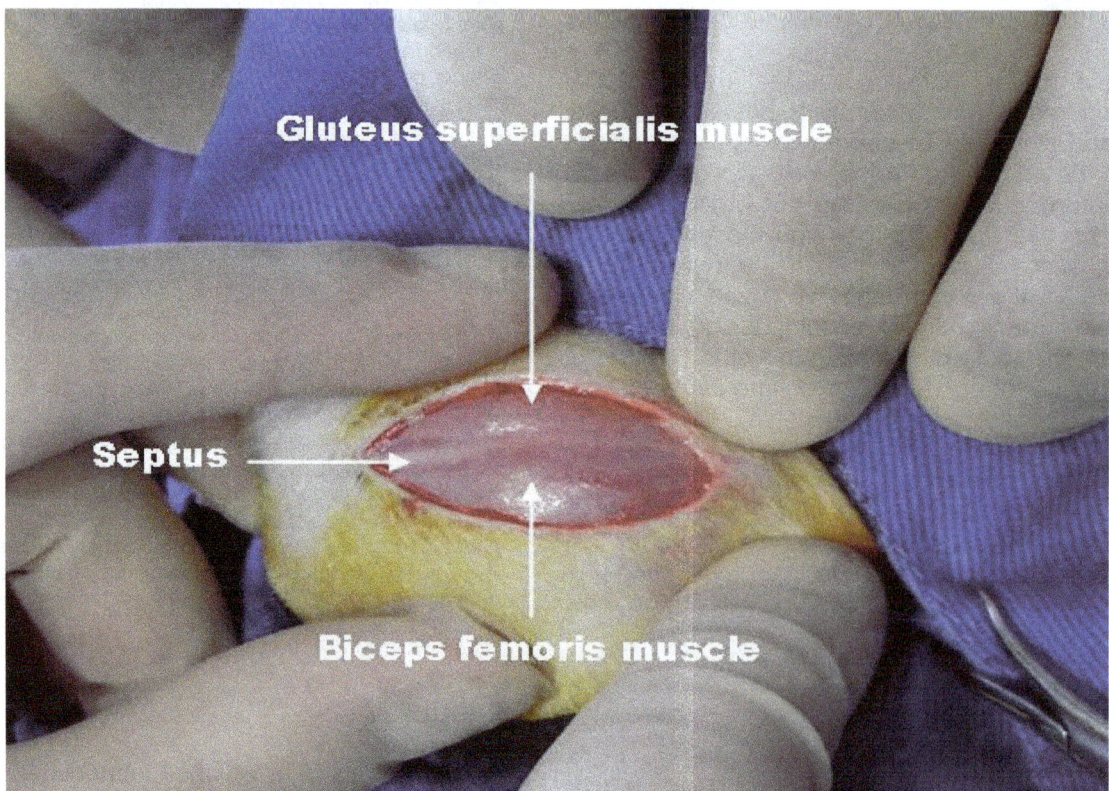

Figure 2. Intermuscular septus that separates the superficial gluteal muscle (above) from the biceps femoral muscle (Hebel and Stromberg, 1976; Hunt, 1924; Rowett, 1957).

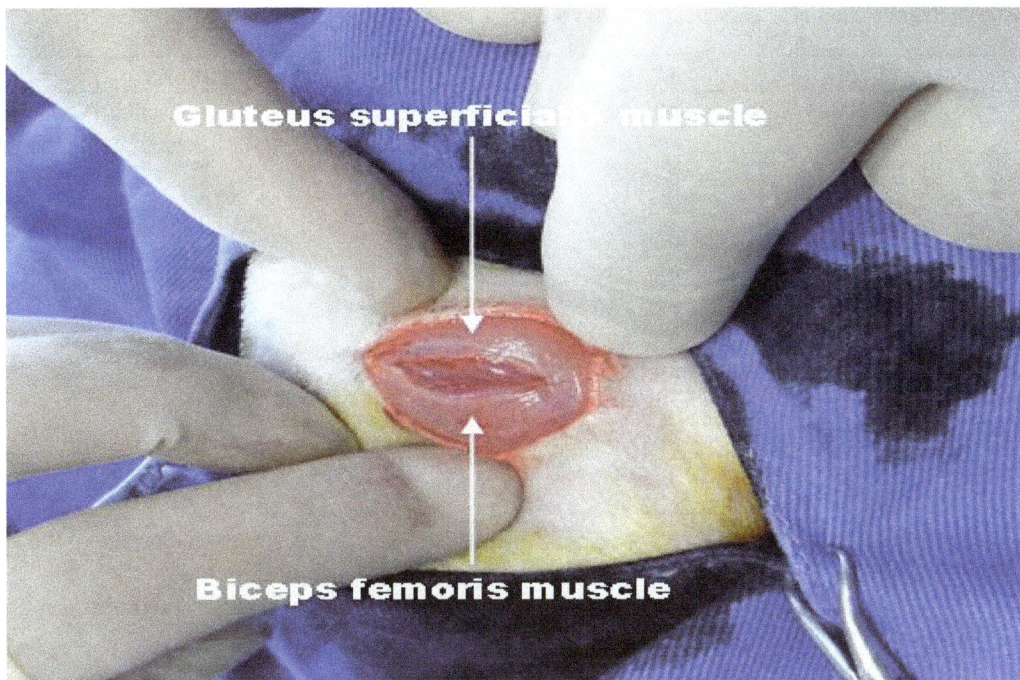

Figure 3. Details of muscular separation.

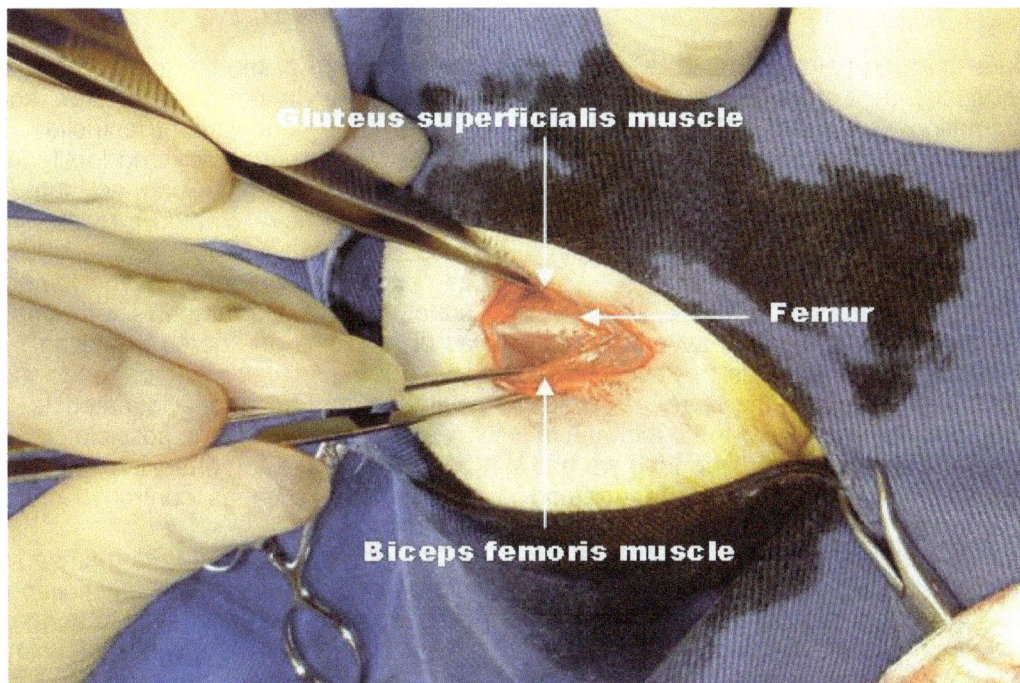

Figure 4. Access to the entire length of the bone.

and those sacrificed after 60 days had a gain of 1.34%. There was no significant change in the mass of the animals analyzed during the observation period. This experiment revealed that a small external rotation allowed for the visualization of the entire femur, including the femoral sulcus, which was relatively short. Prophylactic antibiotic therapy was not undertaken during the study and no skin infection was seen. This was attributed to the use of skilled surgical technique in addition to the aseptic conditions of the surgical space.

Table 1. Follow-up of the evolution of animal weight during the experiment.

Rat	M_i (g)	M_f (g)	T_a (days)	Observations
1	481.00	480.30	60	
2	494.00	**		Femur fracture during the procedure
3	373.90	336.60	60	
4	456.06	458.40	60	
5	477.55	460.48	60	
6	453.38	481.06	60	
7	385.00	422.29	60	
8	361.55	389.42	60	
9	457.15	485.00	30	
10	410.76	420.50	30	
11	566.95	594.00	30	
12	428.69	424.40	30	
13	456.11	433.46	30	
14	490.00	**		Femur fracture during the procedure
15	496,20	505,00	30	
16	465,90	**		Died after the procedure
17	425.90	425.40	30	
18	564.10	572.80	30	
Mean	452.95	459.27		1.34% weight gain

M_i – Weight at the start of the experiment (in grams); M_f – weight at the end of the experiment (in grams); T_a – time until induced painless death (in days); ** - eliminated from the experiment.

After the procedure inducing painless death, withdrawal was performed for subsequent examination of the femur. No signs of inflammation were seen at the surgical site. Both the muscle layer and the bone plane showed a normal macroscopic appearance.

DISCUSSION

This research was conducted because of several deficits in the current knowledge. The first was the absence of previously published parameters in the literature, which led to the trial-and-error construction of the experimental model presented here. This took more time than we had expected it to take. The attempt to transpose the method to larger animals such as pigs proved impractical because of the cost as well as the difficulty of maintaining the animals in the laboratory. Local law prohibits experiments with dogs. Thus, we conclude that the rat would be the animal of choice for this type of experimental procedure in which access to the femur for posterior composite implantation is required.

The experiment will be useful primarily to craniofacial surgeons, orthopedic specialists and dentists, who can use all the method for easy access to the rat femur. This experimental model gives them the possibility to test a number of biomaterials that, in the future, could replace bone with the quality and safety standards required for use in humans.

The purpose of this research was to find a proper

access route for the complete, quick and easy exposure of the femur in rats and to allow access to the entire length of the bone without restriction. This access allows for the viewing of a wide and totally free bone surface in which the researcher can act without any restriction. Because we have created a reliable method for surgical access, we can now test a wide field of implants as well as conduct experiments on the femoral plane, including puncture, and use materials with small dimensions such as plates, pins and screws, making it possible to test *in vivo* biomaterials, scaffolds and engineered products manufactured from specific materials. The ultimate aim of these experiments is the possible creation of replacement bone that could integrate into the tissue or therapeutic applications, such as controlled drug delivery thrrough scaffold.

The rat was chosen for its many advantages, including the ease of obtaining and maintaining the animal in the laboratory due to its small size and space requirements. In addition, a large number of animals can be used, allowing for standardization and validation of the technique for future work. After surgery, the animals were observed postoperatively three times a day by the same investigators at three different time points. No wound infection or occurrence of abnormalities such as decreased mobility, bristling hair, decrease in search of food or water or the occurrence of death within 60 days. After the observation period, the animals were subjected to painless assisted death under general anesthesia.

No infection of surgical wound was observed, and only

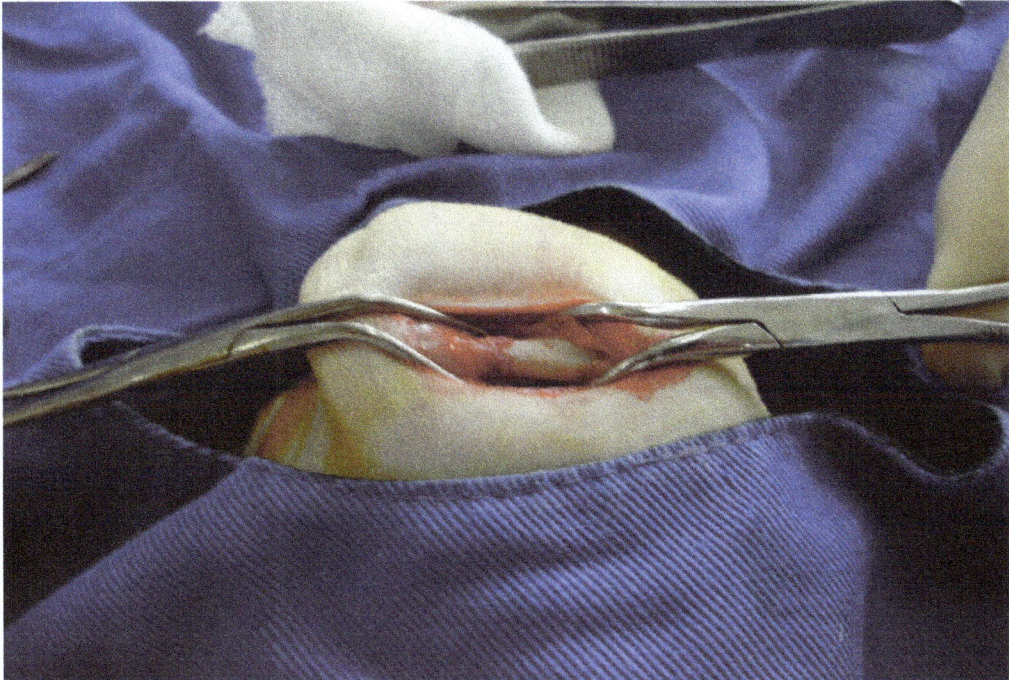

Figure 5. Stabilization of the femur by means of *Backhauss* tweezers.

one immediate death occurred in the postoperative period. There was no replacement of animals from the original group of 18. Only one animal died from hypothermia after the procedure and two animals experienced a fracture of the femur during handling with inadequate instrumentation. Thus, the experimental group was made up of a final number of 15 animals.

All researchers in the group were able to easily repeat the technique of exposure of the bone after it was standardized. The easy access provided by this path was reproducible and the learning curve was short. After incision at the marked site, between the greater trochanter and the femorotibiopatelar articulation of the animal (Figure 1), it became easy to identify the muscle layer of the superficial gluteal muscle and the biceps femoris where the septum between the two muscles can be observed (Figure 2).

When making an incision over the intermuscular septum, the femur of the rat is exposed (Figure 4) and completely isolated through muscle divulsion (Figure 5).

The premise of this study was to demonstrate a quick path and easy access route to the Wistar rat femur. The research team believes that this method is the appropriate way to conduct experiments involving the femur of the Wistar rat, as it leads to easy bone access with a short learning curve using a convenient research animal.

Conclusion

The approach proposed is suitable for full exposure of the femur of Wistar rats.

REFERENCES

d'Acampora AJ, Rossi LF, Ely JB, Vasconcellos ZA (2008). Is animal experimentation fundamental? Acta. Cir. Bras., 24(5): 42342-5.

Ethical principles for animal experimentation. Brazilian College of Animal Experimentation (1991). Festing S, Wilkinson R (2007). The ethics of animal research. Talking Point on the use of animals in scientific research. EMBO Rep., 8(6): 526-530.

National Centre for Replacement, Refinement and Reduction of animals in research (NC3RS). Responsibility in the use of animals in bioscience research: Expectations of the major research council and charitable funding bodies. 2010. http://www.nc3rs.org.uk/document.asp?id=1319.

National Research Council, Institute for Laboratory Animal Research (1996). Guide for the care and use of laboratory animals. National Academy Press, Washington, DC.

Westmoreland C, Holmes AM (2009). Assuring consumer safety without animals. Organogenesis, 5(2): 72-67

Schossler JE (1993). A escolha, contenção e manuseio de animais de experimentação. Acta Cir Bras., 8(4): 166-169.

Vialle E, Vialle LR, Boechat R, Bley JP, Scussiato R, Busato T, Carvalho, D., Fedato, F, Fernandes, B., Torres, R (2004). Produção de fratura padronizada de fêmur em ratos. Rev Bras Ortop., 39(6): 323-329.

Giordano V, Rezende R, Senna LF, Pompei A, Amaral NP, Albuquerque RP, Giordano M, Apfel MIR, Bastos JSA (2006). Estudo histomorfológico da incorporação de aloenxerto fresco e da hidroxiapatita de alta porosidade em defeito ósseo produzido em fêmures de ratos. Rev Bras Ortop., 41(9): 384-6.

National Centre for Replacement, Refinement and Reduction of animals in research (NC3RS): Procedures with care. 2011. http://www.procedureswithcare.org.uk/2010/intramuscular-injection-in-the-rat

De Luca RR, Alexandre SR, Marques T, Souza NL, Merusse JLB, Neves SP (1996). Manual para Técnicos em Bioterismo. São Paulo: H.A. Rothschild, p. 259.

Hebel R, Stromberg MW (1976). Anatomy of the laboratory rat. Baltimore: The Williams & Wilkins Company, p. 169.

Hunt HR (1924). A laboratory manual of the anatomy of the rat. New York: The Macmillan Company, p. 169.

Rowett HGQ (1957). Dissection Guides: III The Rat. New York: Holf, Rinehart and Winsto, p. 64.

6

Transient receptor potential channels interference has broader implications in safeguarding against genomic instability

Ammad Ahmad Farooqi[1]*, Qaisar mansoor[2], Aamir Rana[1], Zeeshan Javed[1] and Shahzad Bhatti[1]

[1]Institute of Molecular Biology and Biotechnology (IMBB), University of Lahore, Pakistan.
[2]Institute of Biomedical and Genetic Engineering (IBGE), Islamabad, Pakistan.

Prostate cancer is a multistep molecular disorder that arises because of a miscellany of proteins. Calcium channels, hotspots in genome and susceptibility to DNA damage, an illegitimate repair of the genome, all these factors work synchronously at various levels to worsen the clinical management of the disease. Fusion transcripts are recently acclaimed candidates for exacerbation of the disease. Erlotinib has been used for the treatment of various molecular pathologies but it has been ineffective in producing desired level of effects. It is therefore used in conjunction with other therapeutic intercessions. Despite the characterization of a number of fusion genes, therapeutic interventions to address the underlying mechanisms are still insufficient. In this particular study, we have applied a combinatorial approach to silence some calcium channels which have been documented earlier to be contributory in disease progression. The siRNA against TRPM2, TRPV6 and TRPC6 were used to evaluate the striking synergy. Simultaneously, the effects were evaluated for ATM activation and a downregulation of the chimeric gene. The expression of channels was blocked effectively as analyzed by RT PCR and Blotting assays. TRPM2 ablation instigated a DNA damage response which was observed by the blotting assay for phosphorylated ATM. Dampening the expression of TRPV6 and TRPC6 by siRNA concomitantly inhibited the genomic rearrangements. We have evaluated the synergistic impact of Erlotinib along with TRP interference on genomic instability. The observations are indicative of the fact that, silencing of calcium channels offer exciting avenues for getting a step closer to personalized medicine.

Key words: ATM, TMPRSS2-ERG, TRPM2, TRPV6, TRPC6, genomic instability, prostate cancer.

INTRODUCTION

Prostate cancer is a multifactorial disease. A wide range of proteins have been characterized and incriminated to be involved in the disease progression. A confluence of observations indicates that, super-families of transient receptor potential (TRPs) channel are instrumental in prostate carcinogenesis. TRPC6 (Yue et al., 2009) is associated with the disease exacerbation and knocking down of TRPC6 (Wang et al., 2010) and TRPV6 (Zhao et al., 2010) was growth inhibitory in prostate cancer cell lines.

The genomic instability arising because of genomic rearrangements results in the subversion of core biological system. TMPRSS2-ERG is a fusion transcript, that is a well documented example of genomic instability outcomes in prostate epithelium (Tomlins et al., 2005; 2006).

Zeng et al. (2010) has recently documented that, TRPM2 is populated in the nucleus in cancerous cells. The exact mechanism that underlies this trafficking of the protein into the nucleus is still unclear yet some

*Corresponding author. E-mail: ammadahmad638@yahoo.com.

hypothesis can be made. It might be engaged in the initiation of genomic rearrangements. The ataxia telangiectasia mutated (ATM) gene is known to have a central role in sensing general DNA damage and mediating cell-cycle checkpoint. Dynamic multifaceted functions of ATM act as gatekeepers of genomic stability and preventing tumorigenesis, ATM is impaired in the prostate cancer aggressiveness that results in an unfaithful repair of the DNA (Luedeke et al., 2009; Mani et al., 2009). Epidermal growth factor receptor (EGFR) is indispensable for mediation of both proliferative and survival signals to cells. Moreover, in recent years it became noticeable that, in addition to ligand binding-induced activation of the EGFR, ligand-independent receptor activating processes also exist (Boerner et al., 2003). The most harmful DNA damages after treatment with ionizing radiation are double-strand strands breaks, which are preferentially repaired by non homologous end-joining. They concluded that, majority of genomic rearrangements seen in wild-type cells after an acute 80-Gy X-ray exposure via DNA-PK-dependent end-joining pathway (Rothkamm et al., 2001).

Two exciting documentations prompted us to focus our concentration on a probable connection between EGFR and DNA repair. First, Bandyopadhyay et al. (1998) showed a physical co-existence of EGFR and DNA-PK, after incubation with the EGFR-blocking antibody C225. Second, Lin et al. (2001) demonstrated nuclear localization of EGFR which would interdigitate/tether EGFR to sites of DNA damage. Thus, we analyzed the location of EGFR and its molecular interactions after radiation exposure. Erlotinib, gefitinib are the drugs which are designed for inhibition of RTK's. Heterogeneity of the tumor cell population develops chemoresistance and radioresistance. In addition, uncontrolled cellular growth and constitutive activation of survival pathways severely compromise efficacy of erlotinib monotherapy (Qi et al., 2009). The calcium homeostasis triggers a multiplicity of cellular dynamics together with gene transcription, proliferation and apoptosis. Perturbance of Ca^{2+} signaling may stimulate uncontrolled cell proliferation and suppression of apoptosis providing the basis for cancer development. Interestingly, there is an upregulation of specific calcium channels or pumps associated with certain types of cancer. Consistent with the same documentations, upregulation of TRPV6 channel in prostate cancer cells may symbolize an instrument for maintaining a higher proliferation rate, increasing cell survival and apoptosis resistance. Androgen receptor is engaged in TRPV6 regulation in a ligand-independent way. They found that androgen receptor abrogation by siRNA decreased TRPV6 mRNA and protein levels. Discordantly, ligands DHT, an androgen receptor–selective agonist and antagonist, had no noteworthy consequence on TRPV6 mRNA expression (Lehen'kyi et al., 2007; Schwarz et al., 2006). Taking into consideration, TRPM (melastatin-related trp cation channel) and TRPV sub-families of TRP channels, and review of

literature indicates that, TRPM4 was found contributory in cancer. Armisén et al. (2011) stated that, abrogation of TRPM4, resulted in an attenuation of cellular proliferation however reconstituting the cells for TRPM4 resulted in enhanced cellular proliferation. On the contrary, TRPM8, a receptor for cold stimuli and menthol portrays opposite trends. It was considered as an oncogene by Bidaux et al. (2007).

Conflictingly, Yang et al. (2009) appraised it as a tumor suppressor if over-expressed. They documented that PC-3 cell line underwent a decrease in cellular proliferation if this gene was over-expressed. This was further strengthened by a recent study in which Valero et al. (2010) induced an enforced over-expression of TRPM8, that resulted in shuttling of the protein from endoplasmic reticulum to the plasma membrane and retarding cellular proliferation. TRPV6 is strongly expressed in prostate cancer, but not in benign prostate hyperplasia. It is intriguing that, TRPV6 renders higher invasion and metastasizing efficiency to prostate tumors and it is well documented that, drastic calcium influx through TRPC channels, gears up proliferation rate of LNCaP cells. In the present investigation, we adapted a combinatorial targeting of TRPC/TRPM and TRPV along with administration of Erlotinib, to explore probabilistic gene fusions in prostate cancer cell line. Moreover the influence of knock down of the channels and ATM responsiveness was analyzed.

Cell lines and treatments

The prostate cancer cell line, LNCaP, was obtained from the American Type Culture Collection. Human PrECs was maintained in PrEGM media. Transfection of LNCaP and PrECs cells with siRNA was performed with Lipofectamine2000 (Invitrogen). For induction of chromosomal translocation, LNCaP cells or PrECs were grown in charcoal-stripped serum containing media for 48 h followed by mock, DHT (10^{-7} M), g-irradiation (50 Gy) treatment or both. Co-treatment was done because of non-detectable expression of fusion transcripts in single treatments. After treatment, the cells were reincubated for 24 h before being harvested for appropriate assays. Cell viability in cultures before and after irradiation was assessed by trypan blue exclusion. Erlotinib (sc-202154) purchased from santacruz biotech was dissolved in autoclaved water and DMSO (<0.1%) respectively, as stock solutions for in vitro studies. LNCaP cell line was treated with varying doses of the drug, however worked well at 5 μM.

Small interfering RNA assay

To examine the effects of ATM on genomic rearrangements, cells were transfected with TRPV6, TRPM2 and TRPC6 specific or scrambled siRNAs. Briefly, cells were seeded in 24 or 6-well plates till they attained a confluency of 60 to 80%. To achieve the finest evaluations transient transfection of cells in 24-well plates was done with 10, 20, 40 μM/well of specific siRNAs (TRPV6, TRPM2 and TRPC6) or scrambled siRNA, using Lipofectamine™ 2000 (Invitrogen) according to the manufacturer's protocol. After incubation at 37°C in 5% CO_2, cells were then harvested for gene expression and protein analysis after 48 h. siRNA for TRPV6,

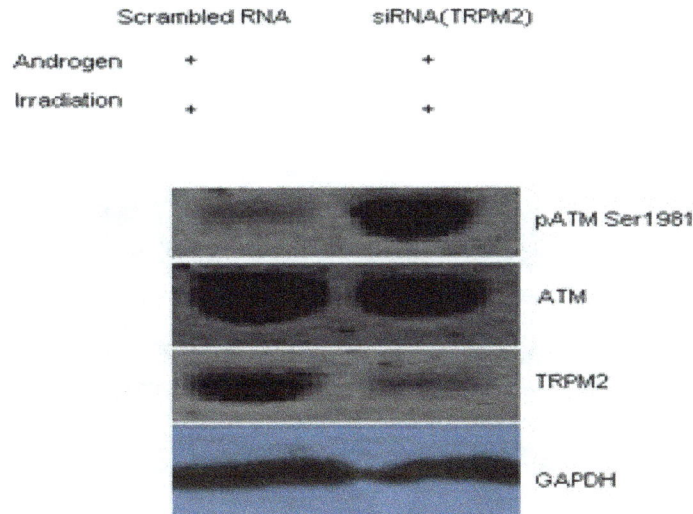

Figure 1. Treatment of androgen treated and irradiated LNCaP cells with scrambled RNA and siRNA (TRPM2).

TRPM2 and TRPC6 were purchased form Santacruz Biotechnology.

Western blot analysis

Cell collection was done by adding trypsin and subsequent centrifugation for 5 min at 13,000 rpm. Pellets obtained after discarding supernatant were treated with 1XPBS. 100 μl of lysis buffer was added incubated in ice for 20 min and subjected to centrifugation for 30 min at 13,000 rpm. Supernatant obtained was shifted carefully to new tube and stored at -20°C. Protein samples were mixed with 5X reducing dye in microtubes and placed in a boiling water bath for 2 min, then placed on ice and again centrifuged. Samples were then carefully loaded in the wells and run at 80 V until the samples passed the stacking gel, with a slight increase in voltage up to 100 V, samples seperated completely. After SDS-PAGE and electrophoretic transfer, the membrane was blocked in 5% skim milk in 1XTBS with 0.05% Tween-20 for 1 h at RT or at 4°C overnight. The antibody for ATM, pATM and TRPM2 were purchased from Santa Cruz biotechnology.

RNA extraction and RT PCR based amplifications

RNA extraction was done from RNeasy Mini kit (Qiagen) from LNCaP cells. There was a collection of cultured cells by preliminary centrifugation for 5 min at 5000 g, 5 min at 5000 g washing in phosphate-buffered saline (PBS). RNAse-free water (Sigma–Aldrich) was used to dissolve all RNA samples. The purity and concentration were evaluated by A260 nm measurement. 2 μg of each sample in a total volume of 50 μl was subjected to reverse transcription using the cDNA Archive Kit. Incubation was given to the mixtures for 10 min at 25°C and for 2 h at 37°C. cDNA products of 2 μl were used as template for polymerase chain reaction (PCR) quantification in all samples. The primers for TMPRSS2-ERG and conditions for PCR amplification were reported earlier by Hessels et al. (2007). The effect of genes and their siRNA were detected by using specific primers of TRPV6 and TRPC6 genes on ABI 7500 real time PCR, using SYBR Green mix (Fermentas) according to manufacture's instructions. Temporarily, 2X SYBR Green RT-PCR Master Mix, template RNA, primers, and RNase-free water were

thawed. After mixing the individual solutions, all regents were placed on ice. In the end, template cDNA of 0.5 μl were added to the individual PCR vessels. The relative gene expression analysis was done by using SDS 3.1 software, provided by ABI. Each real time PCR assay was performed in triplicate. Level of significance and standard error was determined by SPSS software.

RESULTS

Western blot analysis showed that, TRPM2 protein expression was decreased in siRNA TRPM2 treated cells. GAPDH is used as an internal control. There is a remarkable activation of ATM in siRNA treated cells, as evaluated by western blot analysis using ATM and ser 1981 ATM antibodies. RT PCR analysis showed that siRNA treatment efficiently inhibited the expression of TRPV6 and TRPC6. However, scrambled RNA did not inhibit the expression. Down regulation in TMPRSS2-ERG expression is also evident from RT PCR based analysis. GAPDH was used as an internal control.

Explanation

Figure 1 shows the protein expressions of TRPM2 and its influence on ATM activation. The Figure 1 displays that there is a blockade of TRPM2 after pretreatment with siRNA. Sequence specific inhibition was obvious in the TRPM2 RNA interference mediated LNCaP cell line. No results of blockade can be seen in the cells treated with scrambled RNA. This is indicative of the fact that, TRPM2 is actively involved in desensitizing the ATM to DNA damage. ATM once rendered inactive can not contribute in genome repair rather another kinase enzyme (DNA-PK) repairs genome in an unfaithful manner.

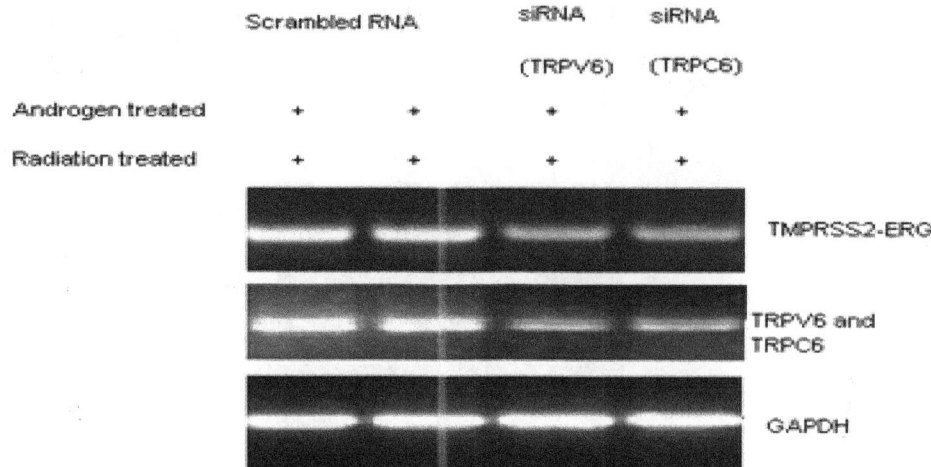

Figure 2. Treatment of androgen and radiation treated LNCaP cells with siRNA (TRPV6) and TRPC6 respectively.

Figure 1 explains that, TRPM2 abolishes the activation of ATM. As the autophosphorylation at serine 1981 is absent in the LNCaP cell line treated with scrambled RNA. This is underlining the aspect that TRPM2 abrogation, stimulates the activation of ATM and there is no illegitimate genome repair.

Another fact that cannot be ruled out is that, TRPM2 activities severely compromise the activation of ATM and incapable of being converted into its autophosphorylated form. Consistent with the same approach in Figure 2, we ablated TRPV6 and TRPC6 and the effect on the genomic instability was studied. In the cells treated with scrambled RNA, there was no down regulation of the fusion transcripts. This genomic rearrangement was induced by pretreatment of cell line with androgen and radiation. Both the treatments induced DNA breaks and subsequent generation of the chimeric transcripts. It seems obvious that DNA repair protein activities are hampered as a result of which, there is no surveillance in terms of genome repair. GAPDH is used as an internal control. Cells treated with TRPV6 specific RNA interference gave remarkable results in terms of attenuation of the genomic instability and genomic rearrangements. This was observed by the fact that, expression of the TMPRSS2-ERG was reduced because of the treatment of the cells with TRPV6 specific siRNA. It seems to be down regulated presumably because of the silencing of the TRPV6 because cells competent for TRPV6, do not show any decrease in the expression profile of the fusion transcripts.

On a similar note, TRPC6 treated cells were also positive for an inhibition of the fusion transcripts. Cells were priorly treated with androgen and radiation and scrambled RNA. The non specific RNA was inefficient in retarding the expression pattern of the chimeric transcripts. However, treatment of the cells with the TRPC6 specific RNA resulted in expression inhibition of the fusion genes. This means that genomic rearrangements are actively mediated by TRPC6 and TRPV6 and both the genes are negatively regulating the activation of ATM. Figure 3 shows remarkable shut down of fusion transcript (TMPRSS2-ERG) expression after pretreatment with siRNA (TRPV6) in lane 2 and with siRNA (TRPC6) in lane 4. Lane 5 depicts expression profile of TMPRSS2-ERG after pretreatment with androgen treatment and irradiation. Moreover, lane 3 shows a co-treatment with siRNA (TRPV6+ TRPC6). Lanes 7 and 8 are treated with scrambled RNA. Statistical analysis shows relative effect of RNA interference. Asterisks indicate significant change. Figures 4a, b and c, shows LNCaP cell line treated with Erlotinib (E) alone and in combination either with siRNA TRPV6(V) or siRNA TRPC6(C) and activation of ATM was evaluated after various time frames.

In Figure 3 we have used an individual and combinatorial blockade. We come to the conclusion that, cotreatment of the cell line with siRNA for TRPV6 and TRPV6 worked with striking synergy. The results displayed in Figure 3 are the RT PCR based amplicons run on agarose gel using suitable ladder. There is a substantial inhibition of the chimeric transcripts after treatment with TRPV6 and TRPC6 individually, yet more pronounced inhibitory effect was observed after synergistic RNA interference. This draws attention towards the application and efficacy of the combinatorial drug design. Henceforth, expression profile of multiple fusion transcripts after treatment with inhibitors of TRPV6 and TRPC6, was evaluated. It is quite impressive that RNA interference of TRP's has wide inhibitory effects. As multiple transcripts generated because of the aberrant genomic rearrangement patterns are inhibited in this particular experiment. It is interesting to note that inhibition/interference of TRPV6 and TRPC6, retards induction of wide range of rearranged genome patterns. It

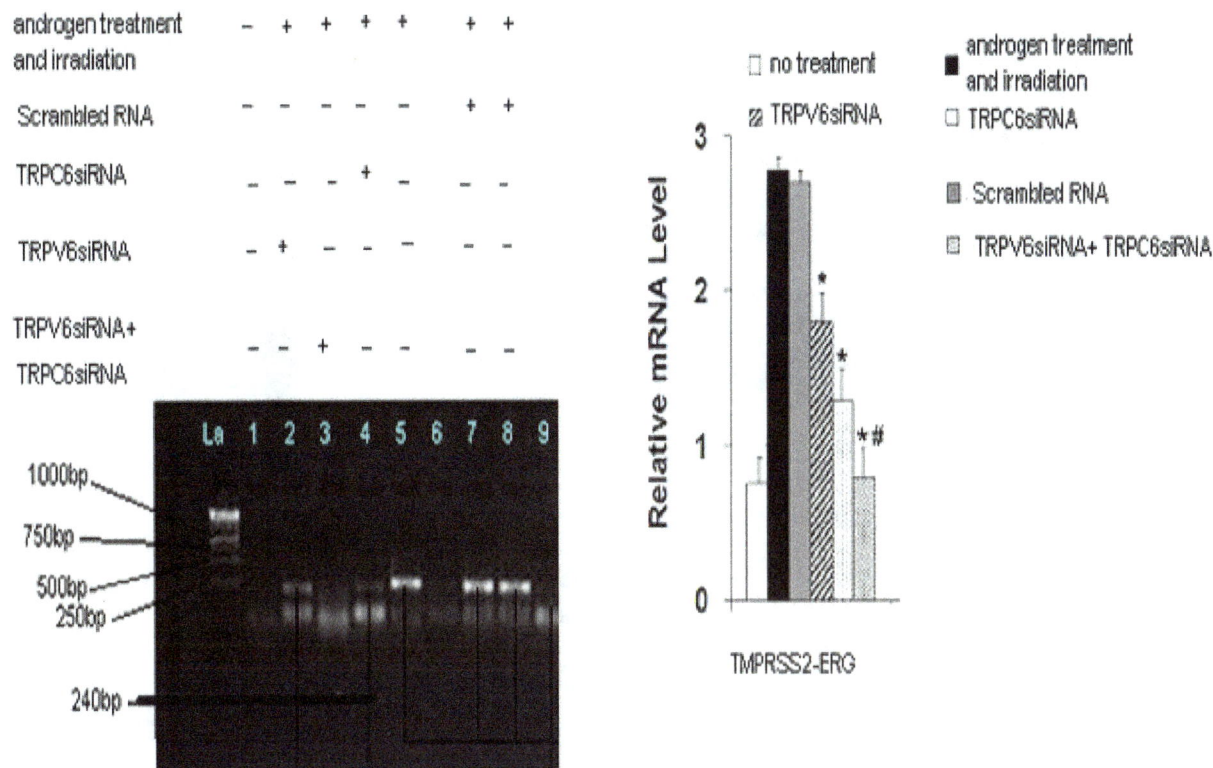

Figure 3. shows expression profile of RT PCR based amplicons of the fusion transcripts after treatment with siRNA's. Statistical analysis displays significant reductions in the fusion transcripts expression. Asterisks indicate.

is also noted in Figure 4 that, Erlotinib treatment acts as a mild stimulus for the activation of ATM. Nonetheless, in combination with siRNA for TRPC or TRPV, the influence is potentiated outstandingly. It is also attention-grabbing that, time span is influential in the activation of ATM, ATM was activated robustly after 72 h. Maximum activation of ATM was achieved after pretreatment with Erlotinib and siRNA TRPV6. It was also observed that, ATM activity in Erlotinib treated cells was similar at 24 and 48 h but there was a sudden increase in the activity at 72 h.

DISCUSSION

Prostate cancer is a multidimensional molecular disease that arises because of deviant activities of broad spectrum of mediators. TRP channels have attracted the attention of researchers. Recently, Amantini et al. (2009) documented a tight association of TRPV with ATM. They registered that activation of TRPV1 with capsaicin, resulted in the activation of ATM along with induction of apoptosis. A closer look at the credentials of ATM indicates that, it has bipartite activities. Either it is involved in the induction of apoptosis or it is engaged in DNA repair in a faithful manner. Despite the growing evidence that these channels and genomic rearrangements have wide contributions in disease

aggressiveness, no study to date addresses the tight association of the abrogation of channels and genomic rearrangements. Presuming the same assumption we designed our experiments that encompassed all these activities.

siRNA-TRPM2 was used to knock down TRPM2 expression by RT PCR. Figure indicates successful blockade of TRPM2 protein by siRNA-TRPM2. TRPM2 protein, as observed in Western blot analysis, was significantly reduced in LNCaP cells treated with siRNA-TRPM2 in comparison with treatments of scrambled siRNA. LNCaP cells treated with scrambled siRNA presented low levels of ATM phosphorylation, which according to Shiotani et al. (2006) is dependent on ADP-ribosylation of DNA. On the contrary, substantial fraction of ATM activation was observed in TRPM2 compromised cells suggesting that, an increase in ADP-ribosylation occurred. We abolished TRPV6 in order to evaluate its effect on the genomic rearrangements. It was observed that, siRNA (TRPV6) stimulated the activation of ATM and suppression of TMPRSS2-ERG. In accordance with the same concept, we treated cells with siRNA (TRPC6). The results were very much concordant in interpretations. There was a substantial downregulation of the fusion transcript expression which underscores the fact that, faithful genomic repair instigate in TRPC6 knocked down cells. It is interesting to note that, ATM activation is

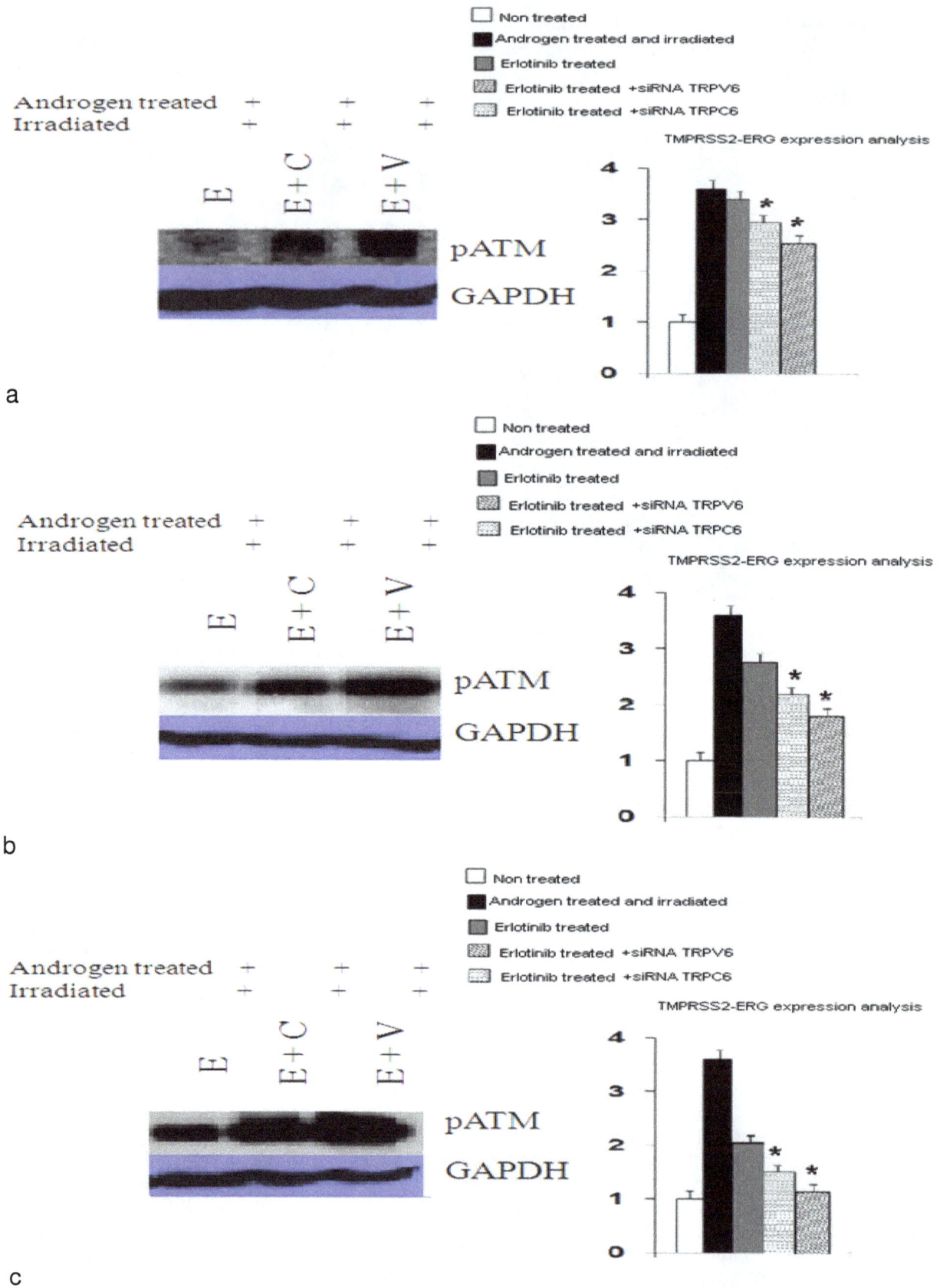

Figure 4. Treatment of LNCaP cell line with Erlotinib alone and in combination with siRNA. (a) After 24 h (b) after 48 h (c) after 72 h. Erlotinib is more effective as displayed by statistical analysis. Asterisks indicate significant change.

impaired in the TRPM2 competent LNCaP cell line. The presence of TRPM2 does not allow the autophosphorylation of ATM at serine 1981. However, treatment of cells with siRNA of TRPM2, recapitulates the activity of ATM and there is a successive retardation of genomic instability. TRPV6 and TRPC6 exclusively and synergistically hold tremendous potential to inhibit the illegitimate genome repair.

Another important finding is that, the silencing of the channels inhibits the expression of a wide range of fusion transcripts. There are tremendous efforts which are being made to induce apoptosis in castration resistant prostate cancer cells. Keeping in view the critical roles of ATM, we have to design target specific interventions. Another thing that can be ruled out is the differential activities of ATM. This study highlights the involvement of ATM in attenuation of genomic rearrangements. Blockade of the channels resulted in the activation of ATM and the DNA damage repair, instead of apoptosis. Study unfolds a different pathway of addressing the genomic insult by abrogation of TRP channels. Yet, there are some outstanding questions which are to be answered. The intermediate players in TRP channels interference mediated activation of ATM have not been identified. We have gone through the DNA repair mode after targeting of calcium channels; however, the predisposition of apoptosis is still controversial.

It is interesting to note that TRPV forms a complex with EGFR, that results in an exacerbated signaling (Cheng et al., 2010). Moreover, TRPV activation results in activation of EGFR, that stimulates downstream signaling (Yang et al., 2010). Erlotinib (OSI-774 or Tarceva), an EGFR tyrosine kinase inhibitor, has strong antineoplastic and chemopreventive efficacies in a variety of cancer types. Erlotinib is involved in the activation of p53. It modulates expression of cell cycle regulatory proteins p21 and p27 and apoptosis-regulatory protein Bim, in a p53-dependent manner. EGCG enhances the proapoptotic potential of erlotinib by inhibiting the expression of p21 and p27 (Amin et al., 2009). This is indicative of the fact that, monotherapeutic approach is inefficient in yielding desired results. Here, we have explored that Erlotinib activates an upstream kinase of DNA damage pathway, ATM because of its antineoplastic activities. Erlotinib is an EGFR inhibitor, that hampers activation of the native receptor. Moreover, Tanaka et al. (2008) did not observe any suppression of ATM after treatment of on-small cell lung cancer cells by gefitinib (EGFR inhibitor). Consistent with the interpretations in this regard are the reports that C225, a humanized antibody (Dittmann et al., 2005 a, b) that binds to the EGFR and blocks relevant downstream transduction cascade, that causes a redistribution of DNA-PK from the nucleus to the cytosol and blocks the transport of EGFR to the nucleus (Friedmann et al., 2006). It is a matter of great significance that exposure to genotoxic stress results in ligand-independent phosphorylation and activation of the EGFR (Boerner et

al., 2003). Our results indicate that, EGFR inhibitor administration along with TRP channel interference mechanism remarkably retarded genomic rearrangements. EGFR is documented to be involved in crosstalk with DNA-PK. DNA PK is the key player, that contributes in genomic rearrangements. DNA-PK overrides the activities of ATM in prostate cancer and chromosomal rearrangements are enhanced in ATM compromised cells. However, recapitulation of ATM dynamics results in faithful repair of the genome (Ammad et al., 2010a). Furthermore, we have previously documented that ATM activities are severely impaired by various key mediators in prostate cancer. Negative regulators of TGF signaling are engaged in hampering ATM autophosphorylation in LNCaP cell line. Yet, silencing of negative regulators of TGF signaling, resulted in reinvigoration of quiescent ATM. (Ammad et al., 2010b). The current study indicates that, TRP channels are undoubtedly the negative regulators of ATM activation. However, silencing of TRP channels resulted in an effective repair of the genome and a marked decrease in genomic rearrangements. It is worth mentioning that, dual inhibition of the channels brought more prominent attenuation of genomic rearrangements. On a similar note, Erlotinib administration alone is inefficient in triggering robust ATM autophosphorylation. But results are quite different if there is a combinatorial administration. Erlotinib in combination with TRPV/TRPC inhibitors, outstandingly increase autophosphorylation activity of ATM after 72 h. It is therefore effective if Erlotinib is administered in combination, as it significantly upregulates ATM dynamics and simultaneously retarding genomic rearrangements.

This is the first documented study that unfolds an interaction between calcium channels and ATM activation. In prostate cancer, these channels are remarkably upregulated and inhibit the activation of ATM after DNA damage. The exact mechanisms underlying ATM inhibition are presently unclear but further studies will add substantial information into the existing pool of negative regulators of DNA damage response. Moreover, phytonutrients based studies on suppressing or promoting calcium channel activities still lack. Hopefully, future studies will converge upon exploring effect of phytonutrients on a broad range of TRP channels and their dynamics.

REFERENCES

Ammad AF, Qaisar M, Mohammad I, Shahzad B (2010). Therapeutic Effect of Epigallocatechin-3-gallate (EGCG) and Silibinin on ATM Dynamics in Prostate Cancer Cell Line LNCaP. World J. Oncol. 1: 242-246.

Ammad AF, Qaisar M, Amir R, Taskeen MM, Maryam I, Syed AN, Zia-ur-Rehman, Shahzad Bhatti (2010). SMURF and NEDD4 interference offers therapeutic potential in chaperoning genome integrity. J. Exp. Integr. Med. DOI, 10: 5455.

Amantini C, Ballarini P, Caprodossi S, Nabissi M, Morelli MB, Lucciarini R, Cardarelli MA, Mammana G, Santoni G (2009). Triggering of of transient receptor potential vanilloid type 1 (TRPV1) by capsaicin induces Fas/CD95-mediated apoptosis of urothelial cancer cells in an ATM-dependent manner. Carcinog. 8: 1320-1329.

Amin AR, Khuri FR, Chen ZG, Shin DM (2009). Synergistic growth inhibition of squamous cell carcinoma of the head and neck by erlotinib and epigallocatechin-3-gallate: the role of p53-dependent inhibition of nuclear factor-kappaB. Cancer Prev. Res. (Phila), 2: 538-545.

Armisén R, Marcelain K, Simon F, Tapia JC, Toro J, Quest AF, Stutzin A (2011). TRPM4 enhances cell proliferation through up-regulation of the β-catenin signaling pathway. J. Cell. Physiol., 226: 103-109.

Bandyopadhyay D, Mandal M, Adam L, Mendelsohn J, Kumar R (1998). Physical interaction between epidermal growth factor receptor and DNA-dependent protein kinase in mammalian cells. J. Biol. Chem. 273: 1568-1573.

Bidaux G, Flourakis M, Thebault S, Zholos A, Beck B, Gkika D, Roudbaraki M, Bonnal JL, Mauroy B, Shuba Y, Skryma R, Prevarskaya N. (2007). Prostate cell differentiation status determines transient receptor potential melastatin member 8 channel subcellular localization and function. J. Clin. Invest., 117: 1647-1657.

Boerner JL, Danielsen A, Maihle NJ (2003). Ligand-independent oncogenic signaling by the epidermal growth factor receptor: v-ErbB as a paradigm. Exp. Cell. Res., 284: 111-121.

Cheng X, Jin J, Hu L, Shen D, Dong XP, Samie MA, Knoff J, Eisinger B, Liu ML, Huang SM, Caterina MJ, Dempsey P, Michael LE, Dlugosz AA, Andrews NC, Clapham DE, Xu H (2010). TRP channel regulates EGFR signaling in hair morphogenesis and skin barrier formation. Cell., 141: 331-343.

Dittmann K, Mayer C, Fehrenbacher B, Schaller M, Raju U, Milas L, Chen DJ, Kehlbach R, Rodemann HP (2005). Radiation-induced epidermal growth factor receptor nuclear import is linked to activation of DNA-dependent protein kinase. J. Biol. Chem. 280: 31182–31189.

Dittmann K, Mayer C, Rodemann HP (2005). Inhibition of radiation-induced EGFR nuclear import by C225 (Cetuximab) suppresses DNA-PK activity. Radiother. Oncol., 76: 157-161.

Friedmann BJ, Caplin M, Savic B, Shah T, Lord CJ, Ashworth A, Hartley JA, Hochhauser D (2006). Interaction of the epidermal growth factor receptor and the DNA-dependent protein kinase pathway following gefitinib treatment. Mol. Cancer Ther., 5: 209–218.

Hessels D, Smit FP, Verhaegh GW, Witjes JA, Cornel EB, Schalken JA (2007). Detection of TMPRSS2-ERG fusion transcripts and prostate cancer antigen 3 in urinary sediments may improve diagnosis of prostate cancer. Clin. Cancer Res., 13: 5103-5108.

Lehen'kyi V, Flourakis M, Skryma R, Prevarskaya N (2007). TRPV6 channel controls prostate cancer cell proliferation via Ca(2+)/NFAT-dependent pathways. Oncogene, 26: 7380-7385.

Lin SY, Makino K, Xia W, Matin A, Wen Y, Kwong KY, Bourguignon L, Hung MC (2001). Nuclear localization of EGF receptor and its potential new role as a transcription factor. Nat. Cell. Biol., pp. 3802-3808.

Luedeke M, Linnert CM, Hofer MD, Surowy HM, Rinckleb AE, Hoegel J, Kuefer R, Rubin MA, Vogel W and C Maier (2009). Predisposition for TMPRSS2-ERG fusion in prostate cancer by variants in DNA repair genes. Cancer Epidemiol. Biomarkers Prev., 18: 3030-3035.

Mani R, Tomlins A, Callahan K, Ghosh A, Nyati M, Varambally S, Palanisamy N and A Chinnaiyan (2009). Induced chromosomal proximity and gene fusions in prostate cancer. Science, 326: 1230

Qi W, Cooke LS, Stejskal A, Riley C, Croce KD, Saldanha JW, Bearss D, Mahadevan D (2009). MP470, a novel receptor tyrosine kinase inhibitor, in combination with Erlotinib inhibits the HER

family/PI3K/Akt pathway and tumor growth in prostate cancer. BMC Cancer, 9: 142-149.

Rothkamm K, Kühne M, Jeggo PA, Löbrich M (2001). Radiation-induced genomic rearrangements formed by nonhomologous end-joining of DNA double-strand breaks. Cancer Res., 61: 3886-3893.

Schwarz EC, Wissenbach U, Niemeyer BA, Strauss B, Philipp SE, Flockerzi V, Hoth M (2006). TRPV6 potentiates calcium-dependent cell proliferation. Cell. Calcium, 39: 163-173.

Shiotani B, Kobayashi M, Watanabe M, Yamamoto K, Sugimura T, Wakabayashi K(2006). Involvement of the ATR- and ATM-dependent checkpoint responses in cell cycle arrest evoked by pierisin-1. Mol. Cancer Res., 4: 125-133.

Tanaka T, Munshi A, Brooks C, Liu J, Hobbs ML, Meyn RE (2008). Gefitinib radiosensitizes non-small cell lung cancer cells by suppressing cellular DNA repair capacity. Clin. Cancer Res., 14: 1266-1273.

Tomlins SA, Mehra R, Rhodes DR, Smith LR, Roulston D, Helgeson BE, Cao X, Wei JT, Rubin MA, Shah RB, Chinnaiyan AM (2006). TMPRSS2:ETV4 gene fusions define a third molecular subtype of prostate cancer. Cancer Res., 66: 3396–3400.

Tomlins SA, Rhodes DR, Perner S, Dhanasekaran SM, Mehra R, Sun XW, Varambally S, Cao X, Tchinda J, Kuefer R, Lee C, Montie JE, Shah RB, Pienta KJ, Rubin MA, Chinnaiyan AM (2005). Recurrent fusion of TMPRSS2 and ETS transcription factor genes in prostate cancer. Science, 310: 644–648.

Valero M, Morenilla-Palao C, Belmonte C, Viana F (2010). Pharmacological and functional properties of TRPM8 channels in prostate tumor cells. Pflugers Arch. AOP., 461: 99-114.

Wang Y, Yue D, Li K, Liu YL, Ren CS, Wang P (2010). The role of TRPC6 in HGF-induced cell proliferation of human prostate cancer DU145 and PC3 cells. Asian J. Androl. AOP., 12: 841-852.

Yang H, Wang Z, Capó-Aponte JE, Zhang F, Pan Z, Reinach PS (2010). Epidermal growth factor receptor transactivation by the cannabinoid receptor (CB1) and transient receptor potential vanilloid 1 (TRPV1) induces differential responses in corneal epithelial cells. Exp. Eye Res., 91: 462-471.

Yang ZH, Wang XH, Wang HP, Hu LQ (2009). Effects of TRPM8 on the proliferation and motility of prostate cancer PC-3 cells. Asian J. Androl., 11: 157-165.

Yue D, Wang Y, Xiao JY, Wang P, Ren CS (2009). Expression of TRPC6 in benign and malignant human prostate tissues. Asian J. Androl., 11: 541-547.

Zeng X, Sikka SC, Huang L, Sun C, Xu C, Jia D, Abdel-Mageed AB, Pottle JE, Taylor JT, Li M (2010). Novel role for the transient receptor potential channel TRPM2 in prostate cancer cell proliferation. Prostate Cancer Prostatic Dis., 13: 195-201.

Zhao XZ, Guo HQ, Liu GX, Ji CW, Zhang SW, Liu TS, Gan WD, Li XG (2010). Inhibitory effect of TRPV6 silencing on prostate cancer cell line LNCaP in vitro. Zhonghua Nan Ke Xue, 16: 423-427.

7

Identification of *Ralstonia solanacearum* using conserved genomic regions

Alka Grover[1]*, Abhinav Grover[2], S. K. Chakrabarti[1], Wamik Azmi[3], Durai Sundar[2] and S. M. P. Khurana[4]

[1]Division of Crop Improvement, Central Potato Research Institute, Shimla 171001, H.P. India.
[2]Department of Biochemical Engineering and Biotechnology, Indian Institute of Technology Delhi, Hauz Khas, New Delhi 110016, India.
[3]Department of Biotechnology, Himachal Pradesh University, Shimla,171005, H.P. India.
[4]Amity Institute of Biotechnology, Amity University, Haryana, 122413, India.

The aim of the present study is to develop a scheme for identification of *Ralstonia solanacearum* with high specificity based on conserved genomic regions. Short Tandem Repeats (STRs) in *R. solanacearum* genome were searched using Tandem Repeat Finder software. A total of 189 and 74 STRs were found in chromosomal and megaplasmid DNA respectively. Sequence homology of these STRs analyzed using BLAST showed that out of total 273 STRs only nine were found unique for *R. solanacearum*. Correspondingly nine pairs of primers were synthesized for the flanking regions of these STRs. Sequence homology of these primer pairs carried using BLAST revealed that out of the nine pairs, one pair uniquely matched only at a single locus in the *Ralstonia* chromosomal DNA. Polymerase chain reaction (PCR) amplification using templates from 44 different isolates of *R. solanacearum* yielding single sized amplicon ascertains the versatility and unambiguousness of the designed primers. The fact that the primer pair did not amplify the genomic DNA of 12 soil bacteria establishes the specificity to *R. solanacearum*. Thus the novel and specific primers designed for *R.solanacearum* would enable fast and definitive identification of the lethal pathogen. The designed primers would be of great importance for detection of *R. solanacearum* in seed tubers, soil and water streams thus helping in establishing preventive measures for checking pathogen spread. This would also allow facilitating epidemiological studies allowing better surveillance of this pathogen.

Key words: *Ralstonia solanacearum*, identification, specific primers, differentiation, polymerase chain reaction.

INTRODUCTION

Brown rot disease caused by *Ralstonia solanacearum* is the second major constraint to potato production in tropical and subtropical regions worldwide after late blight (Hayward, 1991). Worldwide increase in number of infected sites (Castillo and Greenberg, 2007) emphasizes the need for efficient identification tests for material exchanges and epidemiological studies. Polymerase chain reaction (PCR) is the standard and preferred identification technique for *R. solanacearum* (Kutin et al., 2009). Different regions of the genome of *R.*

(Taghavi et al., 1996). These include ribosomal genesviz. 16S rDNA sequences (Seal et al., 1993), 16S rRNA (Pastrik and Maiss, 2000) and internally transcribed spacer region between 16S and 23S (Pastrik et al., 2002). In addition to ribosomal genes other genes of *R. solanacearum* have also been used for specific primers design like tRNA consensus primers (Seal et al., 1992) polygalacturonase gene (Gillings et al., 1993), the *hrp* gene region (Poussier and Luisetti, 2000), insertion sequences (Lee et al., 2001) and targeting the gene coding for the flagella subunit, *"fli C"* (Schonfeld et al., 2003).

Multi-copy target sequences allow greater sensitivity as compared to single or low-copy target sequences. The potential multi-copy target sequences for specific

*Corresponding author. E-mail: alkagrover@hotmail.com.

Table 1. Soil borne bacteria used for checking the specificity of designed PCR primers.

Serial No.	MTCC No.	Name of the bacterial strains
1	532	*Agrobacterium rhizogenes*
2	2274	*Bacillus subtilis*
3	1428	*Erwinia carotovora subsp. Carotovora*
4	134	*Nitrosomonas europea*
5	135	*Nitrobacter winogradskyi*
6	122	*Paenibacillus polymyxa*
7	2406	*Pantoea agglomerans*
8	1748	*Pseudomonas fluorescence*
9	2758	*Pseudomonas marginalis*
10	2475	*Pseudomonas putida*
11	99	*Rhizobium leguminosarum*
12	2286	*Xanthomonas campestris*

amplifications include the short, interspersed tandem repeats (STRs) present in bacterial genome. The presence of prokaryotic STRs is well documented and has been shown to be ideal candidates for use as identification sequences, since they are, in general dispersed around the chromosome and are often non-coding sequences (van Belkum et al., 1998). The sequences bordering the tandem repeats are conserved and unique (Salaun et al., 2006). The genome sequence of *R. solanacearum* has been determined (Salanoubat et al., 2002). It has been shown that the 5.8-Mb genome is organised into two replicons, a 3.7-Mb chromosome and a 2.1-Mb megaplasmid. A total of 2, 21 729 STRs with a motif length between 1 and 10 bp have been reported in the entire *R. solanacearum* genome (Coenye and Vandamme, 2003). In the present study we describe the identification of *R. solanacearum* through specific PCR primers, designed for the conserved sequences which flank the tandem repeat regions found in genome. The differentiation capability and specificity of designed primers was assessed by PCR amplification of genomic DNA of 12 soil borne bacteria.

MATERIALS AND METHODS

Strains and culture conditions

The 44 isolates of *R. solanacearum* were isolated from 16 terraces of a single field of approximately two hectares area of Central Potato Research Institute, Shimla, India (2000 m above mean sea level in the Himalayan mountain range) for our previous study (Grover et al., 2006). Isolates of *R. solanacearum* were isolated from infected potato (*Solanum tuberosum* L.), tomato (*Lycopersicon esculentum* L.) and *Solanum chacoense* plants, by standard procedure on casamino acid peptone glucose (CPG) agar medium (Kelman, 1983). 12 soil-borne bacterial cultures (Table 1) were procured from Microbial Type Culture Collection (MTCC), Institute of Microbial Technology, Chandigarh, India. All the bacterial strains were grown on LB medium at 37ºC (Sambrook et al., 1989) except

R. solanacearum which was grown on TTC agar medium (Kelman, 1983) at 28ºC.

Identification of tandem repeats

The genome sequence of *R. solanacearum* has been determined by (Salanoubat et al., 2002). It has been shown that 5.8 Mb genome is organized into two replicons, a 3.7 Mb chromosome and a 2.1 Mb megaplasmid. The sequences for the chromosomal DNA and megaplasmid DNA are deposited in the EMBL database under accession numbers AL 646052 and AL 646053 respectively. The complete genome of *R. solanacearum* was analyzed to search tandem repeats with the help of Tandem Repeat Finder (TRF) software (Benson, 1999). The genome sequences of the chromosomal (AL 646052) and mega plasmid DNA (AL 646053) of *R. solanacearum* was retrieved from the NCBI (www.ncbi.nlm.nih.gov) website. Input to the TRF consisted of a chromosomal and megaplasmid sequence file in the FASTA format. A total of 189 and 74 repeats were found in chromosomal and megaplasmid DNA respectively. Twelve repeats from chromosomal DNA and 25 repeats from megaplasmid were selected for BLAST search to get *R. solanacearum* specific loci. Out of these 37 loci, nine (five from chromosome and four from megaplasmid) repeat loci were found to be unique to *R. solanacearum* (Table 2).The identified tandem repeats were named according to their location on the *R. solanacearum* genome. 240 bases of upstream and downstream sequences of unique tandem repeats were used for primer designing. Primer 3.0 software (www.primer3.com) was used for producing five primer pairs (primer pairs 0 to 4) for each amplicon. These primer pairs were further used for carrying out BLAST search onto the database- All GenBank+EMBL+DDBJ+PDB. Only those oligonucleotides which uniquely showed similarity with *R. solanacearum* genome were selected and got custom synthesized (Integrated DNA Technologies, IA, USA). The Tm and GC content of these primers were kept on an average 55ºC and 55% respectively.

Genomic DNA extraction and PCR amplification

Genomic DNA of all the bacterial strains, *R. solanacearum* as well as other soilborne bacteria was isolated through alkaline lysis proteinase K method (Sambrook et al., 1989). The quality of the

Table 2. Unique repeat loci in *R. solanacearum* genome.

Name	Consensus pattern (bp)	Period size	Copy No.	Percentage match	Location in genome
CHR 1	TCG CAA	6	7	100	133827-133868
CHR 2	GGA TCG GCA	9	15.9	97	754254-754396
CHR 3	TCG GA	5	12	90	1144916-1144975
CHR 4	CGA CCG ACT	9	3.9	84	1457884-1457918
CHR 5	AAT GGT TG	8	7.3	98	3461892-3461949
MPR 1	GCG ACC GAA	9	5.6	100	266247-266296
MPR 2	GTG CTC	6	9.5	94	761222-761278
MPR 3	CAG AAG	6	16.7	83	1118782-1118881
MPR 4	CAA TGG ACG	9	10.1	95	1549270-1549360

CHR and MPR indicate chromosomal and megaplasmid DNA respectively.

extracted DNA was checked by means of electrophoresis in 0.8% agarose gels, followed by staining with ethidium bromide. The purity of the DNA was estimated from the A260/A280 ratio, whereas the yield was obtained by measuring absorbance at 260 nm with a UV spectrophotometer (BioRad, USA). PCR optimization with selected primers was carried out following the modified Taguchi methods described by Cobb and Clarkson (1994). This involved varying concentrations of DNA (10, 20, 30, 40 and 50 ng), $MgCl_2$ (1, 1.5 and 2 mM) and primer (2, 20, 30, 40 and 50 pM) while keeping the other reaction components constant. Reactions were carried out in a final volume of 25 µl containing 1.0 U Taq DNA polymerase (Applied Biosystems, USA), 1X PCR buffer (Applied Biosystems, USA), and 200 $mmol^{-1}$ dNTPs. Reactions were amplified in a Perkin Elmer thermal cycler (GeneAmp PCR System 9700). The temperature regime used for PCR amplification was 2 min denaturation at 95ºC followed by 40 cycles of 5 s denaturation at 95ºC, 30 s annealing at 60ºC and 30 s elongation at 72ºC and one cycle at 72ºC for 10 min. PCR products were analyzed on 1% agarose gels in TAE buffer at 10 V/cm.

Specificity assessment of the primers

DNA of 44 isolates of *R. solanacearum* and 12 soil borne bacteria (Table 1) were used as templates and a reaction mixture without DNA was used as control.

RESULTS

Identification of tandem repeats

The genome of *R. solanacearum* was analyzed for abundance of tandem repeats and 189 tandem repeats in chromosomal DNA and 74 repeats in megaplasmid were identified. Period size of tandem repeats ranged from 5 to 469 in chromosomal DNA and from 6 to 333 in megaplasmid. The copy number of tandem repeats ranged from 1.9 to 18.8 in chromosomal DNA and from 1.8 to 18.9 in megaplasmid. These loci were distributed around the genome from 1, 33, 827 bp to 3, 630, 007 bp in chromosomal DNA and from 4, 50, 393 bp to 2, 040, 453 bp in megaplasmid. Of the 37 repeats selected for BLAST search, nine (five from chromosome and four

from megaplasmid) were specifically unique to *R. solanacearum* (Table 2). Nine pairs of primers complementary to flanking sequences of each repeat locus were selected (Table 3). After BLAST search it was found that out of these nine primer pairs, eight have sequence similarity to multiple sites of chromosomal and megaplasmid DNA of *R. solanacearum*. Only one primer pair CHR-5 (forward and reverse) uniquely matched at a single locus in the chromosomal DNA (Table 3). The amplification conditions were optimized as: DNA concentration 50 ng, $MgCl_2$ concentration 2.0 $mmol^{-1}$ and primer concentration 30 $pmol^{-1}$.

PCR amplification

After PCR amplification of genomic DNA isolated from 44 isolates of *R. solanacearum* with nine primer pairs, eight primer pairs produced multiple bands whereas only one primer pair CHR-5 amplified a single band of size approximately 730 bp from all the 44 isolates of *R. solanacearum* (Figure 1). The sequencing result of this amplicon when aligned with *R. solanacearum* genome using BLAST matched fairly well with the CHR-5 locus. Analysis demonstrated polymorphism in the size of PCR products at eight loci except CHR-5 (Table 4). At only one locus CHR-5, all the 44 isolates produced the PCR product of same size. Being able to produce only single band the locus CHR-5 was selected and primer pair corresponding to this locus was selected for further analysis.

Specificity of PCR primers

In order to determine whether the chosen locus CHR-5 was specific for *R. solanacearum* and could be used for its identification, CHR-5 primer pair was used for PCR amplification of genomic DNA isolated from 12 other soil borne bacteria and *R. solanacearum*. After PCR

Table 3. Primers for flanking regions of unique repeat loci in *R. solanacearum* genome.

Repeat region		Flanking Primers	Tm (°C)
CHR 1	F	GAT TCG CAT GCC AAG GTC	54.2
	R	GCT GAC GTA CTT CCA GCA GA	56.9
CHR 2	F	ACA GCG ACG TGG TGT CTT C	57.6
	R	CGG GCT GAC TAA CGT ACT GC	57.9
CHR 3	F	GTA GCC CCA CAC GTG GTC	58.4
	R	GCA AGA CGT ATC GCA GCA C	56.7
CHR 4	F	AGA ACA ACA CGC GGG AAG	55.6
	R	CTT GAG GTA CGG ATC GCA CA	56.9
CHR 5	F	TCG TGT GTC GAA AGA GTG CT	56.5
	R	CTT GTC TGC CTC GAG TTG TG	54.2
MPR 1	F	GGT CGC CAT GTG GTA TCC	55.7
	R	TTC CCA GGT CGT CAC ATA CC	56.5
MPR 2	F	CTT GCC CTT GTC GTG CTG	56.9
	R	GCT GTT CTA CCC GTT CAT CC	55.6
MPR 3	F	GTC AGC AGC CCA TTT CAT C	54.1
	R	CTG GTC TCC TCC AGG TGC T	58.9
MPR 4	F	GGC TGA TCT GCG TCG AAT AC	55.8
	R	CGA CCG TAT GCT CCT GCT	57.2

amplification the desired band was absent in soil borne bacteria and present in all the isolates of *R. solanacearum* (Figure 2).

DISCUSSION

In the present study, PCR primers specific to *R. solanacearum* were designed which are complementary to the flanking sequences of the tandem repeats present in the *Ralstonia* genome. A number of studies have supported the notion that tandem repeats reminiscent of mini and microsatellites are likely to be a highly significant source of informative markers for the identification of pathogenic bacteria (Adair et al., 2000; Nascimento et al., 2004; Salaun et al., 2006; Supply et al., 2000). This emphasizes the important contribution of tandem repeats to the adaptation of the pathogen to its host. The availability of whole genome sequences has opened the way to the systematic evaluation of tandem repeats diversity and application to epidemiological studies. There are a number of databases of tandem repeats from publicly available genomes which facilitates the identification and selection of tandem repeats (Chang et al., 2007; Le Flèche et al., 2001). The usage of tandem repeats for epidemiology purposes in bacteria has been reported only scarcely for a few species like *Haemophilus influenza* (van Belkum et al., 1997), *Yersinia pestis*, *Bacillus anthrasis* (Le Flèche et al., 2001) *Mycobacterium tuberculosis* (Supply et al., 2001), *Salmonella enterica* (Witonski et al., 2006) *Paracoccidioides brasiliensis* (Nascimento et al., 2004) and *Leptospirra interrogans* (Salaun et al., 2006) etc.

In this study, we used this molecular approach for identification of *R. solanacearum* by designing novel primer pairs bordering the STR regions. The region bordering the repeats are generally well conserved targets for PCR mediated amplification. Border sequence conservation is sometimes even observed amongdifferent species, allowing a broad spectrum analysis of the nature of the species and subspecific genetic polymorphisms (Schlotterer et al., 1991). As a first step we suggested nine primer pairs for the flanking DNA sequences of tandem repeats. But PCR amplification of eight out of nine primer pairs produced polymorphic banding pattern with 44 isolates of *R. solanacearum*. It was further confirmed during BLAST search that though these primer pairs showed significant identity to *R. solanacearum* bipartite genome, the matching loci were more than one. Polymorphic banding pattern observed in the PCR analysis suggests the homology of these primer sequences to multiple sites of *R. solanacearum* genome. The propensity towards expansion or reduction of the number of repeat units at a given locus through slipped strand mispairing events is emphasized by epidemiological studies. Even genetically

Table 4. Number of bands amplified with 9.

Primers Name	RS 1	RS 2	RS 3	RS 4	RS 5	RS 6	RS 7	RS 8	RS 9	RS 10	RS 11	RS 12	RS 13	RS 14	RS 15	RS 16	RS 17	RS 18	RS 19	RS 20	RS 21	RS 22
CHR-1	6	6	6	8	8	6	6	5	5	5	5	6	9	6	1	8	6	6	6	6	6	6
CHR-2	2	2	3	3	3	3	3	1	3	5	3	2	4	1	4	0	3	5	4	4	5	4
CHR-3	2	2	2	1	2	2	1	2	2	2	2	2	1	2	0	2	2	2	2	1	1	3
CHR-4	1	2	1	1	1	4	4	2	3	1	1	2	1	2	1	1	1	2	2	2	3	3
CHR-5	1	1	1	1	1	1	1	1	1	1	1	1	1	1	1	1	1	1	1	1	1	1
MPR-1	0	0	3	1	0	0	0	0	0	1	1	0	0	0	0	0	1	0	0	0	1	0
MPR-2	1	2	1	2	0	1	2	0	1	1	3	2	0	2	0	1	2	0	3	1	4	4
MPR-3	1	1	0	1	0	0	0	1	0	0	1	1	0	0	0	1	1	0	0	1	0	0
MPR-4	0	1	1	4	0	3	2	1	3	1	1	1	1	3	0	0	2	2	3	1	4	3

	RS 23	RS 24	RS 25	RS26	RS27	RS 28	RS 29	RS 30	RS 31	RS 32	RS 33	RS 34	RS 35	RS 36	RS 37	RS 38	RS 39	RS 40	RS 41	RS 42	RS 43	RS 44	$\overline{\Sigma}$
CHR-1	8	6	5	5	5	5	5	1	6	3	4	6	5	4	6	5	5	5	5	5	4	4	5.43
CHR-2	2	5	5	3	4	5	3	4	4	5	5	4	5	4	4	5	4	5	5	5	3	3	3.61
CHR-3	2	0	2	2	1	1	2	3	1	2	3	1	2	1	1	2	1	1	2	2	2	2	1.66
CHR-4	4	2	3	2	3	2	2	1	2	3	4	3	3	3	3	3	3	2	5	3	3	4	2.36
CHR-5	1	1	1	1	1	1	1	1	1	1	1	1	1	1	1	1	1	1	1	1	1	1	1.00
MPR-1	2	1	1	1	1	0	0	0	0	0	0	0	0	2	1	2	0	0	3	2	0	3	0.61
MPR-2	4	2	1	2	2	1	2	2	0	0	2	2	1	1	0	1	3	3	2	3	1	4	1.64
MPR-3	0	0	0	0	1	0	0	0	0	0	0	0	0	1	0	2	2	2	2	2	1	1	0.55
MPR-4	2	1	2	4	3	1	3	0	1	1	3	2	1	3	1	3	1	1	3	1	1	3	1.80

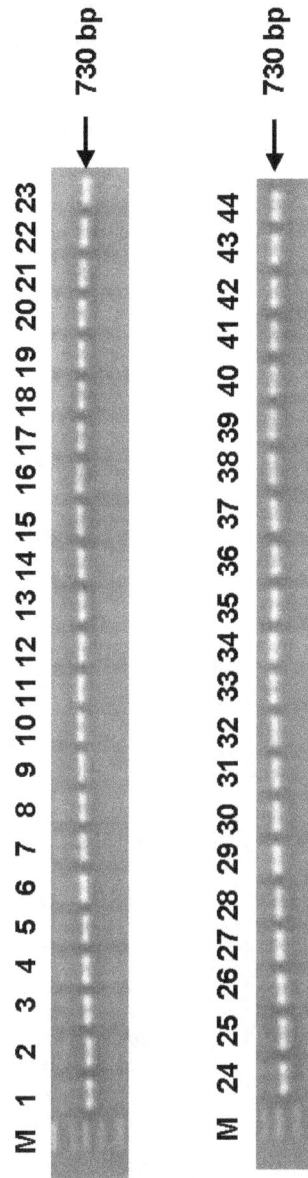

M 1 2 3 4 5 6 7 8 9 10 11 12 13 14 15 16 17 18 19 20 21 22 23

← 730 bp

M 24 25 26 27 28 29 30 31 32 33 34 35 36 37 38 39 40 41 42 43 44

← 730 bp

Figure 1. PCR amplification of genomic DNA of 44 isolates of *R. solanacearum* with primers CHR 5F and CHR 5R. Lanes 1 to 44 represent *R. solanacearum* isolates RS1-RS 44. All the 44 isolates were showing PCR amplification. M represents 1kb DNA ladder (MBI FERMENTAS, USA).

1 2 3 4 5 6 7 8 9 10 11 12 M 13 14 15 16 17 18 19 20 21 22 23 24

730 bp →

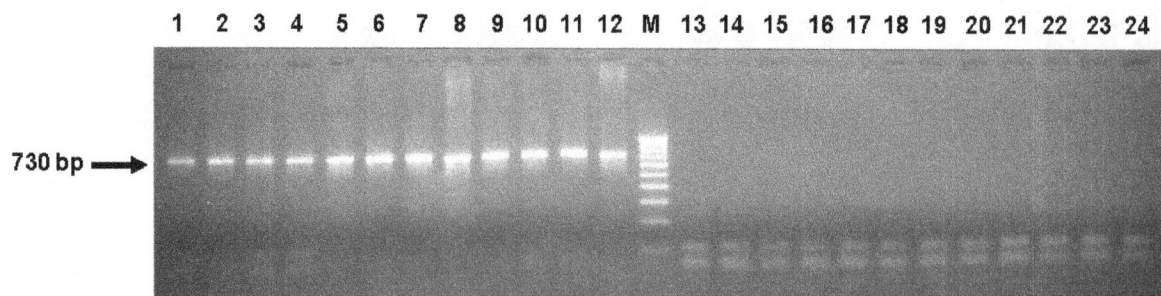

Figure 2. PCR amplification of *R. solanacearum* isolates (RS1-RS12) and different soil borne bacteria with CHR 5 F and CHR 5 R primer pair. Only the genomic DNA of *R. solanacearum* was PCR amplified. Lanes 1-12 *R. solanacearum* isolates. Lane 13: *Agrobacterium rhizogenes*. Lane 14: *Bacillus subtilis*. Lane 15: *Erwinia carotovora subsp. Carotovora*. Lane 16: *Nitrosomonas europea*. Lane 17: *Nitrobacter winogradskyi*. Lane 18: *Paenibacillus polymyxa*. Lane 19: *Pantoea agglomerans*. Lane 20: *Pseudomonas fluorescence* Lane 21: *Pseudomonas marginalis*. Lane 22: *Pseudomonas putida*. Lane 23: *Rhizobium leguminosarum*. Lane 24: *Xanthomonas campestris*.M represents 100 bp DNA ladder (MBI Fermentas; Fermentas Inc., Hanover, MD, USA).

3461892

3461852- CGCGCGTCTGCGACGTCGCCGACACGCTCAGCCGCCGCGA**AA**

TGGTTGAATGGTTGAATGGTTGAATGGTTGAATGGTTGAATGGTTGAATGG

CTGAAGGCGGAGCACCCTGCCCTCAATATGGTCATCAACAATGCCGGC

3461949

Ralstonia solanacearum				
Position	Total length	Unit length	Copy number	% match
3461892 - 3461949	58 bp	8	7.3	98

Figure 3. Schematic presentation of unique locus present in chromosomal DNA of *R. solanacearum*. This locus ranges from 3,461,892 - 3,461,949 bp at chromosome of *R. solanacearum* (Shown in bold letters). Left and right border sequences of length 240 bp were used to design PCR primers

homologous strains may show differences in STR size, even though these may be small as compared to those determined for more distantly related strains (Vanham et al., 1993). In the present study in addition to identify more than one place in bipartite genome, the slipped strand mispairing, polymerase slippage and repeat locus multiplication can also be the reason for polymorphic banding patterns amplified with eight primer pairs.

Since CHR-5 primers amplified a single band from all 44 isolates of *R. solanacearum* and not from any other soil borne bacteria, it is proposed here that this locus is specific to *R. solanacearum*. This locus CHR-5 is located

within the transposase protein and a probable oxidoreductase protein. Left and right border sequences of this locus including 240 bp lengths were used to design PCR primers. When these PCR primers were analyzed by BLAST search, forward primer showed identity with chromosome of *R. solanacearum* from 3,461,516 bp to 3,461,535 bp and reverse primer showed identity with chromosome of *R. solanacearum* from 3,462,391 bp to 3,462,372 bp. It is evident from this data that the primers are homologous to the region (3,461, 892 - 3,461,949) flanking the CHR-5 locus (Figure 3). Thus CHR-5 is one of the marker that could be used for specific identification of *R. solanacearum*. The gene organization around CHR-5 seems to be conserved, irrespective of the isolate and the species considered. Although searching for tandem repeats in *R. solanacearum* genome revealed 289 tandem repeats, but only one locus CHR-5 was identified as a useful marker. The analysis of the isolates of *R. solanacearum* collected from a single field of C.P.R.I. Shimla using the tandem repeat primers designed in the present study also confirmed the genetic diversity which has already been indicated by the RAPD analysis conducted by us previously (Grover et al., 2006) for the same isolates. This indicates the usefulness of tandem repeats for epidemiological investigations of *R. solanacearum*.

ACKNOWLEDGEMENTS

The authors are grateful to the Director and Head of the Division of Crop Improvement, Central Potato Research Institute, Shimla for providing necessary facilities to undertake this study. This work was supported by grants from the Council of Scientific and Industrial Researchunder the sanction no. F.No.2-48/2001 (II) EU.II.

REFERENCES

Adair DM, Worsham PL, Hill KK, Klevytska AM, Jackson PJ, Friedlander AM, Keim P (2000). Diversity in a variable-number tandem repeat from *Yersinia pestis*. J. Clin. Microbiol., 38(4): 1516-1519.

Benson G (1999). Tandem repeats finder: a program to analyze DNA sequences. Nucleic Acids Res., 27(2): 573-580.

Castillo JA, Greenberg JT (2007). Evolutionary dynamics of *Ralstonia solanacearum*. Appl. Environ. Microbiol., 73(4): 1225-1238.

Chang CH, Chang YC, Underwood A, Chiou CS, Kao CY (2007). VNTRDB: A bacterial variable number tandem repeat locus database. Nucleic Acids Res., 35: 416-421.

Cobb BD, Clarkson JM (1994). A Simple Procedure for Optimizing the Polymerase Chain-Reaction (Pcr) Using Modified Taguchi Methods. Nucleic Acids Res., 22(18): 3801-3805.

Coenye T, Vandamme P (2003). Simple sequence repeats and compositional bias in the bipartite *Ralstonia solanacearum* GM11000 genome. Bmc Genomics, p. 4.

Gillings M, Fahy P, Davies C (1993). Restriction Analysis of an Amplified Polygalacturonase Gene Fragment Differentiates Strains of the Phytopathogenic Bacterium *Pseudomonas solanacearum*. Lett. Appl. Microbiol., 17(1): 44-48.

Grover A, Azmi W, Gadewar AV, Pattanayak D, Naik PS, Shekhawat GS, Chakrabarti SK (2006). Genotypic diversity in a localized population of *Ralstonia solanacearum* as revealed by random amplified polymorphic DNA markers. J. Appl. Microbiol., 101(4): 798-806.

Hayward AC (1991). Biology and Epidemiology of Bacterial Wilt Caused by *Pseudomonas solanacearum*. Annu. Rev. Phytopathol., 29: 65-87.

Kelman A (1983). Citation Classic - The Relationship of Pathogenicity in *Pseudomonas solanacearum* to Colony Appearance on a Tetrazolium Medium. Curr. Contents/Agric. Biol. Environ. Sci., (19): 20-20.

Kutin RK, Alvarez A, Jenkins DM (2009). Detection of *Ralstonia solanacearum* in natural substrates using phage amplification integrated with real-time PCR assay. J. Microbiol. Methods, 76(3): 241-246.

Le Flèche P, Hauck Y, Onteniente L, Prieur A, Denoeud F, Ramisse V, Sylvestre P, Benson G, Ramisse F, Vergnaud G (2001). A tandem repeats database for bacterial genomes: Application to the genotyping of *Yersinia pestis* and *Bacillus anthracis*. BMC Microbiol., p. 1.

Lee YA, Fan SC, Chiu LY, Hsia KC (2001). Isolation of an insertion sequence from *Ralstonia solanacearum* race 1 and its potential use for strain characterization and detection. Appl. Environ. Microbiol., 67(9): 3943-3950.

Nascimento T, Martinez R, Lopes AR, Bernardes LAD, Barco CP, Goldman MHS, Taylor JW, McEwen JG, Nobrega MP, Nobrega FG, Goldman GH (2004). Detection and selection of microsatellites in the genome of *Paracoccidioides brasiliensis* as molecular markers for clinical and epidemiological studies. J. Clin. Microbiol., 42(11): 5007-5014.

Pastrik KH, Elphinstone JG, Pukall R (2002). Sequence analysis and detection of *Ralstonia solanacearum* by multiplex PCR amplification of 16S-23S ribosomal intergenic spacer region with internal positive control. Eur. J. Plant Pathol., 108(9): 831-842.

Pastrik KH, Maiss E (2000). Detection of *Ralstonia solanacearum* in potato tubers by polymerase chain reaction. J. Phytopathol., 148(11-12): 619-626.

Poussier S, Luisetti J (2000). Specific detection of biovars of *Ralstonia solanacearum* in plant tissues by Nested-PCR-RFLP. Eur. J. Plant Pathol., 106(3): 255-265.

Salanoubat M, Genin S, Artiguenave F, Gouzy J, Mangenot S, Arlat M, Billault A, Brottier P, Camus JC, Cattolico L, Chandler M, Choisne N, Claudel-Renard C, Cunnac S, Demange N, Gaspin C, Lavie M, Moisan A, Robert C, Saurin W, Schiex T, Siguier P, Thebault P, Whalen M, Wincker P, Levy M, Weissenbach J, Boucher CA (2002). Genome sequence of the plant pathogen *Ralstonia solanacearum*. Nature, 415(6871): 497-502.

Salaun L, Merien F, Gurianova S, Baranton G, Picardeau M (2006). Application of multilocus variable-number tandem-repeat analysis for molecular typing of the agent of leptospirosis. J. Clin. Microbiol., 44(11): 3954-3962.

Sambrook J, Fritsch EF, Maniats TA (1989). Molecular Cloning: A Laboratory Manual 2nd edition. Cold Spring Harbor Laboratory Press, Cold Spring Harbor, New York, pp: 1.21-1.24, 9.14-9.23.

Schlotterer C, Amos B, Tautz D (1991). Conservation of Polymorphic Simple Sequence Loci in Cetacean Species. Nature, 354(6348): 63-65.

Schonfeld J, Heuer H, van Elsas JD, Smalla K (2003). Specific and sensitive detection of *Ralstonia solanacearum* in soil on the basis of PCR amplification of fliC fragments. Appl. Environ. Microbiol., 69(12): 7248-7256.

Seal SE, Jackson LA, Daniels MJ (1992). Use of Transfer-Rna Consensus Primers to Indicate Subgroups of *Pseudomonas solanacearum* by Polymerase Chain-Reaction Amplification. Appl. Environ. Microbiol., 58(11): 3759-3761.

Seal SE, Jackson LA, Young JPW, Daniels MJ (1993). Differentiation of *Pseudomonas solanacearum*, *Pseudomonas syzygii*, *Pseudomonas Pickettii* and the Blood-Disease Bacterium by Partial 16s-Ribosomal-Rna Sequencing - Construction of Oligonucleotide Primers for Sensitive Detection by Polymerase Chain-Reaction. J. Gen. Microbiol., 139: 1587-1594.

Supply P, Lesjean S, Savine E, Kremer K, van Soolingen D, Locht C (2001). Automated high-throughput genotyping for study of global epidemiology of *Mycobacterium tuberculosis* based on mycobacterial interspersed repetitive units. J. Clin. Microbiol., 39(10): 3563-3571.

Supply P, Mazars E, Lesjean S, Vincent V, Gicquel B, Locht C (2000). Variable human mini satellite-like regions in the *Mycobacterium tuberculosis* genome. Mol. Microbiol. 36(3): 762-771.

Taghavi M, Hayward C, Sly LI, Fegan M (1996). Analysis of the phylogenetic relationships of strains of *Burkholderia solanacearum*, *Pseudomonas syzygii*, and the blood disease bacterium of banana based on 16S rRNA gene sequences. Int. J. Syst. Bacteriol., 46(1): 10-15.

van Belkum A, Scherer S, van Alphen L, Verbrugh H (1998). Short-sequence DNA repeats in prokaryotic genomes. Microbiol. Mol. Biol. Rev., 62(2): 275.

vanBelkum A, Scherer S, vanLeeuwen W, Willemse D, vanAlphen L,

Verbrugh H (1997). Variable number of tandem repeats in clinical strains of *Haemophilus influenzae*. Infect. Immun., 65(12): 5017-5027.

Vanham SM, Vanalphen L, Mooi FR, Vanputten JPM (1993). Phase Variation of *Haemophilus influenzae* Fimbriae - Transcriptional Control of 2 Divergent Genes through a Variable Combined Promoter Region. Cell, 73(6): 1187-1196.

Witonski D, Stefanova R, Ranganathan A, Schutze GE, Eisenach KD, Cave MD (2006). Variable-number tandem repeats that are useful in genotyping isolates of *Salmonella enterica* subsp enterica serovars Typhimurium and Newport. J. Clin. Microbiol., 44(11): 3849-3854.

8

Nutrigenomic analysis of mulberry silkworm (*Bombyx mori* L.) strains using polymerase chain reaction - simple sequence repeats (PCR-SSR)

Chinnaswamy Ramesha[1*], Savarapu Sugnana Kumari[2], Chebba Moremnagari Anuradha[3], Hothur Lakshmi[4] and Chitta Suresh Kumar[5]

[1]Silkworm Breeding and Molecular Genetics Laboratory, Andhra Pradesh State Sericulture   Research and Development Institute (APSSRDI), Kirikera-515 211, Hindupur, AP, India.
[2]Biology Division, Indian Institute of Chemical Technology, Tarnaka, Hyderabad - 500 007, AP, India.
[3]Department of Biotechnology, College of Engineering and Technology, Sri Krishnadevaraya University, Anantapur-515 003, AP, India.
[4]Silkworm Genetics and Breeding Laboratory, Central Sericultural Research and Training, Institute (CSR and TI), Central Silk Board, Berhmpore - 742 101, West Bengal, India.
[5]Bioinformatics Centre, Department of Biochemistry, Sri Krishnadevaraya University, Anantapur- 515 003, AP, India.

The DNA marker use in assisting selection are safe method in breeding process and it is an important tool for authentication of new gene cascade in genome. In mulberry silkworm, the major economic and nutrigenomic traits are polygenic in nature. In the present study, we have utilized ten PCR-SSR microsatellite markers to gain better understanding on genotyping of certain nutrigenomic gene loci in nutritionally efficient silkworm breeds / hybrids. Results showed that a single yet varying size amplified band in all four parental silkworm strains (RMG$_4$, RMW$_2$, RBD$_1$ and RBO$_2$) and two clear amplified bands in the hybrids (RMG$_4$ × RBD$_1$ and RMW$_2$ × RBO$_2$) with different molecular weight from three PCR-SSR primers loci *viz.*, F11139, F10429 and F10705. The PCR-SSR results demonstrated that homozygosity in newly evaluated nutritionally efficient parental silkworm strains and heterozygosity in hybrid. These investigations authentically confirmed the previous findings of heterotic nutritionally efficient silkworm hybrids with superior nutrigenomic traits. The developed molecular analysis in silkworm could be utilized for the benefit of the farmers in sericulture industry. In conclusion, these results would be useful in identification of nutrigenomic cascade of genes in silkworm and also emphasize the future prospects of silkworm functional mechanism in nutrigenomic studies.

Key words: Silkworm, breed, hybrid, nutrigenomics, PCR-SSR marker, homozygosity, heterozygosity, cascade of genes.

INTRODUCTION

The silkworm, *Bombyx mori* L. is the well-studied central Lepidopteron model system because of its rich repertoire, and well characterized mutations. It is an important source of livelihood for subsistence of farmers engaged in commercial silk production in many countries. Silkworm is a major insect model for research and the

first Lepidopteron for which draft genome sequences became available (Tazima, 1978; Xia et al., 2004; Nguu et al., 2005). Silkworm with haploid nuclear genome size of 530 Mb broken into 28 chromosomes utilized as a framework genome to organize information for other Lepidopteron. Silkworm is known for its vast genetical, geographical diversity and the model animal for the utilization in the study of heterosis by crossing with deferent parents for commercial exploitation. Hundreds of such different geographical and mutant strains have been preserved in Japan, China, Korea, India, Italy, France and other countries (Goldsmith et al., 2005; Archak et al.,

*Correspondening   author.   E-mail: ramesh_silk53@rediffmail.com.

2007).

The microsatellite markers consist of an array of simple tandemly repeated mono, di-, tri- tetra-, penta or hexanucleotide repeats such as $(A)_n$, $(CA)_n$, $(GA)_n$, $(GTA)_n$, $(ATT)_n$, $(GATA)_n$, $(ATTTT)_n$, $(ACGTCG)_n$ which distributed across the genomes. These are ubiquitous in prokaryotic and eukaryotic genomes, randomly distributed both in protein coding and non-coding regions; these become highly informative and versatile classes of genetic markers (Tautz, 1989). The advantages of SSRs over other molecular markers: (i) multiple SSR alleles may be detected, (ii) SSRs are evenly distributed all over the genome, (iii) they are co-dominant, (iv) very small quantities of DNA are required for screening and (v) analysis may be semi-automated. Therefore, it is an imperative that these molecular markers are utilized for broad range of applications, such as genome mapping and characterization, phenotype at a single gene locus using a simple PCR-based mapping (Powell et al., 1996; Robinson et al., 2004). Recently, these microsatellite PCR-SSRs and ISSR markers has enabled researchers to investigate genetic linkage, polymorphism, homozygosity and heterozygosity analysis in gene loci among silkworm populations (Reddy et al., 1999a, b; Rao and Chandrashekharaiah, 2003; Shen et al., 2004; Li et al., 2005, 2006). Recently, RAPD markers linked to nsd-Z had been screened (Li et al., 2001) and the molecular linkage map of nsd-Z by SSR markers had been constructed in silkworm on monogenic trait against densonucleosis virus (Li et al., 2006; Muwang et al., 2007). However, not much is known about the pattern of gene expression, genomic organization, or molecular evolution of the silkworm breeds / hybrids on nutrition consumption and its metabolism in the recent trust area of nutrigenomics as these are the most important commercial parameter in sericulture industry.

The foremost component of the present study to comprehend nutrigenomic analysis for homozygosity and heterozygosity on certain gene loci in newly developed nutritionally efficient silkworm strains. Parental polyvoltine (RMW_2 and RMG_4), bivoltine (RBO_2 and RBD_1) and the hybrid combinations such as $RMW_2 \times RBO_2$ and $RMG_4 \times RBD_1$ were analyzed by DNA fingerprinting with the assistance of ten different PCR-SSR primers for identification of new nutrigenomic cascade of genes in silkworm.

MATERIALS AND METHODS

Silkworm strains and rearing

The two newly identified nutritionally efficient silkworm hybrids viz., $RMW_2 \times RBO_2$ and $RMG_4 \times RBD_1$ and the parental breeds, polyvoltine (RMW_2 and RMG_4) and bivoltine (RBO_2 and RBD_1) were utilized for the study. All these silkworm strains were brushed and reared according to standard rearing methods adopted by Datta (1992). After cocooning, the healthy silkworm pupae were utilized for preparation of DNA for genomic analysis with PCR-SSR primers.

DNA extraction

Ten pooled silkworm pupae of two newly identified nutritionally efficient silkworm hybrids ($RMW_2 \times RBO_2$ and $RMG_4 \times RBD_1$) with its respective four parental breeds viz., RMW_2, RMG_4 (polyvoltine) and RBO2, RBD_1 (bivoltine) were frozen in liquid nitrogen and the genomic DNA was prepared (Suzuki et al., 1972; Sambrook et al., 1989).

These frozen pupa was pulverized with a mechanical homogenizer in a microcentrifuge tube and suspended in DNA extraction buffer (50 mmol/L Tris–HCl (pH 8.0), 100 mmol/L NaCl, 20 mmol/L EDTA) that contained 150 µg/ml proteinase K. After digestion with the proteinase K at 50°C for 8 – 10 h, phenol–chloroform extraction was carried out. The DNA was recovered by isopropanol precipitation and purified DNA molecule was dissolved in Tris-EDTA buffer (pH 8.0) and concentration was measured spectrophotometrically (Li et al., 2005, 2006).

Microsatellite SSR primers

Ten primer sequences of PCR-SSR repeat motif in silkworm were selected from the previously well characterized microsatellite repeats represented different gene loci viz., Fl0429, Fl1139, Fl0548, Fl0316, Fl0568, Fl0705, Fl0650, Fl0665, Fl0664, Fl0537 (Li et al., 2005, 2006).

PCR amplification

The basic program used to amplify PCR-SSR DNA was performed on a thermal cycler PTC 100 (MJ Research). Polymerase chain reaction cycles for the SSR microsatellite loci included (i) an initial denaturation step at 95°C for 3 min, an annealing step at 63°C for 1 min and an extension step at 72°C for 1 min followed by (ii)14 cycles of 94°C for 30 s denaturation, a 14-step touchdown decreasing by 0.5°C at each step to 56°C (30 s) and an extension step at 72°C for 1 min. (iii) conditions for the last 24 cycles were 94°C for 0.5 min, 56°C for 30 s, and 72°C for 1 min followed by (iv) a final elongation step at 72°C for 10 min extension. The PCR was performed in a final volume of 15 µL containing 10 mmol Tris–HCl/L (pH 8.4), 50 mmol KCl/L, 1.5 mmol MgCl2/L, 0.2 mmol each dNTP/L, 0.2 µmol each primer/L, approximately 20 ng of each silkworm parental breeds / hybrids genomic DNA, 0.5 U of Taq polymerase and distilled de-ionized water.

Electrophoresis of PCR products

The following PCR amplified product was mixed with 5 µl TE buffer and 2 µl 40% sucrose containing 0.5% bromophenol blue was loaded on to a 1.5% agarose gel along with the 100 bp ladder molecular weight marker (Promega) and run at constant voltage (150 V) for 2 h. After electrophoresis, the gels were stained with Ethidium Bromide, clear bands were UV visualized and photographed with digital scientific camera in gel documentation system (Hou et al., 2005). To authenticate repeatability of the results obtained, the PCR amplification and gel electrophoresis was repeated twice.

RESULTS

Morphological features nutritionally efficient silkworm strains

The newly identified nutritionally efficient silkworm strains

Table 1. Nutritionally efficient silkworm strains with origin and cocoon traits utilized for genomic analysis.

Parental breeds / Hybrids	Origin	Cocoon color	Cocoon shape	Cocoon built	Cocoon grains
RMW_2	Madagascar	White	Oval	Medium	Fine
RMG_4	Madagascar	Greenish Yellow	Oval	Medium	Medium
RBO_2	Exogenous	White	Oval	Medium	Fine
RBD_1	Exogenous	White	Dumbbell	Medium	Fine
RMW_2 x RBO_2	Hybrid	White	Oval	Hard	Fine
RMG_4 x RBD_1	Hybrid	Greenish Yellow	Hybrid	Hard	Fine

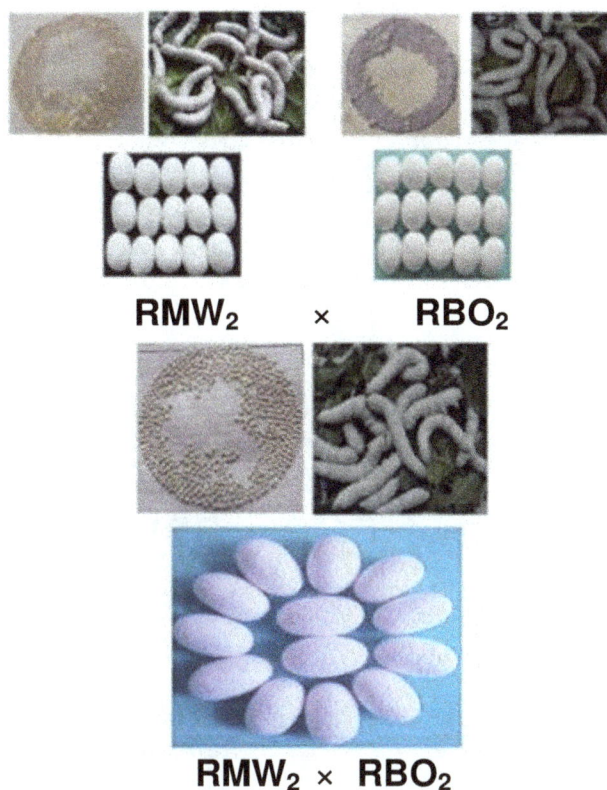

Figure 1. Nutritionally efficient silkworm hybrid RMW_2 × RBO_2 with its parental strains.

and its phenotypic salient features were showed in Table 1. Both polyvoltine non-hibernating parental strains (RMW_2 and RMG_4) were of Madagascar origin with oval shape cocoon, RMW_2 with white color cocoon and RMG_4 with greenish yellow cocoon color. Both the breeds were with pigmented egg color and slender bluish larval color. Bivoltine parental breeds were hibernating in nature (RBO_2 and RBD_1) and exogenous origin with white color cocoon but one with oval shape and another with dumbbell shape cocoon respectively. The egg color of the both the bivoltine were grayish color. The cocoon built and grains of all parental breeds were fine except in RMG_4 where in medium cocoon grains were observed. The newly developed nutritionally efficient hybrid RMW_2 × RBO_2 was characterized with white color oval cocoon,

pigmented egg color and robust bluish white larval color (Figure 1). Another hybrid, RMG_4 × RBD_1 was characterized with greenish yellow color cocoon with hybrid shape, pigmented egg color and bluish larval color (Figure 2). This selected silkworm strains were identified as nutritionally efficient based on nutrigenetic traits of few strains revealed from earlier study of Ramesha et al. (2010).

Characterization of SSR primers

Out of ten SSR primers selected and utilized, seven were $(CT)_n$ and three $(CA)_n$ repeat motifs with varying length. The PCR amplification were found in two $(CA)_n$ repeat

RMG$_4$ × RBD$_1$

RMG$_4$ × RBD$_1$

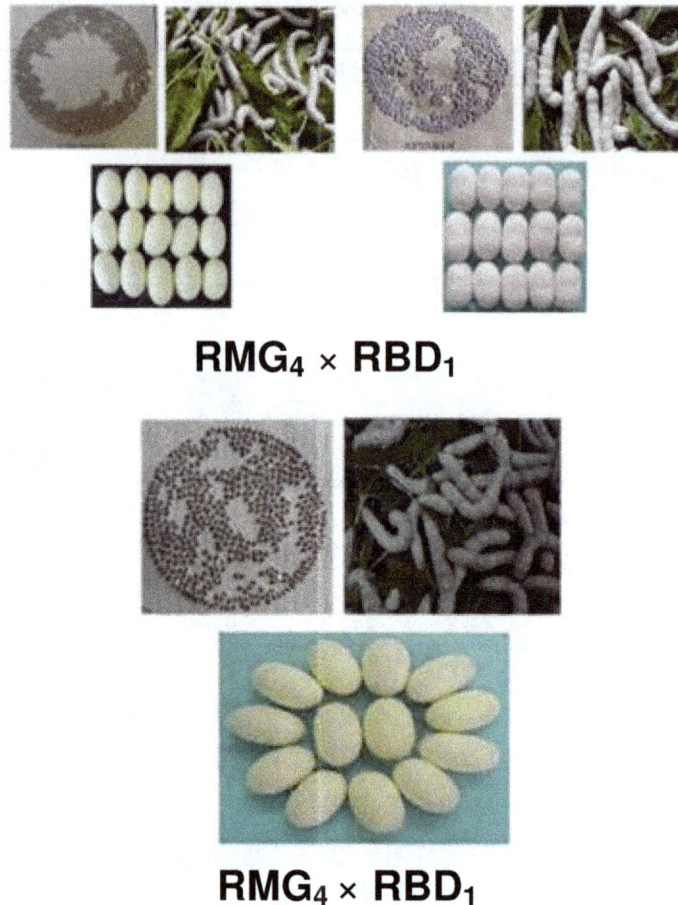

Figure 2. Nutritionally efficient silkworm hybrid RMG$_4$ × RBD$_1$ with its parental strains.

motif viz., F10429 and F11139 and one (CT)$_n$ repeat motifs of F10705 in this study (Table 2).

Optimization of PCR-SSR amplification

The pooled DNA sample from ten - fifteen silkworm pupae from newly identified nutritionally efficient polyvoltine silkworm breed, RMW$_2$ was amplified from randomly selected SSR primer, F10429 and resulted in a series of discrete bands with varying intensity. The genomic DNA template concentration assays were carried out over a range from 5 - 50 ng (5, 10, 15, 20, 25, 35, and 50 ng) in a total volume 15 µl final PCR reaction mixture (Figure 3; Lanes A - G). Template DNA concentration of 15 - 35 ng (Lane B, C, D, E and F) was found to generate a consistent amplification profile but lower (Figure 3; Lane 'A' with 5 ng) and higher (Figure 3; Lane 'G' with 50 ng) concentration appeared to be unenthusiastic amplification. At optimum template DNA concentration (15 - 25 ng/15 µl of final reaction volume), the PCR-SSR DNA profile remained consistent and repeatable (Figure 3).

Homozygosity and heterozygosity specific gene amplification

Three primers yielded PCR-SSR amplification products in all six silkworm strains. The number and size of the amplified products varied depending upon the sequence of SSR primers and silkworm breeds utilized. Size of the amplified products ranged from 250 bp to 1 kb in the monomorphic loci (Figure 4). It was observed that primers F11139 gave 400 and 300 bp amplified products in two set of parents, RMW$_2$ and RMG$_4$ and RBO$_2$ and RBD$_1$ respectively. Whereas in hybrids (RMW$_2$ × RBO$_2$ and RMG$_4$ × RBD$_1$), both 300 and 400 bp products were observed (Figure 4A). Primer F10429 amplified a 250 bp product in all six silkworm breeds/hybrids (Figure 4B) in addition to 1 kb band in hybrids. Approximately 1 kb PCR amplified band was observed in all six silkworm strains from the primer F10705 and 300 bp product in hybrids in addition (Figure 4C). The PCR-SSR results revealed one band in parents and two different size bands in hybrid combinations from all three primers. The amplification reactions of PCR-SSR patterns were consistent and reproducible for the silkworm strains and PCR-SSR

Table 2. The repeat motif with primer sequence on SSR loci.

Locus symbol	Repeat motif	Primer sequence (5'→3')
FI0316	$(CT)_9$	GCGATAAGACCGCCTATTGAAC GTGTATTAGGCACGAGAACTGACG
FI0429	$(CA)_{11}$	AAGGGATTCTCTACCAGTCAACCA TTTGACGCTGGCTTATAAATACTGTAT
FI0548	$(CT)_{12}$	ACAAAGTTCCCCAAAACGCTC TTCGGAATGAAACATCCTCAACTA
FI0568	$(CT)_7$	TCGTCCTACACTTGCGGGTT TGTTTCGTCAAGTCTGCTCGGT
FI0705	$(CT)_5$	GGGATAAGTGGGTCGTTTTGATT TGAGACCCAATAATGTCCCGAG
FI0650	$(CA)_6$	GAAAGCGGATGGTCCTACTCTG CTAAGTAAGAACCCAAGCTACACGA
FI0665	$(CT)_8$	TCCAAATGATTCTTGTCCACCTG TTCCTTTCTTTCAATTCTTCTGT
FI0664	$(CT)_{10}$	AAATTTCATACTCCTCCGTCCG CATTTCTACCACAGCCAAACGAT
FI0537	$(CT)_{15}$	CCATTTACAGGCTGGTATCCAT TAGCGATAAGACCGCCTATTGTA
FI1139	$(CA)_7$	CGGCACTTAAAAGTTTTCATATCAATC CTGACAGTGGTGAGTTAATAAAACAAAA

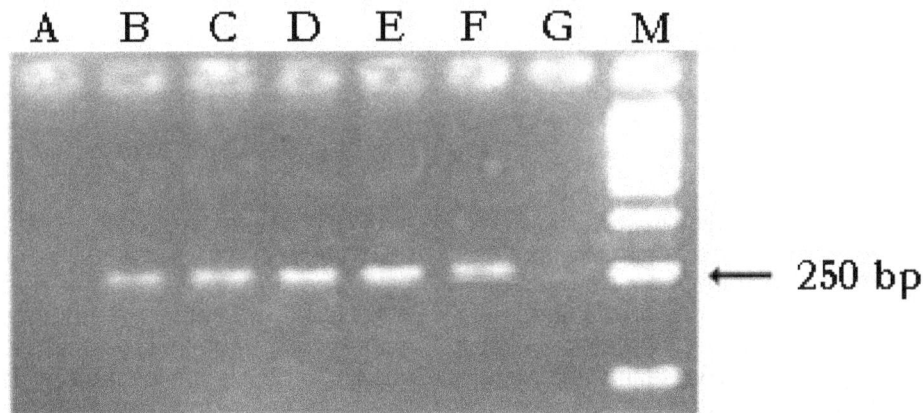

Figure 3. Optimization of PCR-SSR amplification with different concentration of of template DNA (Lane: A-G contains 5, 10, 15, 20, 25, 35 and 50 ng genome template DNA: M-Marker).

primer combinations.

DISCUSSION

The silkworm strains used for the current study were chosen based on the silkworm ten important commercial traits after been analyzed through general combining ability (GCA), specific combining ability (SCA), mid parent heterosis (MPH) and better parent heterosis (BPH) followed by nineteen nutrigenetic traits analysis. Dietary or nutritional factors and related metabolic interactions have direct and indirect influence on specific gene regulation and expression (Walker and Blackburn, 2004)

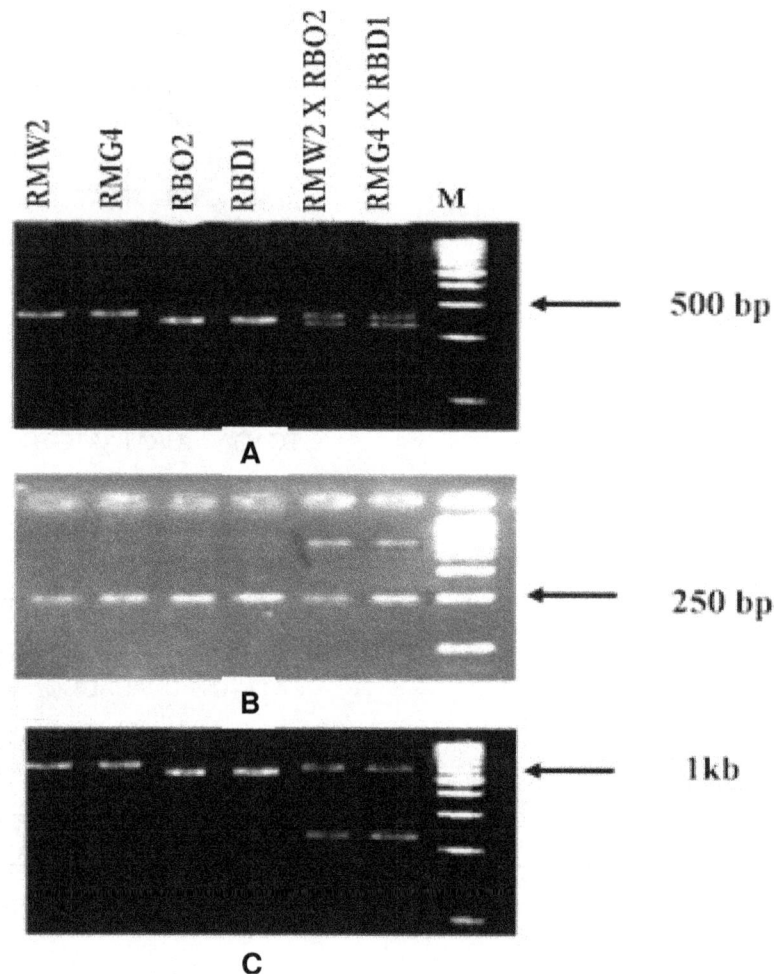

Figure 4. PCR-SSR profile on certain gene loci of homozygosity and heterozygosity in nutritionally efficient silkworm parental breeds and hybrids (A. F11139; B. F10429; C. F10705).

and such interactions and variation in the field of nutrigenetics applied to choose the silkworm strains based on their nutritional efficiency traits as 'biomarkers'. Our previous work has revealed insight on nutrigenetic traits difference amongst silkworm strains while suggesting as nutritionally efficient (Ramesha and Raju, 2009; Ramesha et al., 2010).

Though tremendous achievements have been made in silkworm improvement by exploring the genome at exponential growth in the field of molecular biology envisage the pattern of gene expression in silkworm. PCR-SSR technique proved to be valuable in uncovering genomic variability in silkworm, since past such variability was estimated only through morphological, biochemical and yield attributes (Chatterjee and Datta, 1992). In the past 20 years, the major effort in animal breeding has changed from quantitative to molecular genetics with emphasis on quantitative trait loci (QTL) identification and marker assisted selection (MAS). With the recent advent

of PCR based approach, gel free visualization with automation in genomic analysis and genotyping methods have resulted in a rapid expansion of the power of molecular markers to address individual's identity authentically. The microsatellites emerged as the most popular and versatile marker for markers assistance selection (MAS) in silkworm breeding technology. It was observed that these markers revealed genomic variation among silkworm strains and results obtained in the present study were in accordance with the principle that the number of individuals used to estimate average heterozygosity could be very small if a large number of loci studied (Nei, 1978; Lewontin and Hubby, 1996). Two SSR markers, Fl0316 and Fl0568 showed from *nsd-Z* gene expression (Li et al., 2006) and Muwang et al. (2007) detected the homozygosity in the NILs of *nsd-Z* to inspect the replacement of the linkage groups of the donor parent on SSR markers in silkworm densonuleosis virus. Our results also suggested homozygosity in parents and heterozygosity for hybrid by three viz., F11139, F10429 and F10705 out of ten SSR primers utilized on certain gene loci.

The silkworm breeds/hybrids with superior quantitative traits are by utilizing genetically divergent breeds through breeding programs. Most of the inbred lines or stocks of silkworm available today were generated by successive selection for specific traits of economic importance. Accurate estimation of genetic traits within inbred lines and relativity between them is important for both maintenance of inbred lines and hybrid preparation (Ramesha et al., 2009). Such estimates could be obtained by exploring the power of DNA fingerprinting or molecular analysis of silkworm genomic. In the present study, primers for di-nucleotide repeat motif of PCR-SSR markers were selected because of its abundance and estimated that the silkworm genome of 530 Mb accounted for 1.63 Mb of microsatellite repeats, equivalent to 0.31% of the genome (Dharma et al., 2005). In the achievement of reproducibility and strong signal in the PCR-SSR analysis, the significant parameter was the genomic template DNA concentration. PCR-SSR optimization procedures were carried out with varying concentration of silkworm genomic template DNA in order to ease the analysis on PCR-SSR amplification by only visible clear DNA bands on the gel and observed a reliable amplification with template DNA concentration between 10 - 35 ng (Figure 3). In this study, 20 ng of silkworm genomic template DNA was utilized for efficient PCR amplification and is in concurrence with the earlier studies. Among these selected ten SSR primers utilized for the study, seven $(CT)_n$ and three $(CA)_n$ microsatellite repeat motifs sequence (Rozen and Skaletsky, 2000).

The study examined genetic variability within and among the nutritionally efficient silkworm strains by DNA fingerprinting and the result obtained was consistent with voltinism and morphological differences. It is noticed that in silkworm, both dominance and epistasis are important for obtaining heterosis in traits for productivity of silk (Nagaraju and Kumar, 1995). When compared to closely related strains, genetically distant strains were more likely to have different fixed alleles at the same loci and hence their crosses should give higher degree of heterosis. Furthermore, heterosis was higher if both the parental strains are homozygous. The recent trend of genomic analysis led to the development of the PCR-SSR analysis promise to become a valuable tool for genomic analysis, led to generating molecular markers for specific trait's heterozygosity and homozygosity at allelic nature in gene locus than any other marker system. Apart from being the source of informative genetic markers, microsatellites attracted a lot of attention with respect to their origin, distribution, expansion, mutation and disintegration. The highest microsatellite content in genomic size of 397.71 Mb in B. mori among other full genome sequenced five insect species viz., Drosophila melanogaster, Anopheles gambiae, Apis mellifera and Tribolium castaneum. It

represented about 0.72% of total silkworm genome with a maximum number of 111.006 repeats and of about 280 numbers of microsatellites per Mb genome (Dharma et al., 2005; Archak et al., 2007). Moreover, it was well established that of hereditary properties of PCR-SSR marker in silkworm and demonstrated that the PCR-SSR markers techniques could be successfully applied to the silkworm to reveal gene loci with many PCR-SSR primers producing patterns and potential to serve as molecular markers (Reddy et al., 1999a; Robinson et al., 2004; Li et al., 2006; Miao et al., 2005, 2007). The well characterized and accessible information on ten PCR-SSR markers in silkworm was utilized to investigate its utility in the molecular nutrigenomic analysis for the homozygosity and heterozygosity in newly identified nutritionally efficient silkworm parental strains of polyvoltine (RMW_2, RMG_4) and bivoltine (RBO_2, RBD_1) with its hybrids viz., $RMW_2 \times RBO_2$ and $RMG_4 \times RBD_1$ for the certain gene loci. PCR amplicon was found in two $(CA)_n$ repeat motif viz., F10429 and F11139 and one $(CT)_n$ repeat motifs of F10705. These observations strongly suggest $(CA)_n$ repeat motifs are closely linked in heterozygosity and homozygosity than $(CT)_n$. It was revealed that single varying size amplified DNA band in parental silkworm strains and two clear amplified bands in hybrids from three different gene loci (Figure 4). The result obtained in the present study was in concurrence with PCR-SSR primers utilized in the homozygosity and heterozygosity profile analysis for the nsd - Z gene (Chandrashekharaiah et al., 2006; Li et al., 2006). The PCR-SSR results suggested that the newly identified nutritionally efficient silkworm hybrids were genotypically heterozygous for certain nutritional gene loci. This heterozygosity nature in silkworm hybrid in contrast to the homozygous nature of parental silkworm breed was important for commercial exploitation of heterosis in sericulture industry (Nagaraju and Kumar, 1995; Ramesha et al., 2009).

We noticed that the silkworm breeds/hybrids chosen for the present study reveal marked differences for genetical traits such as origin, voltinism, fecundity, body weight, larval duration, cocoon and shell color and weight, silk filament length etc. in addition to nineteen nutrigenomic traits such as ingesta, digesta, excreta, approximate digestibility (AD), reference ratio (RR), consumption indices (CI), relative growth rate (RGR), respiration and metabolic rate (MR), efficiency conversion of ingesta (ECI) and efficiency conversion of digesta (ECD) for larva, cocoon and shell, ingesta per gram (I/g) and digesta per gram (D/g) for cocoon and shell (Ramesha et al., 2010). The PCR-SSR results in the present study on nutritionally efficient silkworm are similar to the observation made in various other systems (Pejic et al., 1998; Lima et al., 2006; Moyib et al., 2007). Microsatellite molecular analysis reflected the geographical and morphological relations of nutritionally efficient silkworm strains.

The two newly identified nutritionally efficient hybrids

were distinctively high yielding than its relative parental silkworm breeds. The unique PCR-SSR profile in the hybrids led to identification of nutritional related cascade of genes families. Several such different gene cascades were identified, categorized and cloned in silkworm through molecular approaches (Li et al., 2006; Muwang et al., 2007; Kanginakudru et al., 2007; Roller et al., 2008; Chai et al., 2008; Pan et al., 2009).

From the present microsatellite nutrigenomic analysis on molecular markers (PCR-SSR) was noticed to be an ideal tool and play an imperative role for marker-assisted selection (MAS) in silkworm breeding programs. This study also indicated that marker assisted selection could be actualized to new nutritionally efficient silkworm breeds/ hybrids. Simple sequence repeats often have flanking regions highly conserved in related species, which allows the use of the same primer pairs in related genomes.

These linked molecular markers would be very useful in marker-assisted screening, because the co-dominant markers offer information on homozygous and heterozygous genotypes and save time in the breeding program of silkworms with authenticity. Furthermore, not much is known about the pattern of gene expression, genomic organization or molecular evolution of the silkworm breeds / hybrids on nutrition consumption and its metabolism as it is the most important commercial parameter in sericulture industry involving over 60% of the cost for production of mulberry leaf. Therefore, these results led to identification of nutrigenomic related cascade of genes in silkworm and more emphasize the future prospects of silkworm functional mechanism in nutrigenomic studies after recent completion of total silkworm genome sequencing.

Conclusion

This study demonstrated a single amplified band in four parental silkworm strains of RMG$_4$, RMW$_2$, RBD$_1$ and RBO$_2$ whereas two clear amplified bands in the hybrids of RMG$_4$ × RBD$_1$ and RMW$_2$ × RBO$_2$ with varying molecular size from three PCR-SSR primers gene loci viz., F11139, F10429 and F10705. Therefore, it was predicted strongly that the indication of homozygosity in newly evaluated nutritionally efficient parental silkworm strains and heterozygosity in its hybrid. Ultimately, these results can be used in marker assisted selection or gene transmission in silkworm breeding programs and emphasizes the future prospects of silkworm functional mechanism in nutrigenomic studies.

ACKNOWLEDGEMENTS

The authors wish to thank the authorities of Andhra Pradesh State Sericulture Research and Development Institute (APSSRDI), Kirikera- 515 211, Hindupur, AP., India for their encouragement to undertake the work. The authors also express deep regards to Department of Biotechnology, Sri Krishnadevaraya University, Anantapur-515 003, AP, India for their support.

Abbreviations: PCR-SSR, Polymerase chain reaction-simple sequence repeats; **bp,** base pairs; **TE,** tris- EDTA; **DNA,** deoxyribonucleic acid; **EDTA,** ethylene diamine trichloro acetic acid; **MAS,** markers assistance selection; **GCA,** general combining ability; **SCA,** specific combining ability; **MPH,** Mid parent heterosis; **BPH,** better parent heterosis, **AD,** Approximate digestibility; **CI,** consumption index; **RGR,** relative growth rate; **MR,** metabolic rate; **ECI,** efficiency conversion of ingesta; **ECD,** efficiency conversion of digesta.

REFERENCES

Archak S, Meduri E, Sravana Kumar P, Nagaraju J (2007). InSatDb: a microsatellite database of fully sequenced insect genomes. Nucleic Acid Res., 35: 36-39.

Chai CL, Zhang Z, Huang FF, Wang XY, Yu QY, Liu BB, Tian T (2008). A genome wide survey of homeobox genes and identification of novel structure of the Hox cluster in the silkworm, *Bombyx mori*. Insect Biochem. Mol. Biol., 38: 1111-1120.

Chandrashekharaiah, Rao CGP, Nagaraju J (2006). Synthesis of polyvoltine silkworm hybrid based on hybrid performance and heterosis and molecular markers heterozygosity. In: Asia pacific conference of sericulture and insect biotechnology, Samgju, Korea, p. 69.

Chatterjee SN, Datta RK (1992). Hierarchical clustering of 54 races and strains of mulberry silkworm, *Bombyx mori* significance of biochemical parameters. Theor. Appl. Genet., 185: 394-402.

Datta RK (1992). Guidelines for Bivoltine Rearing, Central Silk Board, Bangalore, India, p. 24.

Dharma Prasad M, Muthulakshmi M, Madhu M, Sunil Archak, Mita K, Nagaraju J (2005). Survey and analysis of microsatellites in the silkworm, *Bombyx mori*: frequency, distribution, mutations, marker potential and their conservation in heterologous species. Genetics, 169: 197-214.

Goldsmith MR, Shimada T, Abe H (2005). The genetics and genomics of the silkworm, *Bombyx mori* L. Annual Rev. Entomol. 50: 71-100.

Hou C, Li M, Gui Z, Xu A, Guo X (2005). PCR-based detection of densovirus infection in silkworm (*Bombyx mori* L.). Int. J. Indust. Entomol., 11: 135-138.

Kanginakudru S, Royer C, Edupalli SV, Jalabert A, Mauchamp B, Chandrashekaraiah, Prasad SV, Chavancy G, Couble P, Nagaraju J (2007). Targeting *ie-1* gene by RNAi induces baculoviral resistance in lepidopteran cell lines and in transgenic silkworms. Insect Mol. Biol., 16: 635 - 644.

Lewontin RC, Hubby JL (1996). A molecular approach to the study of genetic heterozygosity in natural populations. II. Amount of variation and degree of heterozygosity in natural population of *Drosophila pseudoobscura*. Genetics, 54: 595-609.

Li MW, Yao Q, Hou CX, Lu C, Chen KP (2001). Studies on RAPD markers linked to the densonucleosis refractoriness gene, *nsd-Z* in silkworm, *Bombyx mori* L. Sericologia, 41: 409-415.

Li M, Shen L, Xu, A, Miao X, Hou C, Sun J, Zhang Y, Huang Y (2005). Genetic diversity among silkworm (*Bombyx mori* L. Lep. Bombycidae) germplasm revealed by microsatellites. Genome 48: 802-810.

Li M, Guo Q, Hou C, Miao X, Xu A, Guo X, Huang Y (2006). Linkage and mapping analyses of the densonucleosis non-susceptible gene *nsd-Z* in the silkworm *Bombyx mori* using SSR markers. Genome, 49: 397-402.

Lima Brito J, Carvallho A, Martin A, Heslop Harrison JS, Guedes Pinto H (2006). Morphological, yield, cytological and molecular characterization of a bread wheat x tritordeum F₁ hybrid. J. Genet., 85: 123-131.

Miao XX, Xu SJ, Lia MH, Lia MW, Huang JH, Dai FY, Marino SW, Mills DR, Zeng P, Mita K, Jia SH, Zhng Y, Liu WB, Xiang H, Guo QH, Xu AY, Kong XY, Lin HX, Shi YZ, Lu G, Zhng X, Huang W, Yasukochi Y, Sugasaki T, Shimada T, Nagaraju J, Xiang ZH, Wang SY, Goldsmith MR, Lu C, Zho GP, Huang YP (2005). Simple sequence repeat based consensus linkage map of Bombyx mori. Proc. Natl. Acad. Sci., 102: 16303-16308.

Miao X, Li M, Dai F, Lu C, Goldsmith MR, Huang Y (2007). Linkage analysis of the visible mutations Sel and Xan of Bombyx mori (Lepidoptera: Bombycidae) using SSR markers. Eur. J. Entomol., 104: 647-652.

Moyib OK, Odunola OA, Dixon AGO (2007). SSR markers reveal genetic variation between improved cassava cultivars and landraces within a collection of Nigerian cassava germplasm. Afr. J. Biotech., 6: 2666-2674.

Muwang L, Chengxiang H, Yunpo Z, Anying, X, Xijie G, Yongping H (2007). Detection of homozygosity in near isogenic lines of non-susceptible to Zhenjiang strain of densonucleosis virus in silkworm. Afr. J. Biotech., 6: 1629-1633.

Nagaraju J, Kumar T (1995). Effects of selection on cocoon filament length in divergently selected lines of the silkworm, Bombyx mori. J. Seri. Sci. Jpn., 64: 103-109.

Nei M (1978). Estimation of average heterozygosity and genetic distance from a small number of individuals. Genetics 89: 583-590.

Nguu EK, Kadono-Okuda K, Mase K, Kosegawa E, Hra W (2005). Molecular linkage map for the silkworm, Bombyx mori, based on restriction fragment lenghth polymorphism of cDNA colones. J. Insect Biotechnol. Sericol., 74: 5-13.

Pan MH, Wang XY, Chai CL, Zhang CD, Lu C, Xiang ZH (2009). Identification and function of Abdominal-A in the silkworm, Bombyx mori. Insect Mol. Biol., 18: 155 -160.

Pejic I, Ajmone Marsan P, Morgante M, Kozumplick V, Castiglioni P, Taramino G, Motto M (1998). Comparative analysis of genetic similarity among maize inbred lines detected by RFLPs, RAPDs, SSRs and AFLPs. Theor. Appl. Genet., 102: 440-449.

Powell W, Morgante M, Andre C, Hanafey M, Vogel J, Tingey S, Rafalski A (1996). The comparison of RFLP, RAPD, AFLP and SSR (microsatellite) markers for germplasm analysis. Mol. Breed., 2: 225-238.

Ramesha C, Seshagiri SV, Rao CGP (2009). Evaluation and identification of superior polyvoltine crossbreeds of mulberry silkworm, Bombyx mori L. J. Entomol., 6: 179-188.

Ramesha C, Raju PJ (2009). Analysis of nutrigenetic traits for identification of nutritionally efficient germplasm breeds of bivoltine silkworm, Bombyx mori L. National Conference on Recent Trends in Animal Physiology, Mysore. India, p. 27.

Ramesha C, Anuradha CM, Lakshmi H, Sugnana Kumari S, Seshagiri SV, AK Goel, Suresh Kumar C (2010). Nutrigenetic traits analysis for identification of nutritionally efficient silkworm germplasm breeds. Biotechnology, 9: 131-140.

Rao CGP, Chandrashekharaiah (2003). Molecular markers assisted breeding in silkworm. A concept paper presented in mulberry silkworm breeders summit. Hindupur, AP, India, pp. 80-85.

Reddy KD, Abraham EG, and Nagaraju J (1999a). Microsatellites in the silkworm, Bombyx mori: abundance, polymorphism, and strain characterization. Genome, 42: 1057-1065.

Reddy KD, Nagaraju J, Abraham EG (1999b). Genetic characterization of silkworm Bombyx mori by Simple Sequence Repeat (SSR) - anchored PCR. Heredity, 83: 681-687.

Robinson AJ, Love CG, Batley J, Barker G, Edwards D (2004). Simple sequences repeat marker loci discovery using SSR primer. Bioinformatics, 20: 475-1476.

Roller L, Yamanaka N, Watanabe K, Daubnerova I, Zitnan D, Kataoka H (2008). The unique evolution of neuropeptide genes in the silkworm Bombyx mori. Insect Biochem. Mol. Biol., 38: 1147–1157.

Rozen S, Skaletsky HJ (2000). Primer3 on the WWW for general users and for biologist programmers. In: Krawetz, S. and Misener, S. (eds), Bioinformatics Methods and Protocols: Methods Mol. Biol. Humana Press, Totowa, N.J.: pp: 365-386.

Sambrook J, Fritsch EF, Maniatis TC (1989). Molecular cloning: A laboratory manual (2nd ed.) Cold spring Harbor Laboratory Press, Cold spring Harbor, New York, 1(3): 253.

Shen L, Li M, Li M, Miao X, Lu C, Huang Y (2004). Development and genetic diversity analysis of microsatellite markers in silkworm, Bombyx mori L. Can Ye Ke Xue, 30: 230-236.

Suzuki Y, Gage LP, Brown DD (1972). The gene for silk fibroin in Bombyx mori. J. Mol. Biol., 70: 637-649.

Tautz D (1989). Hypervariability of simple sequences as a general source for polymorphic DNA markers. Nucleic Acids Res., 17: 6463-6471.

Tazima Y (2001). Silkworm, an important laboratory tool. Kodash LTD. Tokyo. pp: 53-81.

Walker WA, Blackburn G (2004). Symposium introduction: nutrition and gene regulation. J. Nutr., 134: 2434-2436.

Xia Q, Zhou Z, Lu C (2004). A draft sequence for the genome of the domesticated silkworm (Bombyx mori). Science, 306: 1937-1940.

9

HLA-DQA1 and –DQB1 genotyping in individuals with family history of gastritis

Nibras S. Al-Ammar[1], Ihsan Al-Saimary[1]*, Saad Sh. Hamadi[2] and Ma Luo[2]

[1]Department of Microbiolody, College of Medicine, University of Basrah, Basrah, Iraq.
[2]Department of Medicine, College of Medicine, University of Basrah, Basrah, Iraq.

The objective was to study HLA-DQA1 and HLA-DQB1 genotyping in individuals (patients and controls) with family history of gastritis. This study was carried out in College of Medicine, University of Basrah. HLA-DQA1 and HLA-DQB1 genotyping was done in College of Medicine, University of Manitoba, and Winnpeg, Canada during the period from 17th of April 2009 to 15th of July 2010. A total of 100 patients (41 males and 59 females) and a total of 30 controls (18 males and 12 females) were included in this study. A significant decreased frequency of DQA1*0201 allele was found in individuals (patients + controls) with family history of gastritis with a strong association (odds ratio = 4.57), as compared with individuals without family history of gastritis. Significant increased alleles frequencies of DQA1*0402 and DQB1*0402 were found in individuals with family history of gastritis, but with weak association (odds ratios, 0.16 and 0.20 respectively), as compared with individuals without family history of gastritis.

Key words: HLA-DQA1, HLA-DQB1, genotyping and gastritis.

INTRODUCTION

Human leukocyte antigens (HLA) are an inherent system of alloantigens, which are the products of genes of the major histocompatibility complex (MHC). These genes span a region of approximately 4 centimorgans on the short arm of human chromosome 6 at band p21.3 and encode the HLA class I and class II antigens, which play a central role in cell-to-cell-interaction in the immune system (Conrad et al., 2006). They encode peptides involved in host immune response, also they are important in tissue transplantation and are associated with a variety of infectious, autoimmune, and inflammatory diseases (Gregersen et al., 2006; Nair et al., 2006). Moreover, the HLA loci display an unprecedented degree of diversity and the distribution of HLA alleles and haplotypes among different populations is considerably variable (Shao et al., 2004; Blomhoff et al., 2006). The expression of particular HLA alleles may

be associated with the susceptibility or resistance to some diseases (Wang et al., 2006). Heterozygosity within the MHC genomic region provides the immune system with a selective advantage of pathogens (Fu et al., 2003; Kumar et al., 2007). *Helicobacter pylori* infection is in addition to being the main etiologic agent for chronic gastritis, a major cause of peptic ulcer and gastric cancer (Suerbaum and Michetti, 2002). Many studies performed in Iraq about bacteriological and immunological aspects of *H. pylori* (Al-Janabi, 1992; Al-Jalili, 1996; Al-Baldawi, 2001; Al-Dhaher, 2001; Al-Saimary, 2008), but no study was performed yet on HLA genotyping, hence, the results of the present study is compared with studies done in other countries. In developing countries, prevalence of *H. pylori* infection is > 80% among middle-aged adults, whereas in developed countries prevalence ranges from 20 to 50%. Approximately 10 to 15% of infected individuals will develop peptic disease and 3% a gastric neoplasm (Torres et al., 2005). Therefore, *H. pylori* infection is a necessary but not a sufficient cause of severe forms of gastric disease. *H. pylori* induce a host

---

*Corresponding author. E-mail: ihsanalsaimary@yahoo.com.

Table 1. Frequencies of males and females (Patients + controls) with and without family history of gastritis.

Gender	With family history of gastritis {N= 15 (%)}	Without family history of gastritis {N=85 (%)}	Total N = 100
Male	5 (10.64)	42 (89.36)	47
Female	10 (18.87)	43 (81.13)	53

(χ^2 = 1.32; P = NS; OR= 0.51; 95% CI = 0.16 to 1.62).

immune response, but the persistence of the infection suggests that the response is not effective in eliminating the infection. Furthermore, multiple lines of evidence suggest that the immune response contributes to the pathogenesis associated with the infection. As a result, the immune response induced by *H. pylori* is a subject of continuous study that has encouraged numerous questions (Azem et al., 2006). The inability of the host response to clear infections with *H. pylori* could reflect down-regulatory mechanisms that limit the resulting immune responses to prevent harmful inflammation as a means to protect the host (Yoshikawa and Naito, 2000).

METHODS

A total of 100 patients (41 males and 59 females with age groups from (15 to 66) years, with various gastritis symptoms attending endoscopy unit at Al-Sadder Teaching Hospital in Basrah and a total of 30 controls (18 males and 12 females), with age groups from (15 to 61) years, without any symptoms of gastritis were included in the present study. Blood samples were drawn from gastritis patients and subjected to HLA-DQ genotyping. The study was carried out during the period from (17th of April 2009 to 15th of July, 2010). DNA was isolated from the blood samples, using wizard genomic DNA purification Kit, Promega Corporation, USA; Protocol (Beutler et al., 1990).

HLA-DQA1 and –DQB1 genotyping

HLA-DQA1 and –DQB1 genotyping protocol was done according to Sequence-Based-Typing (SBT), which had been developed in National Microbiology Laboratories (NML), Winnipeg, Canada (Luo et al., 1999). All the steps of HLA-DQA1 and –DQB1 genotyping were done under supervision of Dr. Ma Luo in Medical Microbiology Laboratory, College of Medicine, and University of Manitoba and in Dr. Ma Luo Laboratory in National Microbiology Laboratories (NML). Agarose Gel Electrophoresis (Fisher-Biotech FBSB-710; Bio RAD).

PCR amplification of HLA-DQA1 and –DQB1 gene

The PCR amplification of HLA-DQA1 and –DQB1 gene was done in Medical Microbiology Laboratories in College of Medicine, Manitoba University, Winnipeg, Canada.

DNA purification

The purification of the amplified HLA-DQA1 and –DQB1 gene was done in National Microbiology Laboratories (NML), in Dr. Ma Luo

Laboratory, College of Medicine, Manitoba University, Winnipeg, Canada.

Three methods had been used for purification of the amplified PCR DNA samples:

1. DNA purification by using vacuum.
2. DNA purification by using GenElute™ PCR clean-Up Kit (Sigma-Aldrich, Inc. USA). GenElute™ PCR Clean-Up Kit.
3. Purification in DNA core section in NML (NML, Canada).

The amplified PCR DNA was purified in DNA core laboratory in National Microbiology Laboratories (NML), Winnipeg, Canada.

Sequencing–PCR

Sequencing–PCR was done in National Microbiology Laboratories (NML), under supervision of Dr. Ma Luo.

Ethanol precipitation

Ethanol precipitation was done under supervision of Dr. Ma Luo in National Microbiology Laboratories (NML), College of Medicine, Manitoba University, Winnipeg, Canada.

Sequencing-using the (3100 Genetic Analyzer, USA)

HLA-DQA1 and –DQB1 genotyping protocol was done according to Sequence-Based-Typing (SBT), which had been developed in National Microbiology Laboratories (NML), Winnipeg, Canada (Luo et al., 1999).

Statistical analysis

For qualitative variables, frequency data were summarized as percentage. Statistical significant of the differences between two groups was tested by Pearson Chi-square (χ^2) with Yates' continuity correction. Risk was estimated using odds ratio (OR) and 95% confidence interval (95% CI). P-value was determined by Fisher's exact test, P- value of (< 0.05) was considered statistically significant. Data were analyzed using SPSS program for window (Version 10).

RESULTS

Distribution of individuals (patients+controls) with family history of gastritis according to gender

As shown in Table 1, out of 47 males, 5 (10.64%) were with family history of gastritis and 42 (89.36%) were without

Table 2. Distribution of individuals (patients + controls) with and without family history of gastritis according to age groups.

Age group	With family history of gastritis {N= 15 (%)}	Without family history of gastritis {N=85 (%)}	Total N = 100
15 > 45	11 (15.49)	60 (84.51)	71
> 45	4 (13.79)	25 (86.21)	29

(x^2 = 0.05; P=NS; OR= 1.15; 95% CI = 0.33 to 3.94).

Table 3. HLA-DQA1 Genotype frequency in individuals with family history of gastritis and without family history of gastritis.

HLA-DQA1 allele	Individuals with family history of gastritis		Individuals without family history of gastritis		X^2	P	OR	95% CI
	N = 14	%	N = 73	%				
010101/010102/0104 01/010402/0105	2	14.28	12	16.44	0.04	NS	1.18	0.23-5.96
010201/010202/0102 03/010204	5	35.71	21	28.77	0.27	NS	0.73	0.22-2.43
0103	3	21.42	13	17.81	0.10	NS	0.79	0.19-3.26
0201	1	7.14	19	26.03	5.37	<0.05	4.57	0.56-37.36
030101/0302/0303	3	21.42	17	23.29	0.02	NS	1.11	0.28-4.46
040101/040102/0402/ 0404	3	21.42	3	4.11	5.49	<0.05	0.16	0.03-0.88
050101/0503/0505/05 06/0507/0508/ 0509	9	64.28	45	61.64	0.89	NS	0.89	0.27-2.94

family history of gastritis. Also Table 1 indicated that, out of 53 females, 10 (18.87%) were with family history of gastritis and 43 (81.13%) were without family history of gastritis. These results showed no significant differences between males and females with and without family history of gastritis (x^2 = 1.32; P = NS; OR = 0.51; 95% CI = 0.16 to 1.62).

Distribution of individuals (patients+controls) with family history of gastritis according to age groups

Results shown in Table 2, indicated that out of 71 individuals from 15> 45 age group, 11 (15.49%) were with family history of gastritis and 60 (84.51%) were without family history of gastritis. Also results in Table 2 showed that, out of 29 individuals from > 45 age group, 4 (13.79%) were with family history of gastritis and 25 (86.21%) were without family history of gastritis. These results showed no significant differences between individuals with or without family history of gastritis from these two age groups (x^2 = 0.05; P=NS; OR = 1.15; 95%

CI = 0.33 to 3.94).

Genotype frequencies of HLA-DQ of individuals (patients + controls) with and without family history of gastritis

Genotyping of HLA-DQA1 was studied in individuals with family history of gastritis and compared with individuals without family history of gastritis (patients + controls). Results shown in Table 3, indicated that HLA-DQA1*0201 allele was present in 1 out of 14 individuals with family history of gastritis and in 19 out of 73 individuals without family history of gastritis, with allele frequencies of 7.14 and 26.03 respectively. The decreased frequency of HLA-DQA1*0201 allele in individual with family history of gastritis was statistically significant and showed very strong association (x^2 = 5.37, P< 0.05, OR = 4.57, 95% CI = 0.56 to 37.46) as compared with individuals without family history of gastritis. Results shown in Table 3 also indicated that HLA-DQA1*0402 allele was present in 3 out of 14

Table 4. HLA-DQB1 Genotype frequencies of individuals with and without family history of gastritis.

HLA-DQB1 allele	Individuals with family history of gastritis		Individuals without family history of gastritis		X^2	P	OR	95% CI
	N = 14	%	N = 79	%				
020101/0202/0204	7	50.00	39	49.37	0.01	NS	0.98	0.31-3.04
030101/030104/0309/03 21/0322/0324/030302	5	35.00	32	40.37	0.11	NS	1.23	0.38-3.99
030201	1	7.14	11	13.92	0.49	NS	2.10	0.30-17.72
030302	0	0.00	1	1.27	0.94	NS	1.01	0.99-1.04
0402	3	21.43	4	5.06	4.58	<0.05	0.20	0.04-0.99
050101	1	7.14	11	13.92	0.49	NS	2.10	0.30-17.72
050201	2	14.29	12	15.19	0.01	NS	1.08	0.21-5.42
050301	0	0.00	1	1.27	0.18	NS	1.01	0.99-1.04
060101/060103	1	7.14	6	7.59	0.01	NS	1.07	0.12-9.62
060201	2	14.29	3	3.80	2.57	NS	0.24	0.04-1.57
060301/060401	0	0.00	8	10.13	1.55	NS	1.11	1.03-1.20
060401/0634	1	7.14	6	7.59	0.01	NS	1.07	0.12-9.62
060801	1	7.14	0	0.00	5.70	NS	0.93	0.80-1.07
0609	0	0.00	1	0.00	0.18	NS	1.01	0.99-0.85

individuals with family history of gastritis and in 3 out of 73 individuals without family history of gastritis with frequencies of 21.42 and 4.11 respectively. The increased allele frequency of HLA-DQA1*0402 allele in individuals with family history of gastritis was statistically significant but with week association (X^2 = 5.49, P < 0.05, OR = 0.16, 95% CI = 0.03 to 0.88) when compared with individuals without family history of gastritis. Genotype of HLA-DQB1 was studied in 14 individuals with family history of gastritis and in 79 individuals without family history of gastritis. Results shown in Table 4 indicated that HLA-DQB1*0402 allele was present in 3 out of 14 individuals with family history of gastritis and in 4 out of 79 individuals without family history of gastritis with frequencies of 21.43 and 5.06 respectively. The increased frequency of HLA-DQB1*0402 allele in individuals with family history of gastritis was statistically significant but with week association (X^2 = 4.58, P < 0.05, OR = 0.20, 95% CI = 0.04 to 0.99) when compared with individuals without family history of gastritis.

Homozygosity of HLA-DQ in individuals (patients + controls) with and without family history of gastritis

HLA-DQ homozygosity was studied in individuals with and without family history of gastritis. Results shown in Table 5, indicated that for HLA-DQA1, out of 14 individuals with family history of gastritis, 2 were homozygous in one or both loci and 16 out of 73 individuals without family history of gastritis, were homozygous in one or both loci, with frequencies of 14.29

and 21.92 respectively. No significant differences were observed in frequencies of homozygous HLA-DQA1 genotype between individuals with and without family history of gastritis (X^2 = 0.42, P = NS, OR = 1.68, 95% CI = 0.34 to 8.31) (Table 5). For HLA-DQB1, 4 out of 14 individuals with family history of gastritis were homozygous in one or both loci, and 18 out of 79 individuals without family history of gastritis, were homozygous in one or both loci, with frequencies of 28.57 and 22.78 respectively. No significant differences were observed in frequencies of homozygous HLA-DQB1 genotype between individuals with and without family history of gastritis (X^2 = 0.22, P = NS, OR = 0.47, 95% CI = 0.21 to 2.64) (Table 5). For HLA- (DQA1 + DQB1), 3 out of 13 individuals with family history of gastritis were homozygous in one or both loci, and 22 out of 69 individuals without family history of gastritis, were homozygous in one or both loci, with frequencies of 23.08 and 31.88 respectively. No significant differences were observed in frequencies of homozygous HLA-(DQA1+DQB1) genotypes between individuals with and without family history of gastritis (X^2 = 0.40, P = NS, OR = 1.56, 95% CI = 0.39 to 6.24) (Table 5).

DISCUSSION

Many studies reported that there is correlation between gastritis and family history of gastritis (Peterson and Graham, 1998; Brown, 2000; Al-Baldawi, 2001). In the present study, HLA-DQA1 and –DQB1 distribution was studied in individuals with family history of gastritis.

Table 5. Homozygousity of HLA-DQ in individuals with and without family history of gastritis.

HLA-DQ Homozygousity*		Cases			
		With family history of gastritis		Without family history of gastritis	
		No	%	No	%
DQA1**	Homozygous	2	14.29	16	21.92
	Heterozygous	12	85.71	57	70.08
	Total	14	100	73	100
DQB1***	Homozygous	4	28.57	18	22.78
	Heterozygous	10	71.43	61	77.22
	Total	14	100	79	100
**** DQA1+DQB1	Homozygous	3	23.08	22	31.88
	Heterozygous	10	76.92	47	68.12
	Total	13	100	69	100

* Homozygous at one or both loci, ** χ^2 = 0.42, P = NS, OR= 1.68, 95% CI = 0.34-8.31, *** χ^2 = 0.22, P =NS, OR= 0.74, 95% CI = 0.21-2.64, **** χ^2 = 0.40, P =NS, OR = 1.56, 95% CI = 0.39-6.24.

Distribution of individuals (patients + controls) with family history of gastritis was studied according to gender. Results showed no significant differences between males and females with and without family history of gastritis (χ^2 = 1.32; P = NS; OR = 0.51; 95% CI = 0.16 to 1.62) (Table 1). Distribution of individuals (patients + controls) with family history of gastritis was studied according to age groups. Results showed no significant differences between individuals with or without family history of gastritis from (15 > 45) and (> 45) age groups (χ^2 = 0.05; P = NS; OR = 1.15; 95% CI = 0.33 to 3.94) (Table 2). Genotype frequency of HLA-DQ was studied in individuals (patients + controls) with and without family history of gastritis. Genotyping of HLA-DQA1 was studied in individuals with family history of gastritis and compared with individuals without family history of gastritis (patients + controls). Results shown in Table 3, indicated that HLA-DQA1*0201 allele was present in 1 out of 14 individuals with family history of gastritis and in 19 out of 73 individuals without family history of gastritis, with allele frequencies of 7.14 and 26.03 respectively. The decreased frequency of HLA-DQA1*0201 allele in individual with family history of gastritis was statistically significant and showed very strong association (χ^2 = 5.37, P< 0.05, OR = 4.57, 95% CI = 0.56 to 37.46) as compared with individuals without family history of gastritis. Results shown in Table 3 also indicated that HLA-DQA1*0402 allele was present in 3 out of 14 individuals with family history of gastritis and in 3 out of 73 individuals without family history of gastritis with frequencies of 21.42 and 4.11 respectively. The increased allele frequency of HLA-DQA1*0402 allele in individuals with family history of gastritis was statistically significant but with week association (χ^2 = 5.49, P < 0.05, OR = 0.16, 95% CI = 0.03 to 0.88) when compared with individuals without family history of gastritis. Genotype of HLA-DQB1 was studied in 14 individuals with family history of gastritis and in 79 individuals without family

history of gastritis. Results shown in Table 4 indicated that HLA-DQB1*0402 allele was present in 3 out of 14 individuals with family history of gastritis and in 4 out of 79 individuals without family history of gastritis with frequencies of 21.43 and 5.06 respectively. The increased frequency of HLA-DQB1*0402 allele in individuals with family history of gastritis was statistically significant but with week association (χ^2 = 4.58, P < 0.05, OR = 0.20, 95% CI = 0.04 to 0.99) when compared with individuals without family history of gastritis. These results agree with a study performed by Herrera-Goepfert et al. (2006) who found a significant increased frequency of HLA-DQB1*0401 allele in H. pylori -positive patients with chronic gastritis when compared with healthy subjects (19 vs 0%, P = 1 × 10-7, odds ratio (OR) = 4.96; 95% confidence interval (95% CI), 3.87 to 6.35). The HLADQB1*0401 allele was found to be associated with atrophic gastritis in H. pylori-infected patients. HLA-DQ homozygosity was studied in individuals with and without family history of gastritis. Results shown in Table 5 indicated that no significant differences were observed in frequencies of homozygous HLA-DQ genotype between individuals with and without family history of gastritis.

REFERENCES

Al-Baldawi MR (2001). Isolation and identification of *Helicobacter pylori* from patients with duodenal ulcer, study of pathogenicity and antibiotic resistance. MSc Thesis. Submitted to College of Science. University of Baghdad.

Al-Dhaher ZAJ (2001). Study of some bacteriological and immunological aspects of *Helicobacter pylori*. MSc Thesis Submitted to College of Science. Al-Mustansirya University.

Al-Jalili FAY (1996). *Helicobacter pylori* peptic ulceration in Iraqi patients, bacteriological and serological study. MSc Thesis submitted to the College of Science. Al-Mustansirya University.

Al-Janabi AAh (1992). *Helicobacter*-associated gastritis, diagnosis and clinicopathological correlation: "A prospective study." MSc Thesis. Submitted to College of Medicine. Al-Mustansirya University.

Al-Saimary AE (2008). The prevalence of *H. pylori* in intra abdominal hydatid disease. Diploma-Dissertation. Submitted to College of

Medicine. Kufa University.
Azem J, Svennerholm AM, Lundin BS (2006). B cells pulsed with

Helicobacter pylori antigen efficiently activate memory CD8[+] T cells from *H. pylori* infected individuals. Clin. Immunol. 158: 962-967.
Beutler E, Gelbart T, Kuhl W (1990). Interference of heparin with the polymerase chain reaction. Biotechniques, 9: 166.
Blomhoff A, Olsson M, Johansson S, Akselsen H E, Pociot F (2006). Linkage diequilibrium and haplotype blocks in the MHC vary in an HLA haplotype specific manner assessed mainly by DRB1*04 haplotypes. Genes Immun., 7: 130-140.
Brown LM (2000). *Helicobacter pylori*: Epidemiology and routes of transmission. Epidemiol. Rev., 22(2): 283-297.
Conrad DF, Andrews TD, Carter NP, Hurles ME, Pritchard JK (2006). A high-resolution survey of deletion polymorphism in the human genome. Nat. Genet., 38: 75-81.
Fu Y, Liu Z, Lin J, Jia Z, Chen W, Pan D (2003). HLA-DRB1, DQB1 and DPB1 polymorphism in the Naxi ethenic group of south-western China. Tissue Antigens, 61: 179-183.
Gregersen JW, Kranc KR, Ke X, Svendsen P, Madsen IS, Thmsen AR (2006). Functional epistasis on a common MHC haplotype associated with multiple sclerosis. Nature, 443: 574-577.
Herrera-Goepfert R, Yamamoto-Furusho JK, Onate-Ocana LF, Camorlinga-Ponce M (2006). Role of the development of chronic gastritis and gastric carcinoma in Maxican patients. World J. Gastroenterol., 28: 7762-7767.
Kumar PP, Bischof O, Purhey PK, Natani D, Urlaub H, Dejean A (2007). Functional interaction between PML and SATB1 regulates chromatin-loop architectur and transcription of the MHC class I locus. Nat. Cell Biol., 9: 45-56.
Luo M, Blanchard J, Pan Y, Brunham K (1999). High resolution sequence typing of HLA-DQA1 and HLA-DQB1 exon 2 DNA with taxonomy-based sequence analysis (TBSA) alleles assignment. Tissue Antigens, 54: 69-82.
Nair RP, Stuart PE, Nistor I, Hiremagalore R, Chia NV, Jenisch S (2006). Sequence and haplotype analysis supports HLA-C as the psoriasis susceptiblilty 1 gene. Am. J. Hum. Genet., 78: 827-851.
Shao W, Tang J, Dorak MT, Song W (2004). Molecular typing of human leukocyte antigen and related polymorphisms following whole genome amplification. Tissue Antigens, 64: 286-292.
Suerbaum S, Michetti P, (2002). *Helicobacter pylori* infection. N. Engl. J. Med. 347(15): 1175-1186.
Torres J, Lopez L, Lazcano E, Camorlinga M, Flores L, Munoz O, (2005).Trends in *Helicobacter pylori* infection and gastric cancer in Mexico. Cancer Epidemiol. Biomark. Prev., 14: 1874-1877.
Wang CX, Wang JF, Liu M, Zou X, Yu XP, Yang XJ (2006) Expression of HLA class I and II on peripheral blood lymphocytes in HBV infection. Chin. Med. J., 119: 753-756.
Yoshikawa T, Naito Y (2000). The role of neutrophils and inflammation in gastric mucosal injury. Free Radic. Res., 33: 785-794.

10

Construction of genome-length cDNA for foot-and-mouth disease virus serotype Asia 1 IND 63/72 vaccine strain

T. Saravanan, C. Ashok Kumar, G. R. Reddy, H. J. Dechamma, G. Nagarajan, P. Ravikumar, G. Srinivas and V. V. S. Suryanarayana*

Molecular Virology Laboratory, Indian Veterinary Research Institute, Bangalore Campus, Hebbal, Bangalore - 560 024, Karnataka, India.

Foot-and-Mouth Disease (FMD) is a highly contagious and economically important viral disease of cloven-hoofed animals. The virus (FMDV) belongs to the genus *Apthovirus* of the family *Picornaviridae*. The FMDV genome consists of a positive sense ssRNA of about 8500 nt encoding a single open reading frame (ORF) which later processed into multiple viral proteins. Although FMDV replication resembles those of other picorna viruses, there are notable differences like broad host range and several unique genetic features. On the Asian continent, FMDV serotype Asia 1 plays a second major role in causing outbreaks after serotype O. The Asia 1 virus displays unique characteristics in its stability, replication kinetics and plaque morphology. In order to study these characteristics we have constructed a genome-length cDNA clone of FMDV Asia 1 vaccine strain IND 63/72. The functionality of the cDNA was checked in BHK 21 cells and it did not yield any viable virus particle. The genome-length cDNA contained a single ORF of 6902 nucleotides terminating at a UAA codon 95 bases from the 3' poly (A) tail sequence. The 8167 base pair sequence and the deduced amino acid sequence (2330 aa) were compared with the published FMDV sequence of Chinese strain YNBS/58 showed 5.3% variation at amino acid level.

Key words: Foot-and-mouth disease virus, Asia 1, *Apthovirus*, genome-length cDNA, *In vitro* transcription and translation.

INTRODUCTION

Foot-and-Mouth Disease (FMD) is the most contagious viral disease of cloven-hoofed wild and domesticated animals and the earliest recorded account was made by Fracastorius in 1546 (Brooksby, 1982). It was the first animal disease demonstrated to be caused by a filterable agent in 1897 by Loeffler and Frosch. The agent has later been identified as an RNA virus (FMDV), belonging to the genus *Apthovirus* of the family *Picornaviridae* (Franki et al., 1991).

The disease is characterized by high morbidity in adults and low mortality in young animals with typical clinical signs of oral and pedal vesicles which can result in significant decline in production of dairy and meat products. The disease is enzootic in many parts of the world including Asia, Africa and South America and the incidence is very low in many of the European countries (Domingo et al., 1990). In India, the disease is endemic and outbreaks occur throughout the year and in all parts of the country. Majority of the outbreaks are caused by type "O", followed by Asia 1, A and C. Type C is the least prevalent and has not been recorded since 1996. The disease is controlled mainly by regular vaccination using inactivated virus as vaccine.

The FMD virion is a non-enveloped icosahedron of 25-30 nm in diameter and the genome consists of an infectious, linear single stranded, positive sense RNA of 8.5 kb. The RNA is uncapped however, virus coded protein VP_g (3B trimer, 3 kDa) is covalently bound to it at 5' end (Grubman, 1980) and 3' end has a poly (A) tail.

*Corresponding author. E-mail: veluvarthy@yahoo.com.

The genome has an ORF flanked on either ends by untranslated regions (UTRs). The 5'UTR of 1300 nucleotides (nts) comprises S fragment (400 nts) followed by poly(C) tract (80 - 200 nts) (Harris and Brown, 1977; De la Torre et al., 1988) and an L fragment (700 nts) which has an Internal Ribosome Entry site (IRES) 435 nts upstream of the start codon. It has a major role in initiation of viral translation in a cap independent fashion (Belsham, 1993; Tratschin et al., 1995). The single viral ORF is of 6996 nts, has two initiation sites both in frame with strong initiation from the downstream start codon and codes for a polyprotein (Fross et al., 1984). Following the ORF is a short 3' UTR (190 nts) and a poly (A) tract (10 - 100 nts) (Chatarjee et al., 1976). The 3' UTR is believed to contain major cis-acting specific sequences signal required for negative-strand RNA synthesis, and is essential to complete a full replication cycle (Saiz et al., 2001). The single ORF consists of 3 polyprotein coding regions, from the 5' end to the 3' end, P1, P2, and P3. The P1 region consists of 1A, 1B, 1C, 1D genes, which codes for the structural proteins VP4, VP2, VP3, and VP1, respectively (Rueckert and Wimmer, 1985). P2 and P3 regions encode non-structural proteins 2A, 2B, 2C and 3A, 3B, 3C, 3D, respectively. The 2A region is highly conserved in all the serotypes (Vakharia et al., 1987). 2C is probably involved in virion RNA synthesis (Saunders et al., 1985). 3B encodes VPg (Fross and Schaller, 1982) and protein 3D functions as the viral RNA polymerase, also known as the Virus Infection Associated Antigen (VIAA) (Polatnick, 1980) used for differentiating infected and vaccinated animals.

The potential of investigating RNA viruses by establishing infectious genome-length cDNA clones has greatly increased after the discovery of the reverse transcriptase enzyme. The availability of cDNA clones makes it possible to analyze and modify genomes at molecular level and helped in the study of gene organization and expression, host-virus interactions, design of antiviral strategies, trans-complementation studies of segmented RNA viruses, development of new viral vectors and for the development of new generation vaccine (Boyer and Haenni, 1994). At present, the infectious cDNA clones of FMDV have been established for O1K from cattle (Zibert et al., 1990), OH99 from swine (Liu et al., 2004), for A12 (Rieder et al., 1993) and for SAT2 vaccine strain ZIM/7/83 (van Rensburg et al., 2004).

Asia 1 remains endemic in India, parts of China, Bangladesh, Israel, Malaysia, Myanmar, Thailand, Laos and Cambodia (Kitching, 1999). The Asia 1 virus is unique in terms of its stability, replication kinetics and plaque morphology (Rajdan et al., 1996). The molecular basis of these characteristics is not understood. Preliminary studies on 5' untranslated region (UTR) (Tratschin et al 1995) showed that variation in internal ribosomal entry site (IRES) might be responsible for variability in host specificity. However, these characteristics can be better understood through infective cDNA clone. So the the present study was undertaken to develop infectious genome-length cDNA clone of FMDV serotype Asia 1 vaccine strain IND63/72 to study its characteristics.

MATERIALS AND METHODS

Cell lines and viruses

Baby hamster kidney (BHK) cells, strain 21, clone 13 maintained in Glasgow Minimal Essential Medium (GMEM) containing 10% fetal bovine serum (FCS) and 10% tryptose phosphate broth were used. Foot and mouth disease virus vaccine strain serotype Asia-1 (IND 63/72) maintained at FMD centre of IVRI, Bangalore, India was once plaque purified passaged once in BHK-21 cells and used as a source of virus.

RNA extraction and cDNA synthesis

The viral RNA extracted directly from infected cell culture supernatant at passage level 6 in BHK21 cell was used for cDNA synthesis using Trizol reagent as per Manufacturer's (Gibco BRL) protocol and used as a template for cDNA synthesis. Superscript (Moloney murine leukemia virus (MMLV) reverse transcriptase, Life Technologies, USA) and gene specific primers corresponding to various regions of the RNA genome were used for the reverse transcription reactions, which were carried out for 90 min. at 37°C. The primer sequences used for the synthesis of cDNA were shown in Table 1.

PCR amplification, construction of genome-length cDNA clone and sequencing

The genome-length cDNA for FMDV serotype Asia 1 (IND63/72) was constructed in the pBSKS+ vector by linking the PCR amplified overlapping fragments. The different regions of the Asia 1 virus were obtained through PCR amplification from cDNA using EXT and Dynazyme DNA polymerase (Finnzyme). The reactions were carried out in the presence of oligonucleotides designed to contain restriction enzyme sites that would facilitate the linking of fragments. The sequences of the primers used for amplification and linking were shown in Table 2. First, the LUTR to 2A fragment was amplified and cloned into pBSKS+ vector in the HindIII and NotI RE sites (pKSLUTR-2A).

The VP1-2A fragment was amplified and cloned sequentially first into pRSETA (BamHI and BglII) and the fragment released with BamHI and EcoRI and cloned into pcDNA3.1+ vector using same sites. Then the fragment was sub-cloned into pBSKS+ vector in HindIII and EcoRI sites (pKSVP12B). The fragment containing LUTR to 2B was obtained by releasing the LUTR-2A fragment in pKSLUTR-2A with HindIII and NheI sites and cloned into the pBSKS+ vector containing VP1-2A fragment and the resulting plasmid named as pKSLUTR2B. The 2BC fragment was amplified using forward and reverse primer containing BglII and EcoRI RE sites respectively and cloned into pKSVP12B, to get the fragment from VP1 to 2BC with partial 3A region (pKSVP12BC3A).

The 3D fragment was amplified and cloned into NheI and NotI sites of pcDNA3.1+ vector. This fragment was released with NheI and PmeI and subcloned into NheI and EcoRV sites of pBR322 vector. Further, the fragment released NheI and HindIII and cloned into pKSVP12B using same RE sites and the resulting plasmid named as pKSVP12B-3D. The 3ABC fragment was amplified and cloned into EcoRI and NheI sites of pKSVP12B-3D plasmid to get pKS3ABCD. The 3ABCD fragment was released from the plasmid pKS3ABCD with BamHI and EcoRI and subcloned into pKSVP12BC3A to get the fragment containing VP1 to 3D region

Table 1. Primers used for first strand complementary DNA synthesis of FMDV genome.

S/No.	Primer	Sequence	Targeted FMDV gene
1	PrepolyCR(SmaI)	5'GGGCCCGGGGGGGGGGTGAAAGGC3'	SUTR-Poly(C) tract
2	LedRm	5'TCGGACGGATCCGGCGTCCA3'	LUTR
3	2BR	5'GCCGTTCTTGAGAATGGC3'	L-protease, P1-2A, 2B
4	3CRNot	5'GGGCGGCCGCCTCGTGTTGTGGTTCAGG3'	2C
5	OligodT	5'TTTTTTTTTTTTTTTTTTT3'	3AB, 3C, 3D
	OligodT(SexA)	5'ACACCAGGTTTTTTTTTTTTTTTTTTTT3'	
6	cDNA(PolyA)NH2-P1	5'GGCCATGGTACGTAGGATCCGGAAAAAAAAAAA3'	3'UTR

Table 2. Primers used for amplification of overlapping fragments of FMDV Asia 1 IND 63/72 genome.

S. No	Gene amplified	Primer sequence
1	T7-Small UTR	STJD+(F): 5'GCGAGTACTGTAATACGACTCACTATAGGTTGAAAGGGGGCGCTAG3' PrepolyCR(Sma)(R): 5'GGGCCCGGGGGGGGGGTGAAAGGC3'
2	Large UTR	5.7UTL(Apa)(F): 5'CCCGGGCCCCCCCCTTTTACCGTCGTTCCCGAC3' 5.7LUTR(R): 5'CCCGGGCCCCCCCCCTTTTACCGTCGTTCCCGAC3'
3	Lpro-P12A	Seq2(F): 5'GAGGTAACACGCGACACT3' 2AR(NotI)(R): 5'GCTAGTGCGGCCGCGAAGGGCCCAGGGTTGGACTC3'
4	Partial VP1-2B	SV1(F): 5'GGGCGGGGATCCGATGGCGCCACACCGTGTGTTGGC3' 2BR(R): 5'GCCGTTCTTGAGAATGGC3'
5	2BC-partial 3A	P2L(F): 5'GAGTCCAACCCTGGGCCCTT3' 3ARM(R): 5'CCATCCCCTCAAAGAATTCAAT3'
6	3ABC-partial3D	3ALn(F): 5'GGTGATTGACCGGGTTGAG3' 3CR(Nhe)(R): 5'AACGGTGGGTGCTAGCTTGGT3'
7	3D-3'UTR-19A	3DL(Nhe)(F): 5'ACCAAGCTAGCACCCACCGT3' OligoT(SexAI)(R): 5'ACACCAGGTTTTTTTTTTTTTTTTTTTT3'
8	3D	3DL(Nhe)(F): 5'ACCAAGCTAGCACCCACCGT3' 3DR(Not)(R): 5'GCTAGTGCGGCCGCTTATGCGTCACCGCACACGG3'

(F) – Forward primer; (R) – Reverse primer.

(pKSVP13D). The fragment VP1-3D was released from the plasmid pKSVP13D using *Bgl*II and *Not*I and subcloned into pKSLUTR2B to get the fragment from LUTR to 3D (pKSLUTR3D). The 3'UTR containing 19 A residues at 3' end were amplified along with 3D and cloned into pcDNA3.1+ vector using *Nhe*I and *Eco*RI RE sites (pc3D-3'UTR). Then the 3'UTR region was released from pc3D-3'UTR using *Pst*I and *Not*I and subcloned into pKSLUTR3D to get the fragment from LUTR to 3'UTR (pKSLUTR-3'UTR).

The 5' oligonucleotide primer containing the T7 promoter and 17 nucleotides of IND63/72 sequence (*Bam*HI) and 3' primer containing *Sma*I was used to amplify the short-untranslated region

(S-UTR). The amplified fragment was blunt ended and inserted into pBSKS+ vector (pKSSUTR). The L-UTR fragment was amplified with the *Apa*I site at 5' end, was digested and blunt ended with T4 DNA polymerase followed by digestion with *Eco*RI at 3' end. This fragment was cloned into pKSSUTR in the *Sma*I and *Eco*RI sites to get the plasmid pKS5'UTR. The plasmid pKS5'UTR was digested with *Bam*HI and *Xba*I and subcloned into the plasmid digested with same RE sites and gel purified vector backbone. The resulting plasmid pKSFAs contained the genome-length cDNA of FMDV serotype Asia 1 (IND63/72). The complete sequence of the assembled genome-length cDNA was submitted to Gene bank with accession no. AY304994 and AY319417.

In vitro transcription

The full-length pFAS plasmid was linearized with *Not* I and extracted with phenol/chloroform, ethanol precipitated and redissolved in RNAse free water. Typically, *in vitro* transcription was performed with 0.5 – 1 μg of linearized full-length cDNA template in a 20 μl reaction using T7 RNA polymerase (Promega) according to the manufacturer's instruction. The transcribed RNA was treated with 1 U of RNAse free RQ1 DNAse (Promega) at 37°C for 30 min. to remove the linearized DNA template. The size and quality of the *in vitro* transcribed RNA was checked by electrophorese in 1.2% agarose gel in Tris-borate-EDTA buffer in the presence of ethidium bromide.

Transfection experiments in BHK21 cells

Transfections of the *in vitro* transcribed RNAs were performed on BHK 21 cells by using Lipofectamine 2000 (Invitrogen). One day before transfection, 0.5–2 × 10⁵ cells in 2 ml of growth medium were seeded in the wells of 6 well plate. At 90 – 95% confluency, the cells were transfected with the RNA and transfection mixture prepared in Opti-MEM medium (Invitrogen) and incubated at 37°C with 5% CO_2 for 4 – 6 h. At the end of incubation, the cells were incubated with fresh Eagle's Minimum Essential Media (EMEM) containing 2% sera for 48 h. The cells were observed for cytopathic effect (CPE) for 48 h and if the CPE is not clear during incubation, either cell supernatant or freeze-thawed cell lysates were blind passaged twice on fresh BHK 21 monolayer cells.

In vitro translation

In vitro translation of the linearized full-length cDNA was performed in rabbit reticulocytes using TNT coupled transcription and translation system (Promega). About 2 μg of DNA was added to an aliquot of the master mix in a 25 μl reaction volume and incubated for 90 min. at 30°C. Luciferase from Renilla and firefly were included in the experiment as a positive control. Synthesized proteins were then analyzed by SDS-PAGE and autoradiography.

RESULTS

Assembly of genome-length cDNA constructs

Sequencing of the overlapping fragments showed unique *NheI* site in the P1-2A gene. Likewise, the *BglII* site identified in the 2B region was used to link the upstream and downstream half of the genome. There is a conserved *PstI* site in the 3D polymerase gene. The restriction site, *XbaI* in the LUTR was identified as unique site. This has been found to be unique for all the FMDV serotypes.

The amplified and cloned fragments were sequenced and confirmed for integrity before being used for final construct. The final construct in pBSKS+ vector was obtained by linking the overlapping fragments and the whole genome was sequenced to confirm the size and coding region of each gene (accession nos. AY304994 and AY319417).

In vitro transcription and infectivity study of RNA into BHK21 cells

RNA was *in vitro* transcribed from the *Not* I linearized plasmid carrying genome-length cDNA (pKSFAs) with T7 RNA polymerase and the quality of the RNA was analysed by gel electrophoresis, which showed the intact and expected size of the transcribed RNA. This RNA contained two non-genomic G residues (added to increase T7 promoter efficiency) at 5' end, a poly(C) tract of 12 nucleotides and 23 non-genomic bases after a 19 base poly (A) tail at 3' end.

To study the infectivity of the genome-length cDNA, the *in vitro* transcribed RNA was transfected into BHK 21 cells using the transfection reagent Lipofectamine 2000 (invitrogen). Apparent CPE is generally observed in case of RNA purified from wild type virus after 24 h, but in the case of *in vitro* transcribed RNA CPE-like appearance was observed after 48 h of incubation (Figure 1A and B). The *in vitro* transcribed RNA-transfected cell supernatant and the freeze thawed cell lysate was passaged twice through BHK 21 cell monolayers. Though initially cells showed rounding and CPE-like appearance, after 48 h the cells recovered and there was no lysis of even after 72 – 96 h of incubation.

Coupled transcription and translation

In order to check the translation and processing of the virus, specific proteins *NotI* linearized plasmid (2 μg) was subjected for coupled transcription and translation in rabbit reticulocytes. After the incubation of sample along with positive control at 30°C, the translated proteins were run in SDS-PAGE and autoradiographed. There was a ladder of proteins observed in the sample lane. It indicated that complete FMDV genome was expressed and processed by viral proteases (like L protease, 2C protease and 3C protease).

So the expressed proteins were cleaved, processed and was seen as multiple bands in the gel (Figure 2, Lane 1). In positive control there was no such ladder, only two bands corresponding to the expressed proteins Firefly and Renilla luciferase, 31 and 65 KDa respectively were noticed (Figure 2, Lane 2). It indicated the integrity, stability and completeness of the FMDV Asia-1 full-length construct (pKSFAs).

DISCUSSION

Foot-and-mouth disease (FMD), a highly contagious and economically important disease is caused by picornavirus which mainly affects wild and domestic cloven-hoofed animals. FMDV populations are genetically hetero-geneous and exhibit an important potential for variation and adaptation (Domingo et al., 2003). Although FMDV replication resembles those of other picorna viruses,

78

Genetics: Analysis and Principles

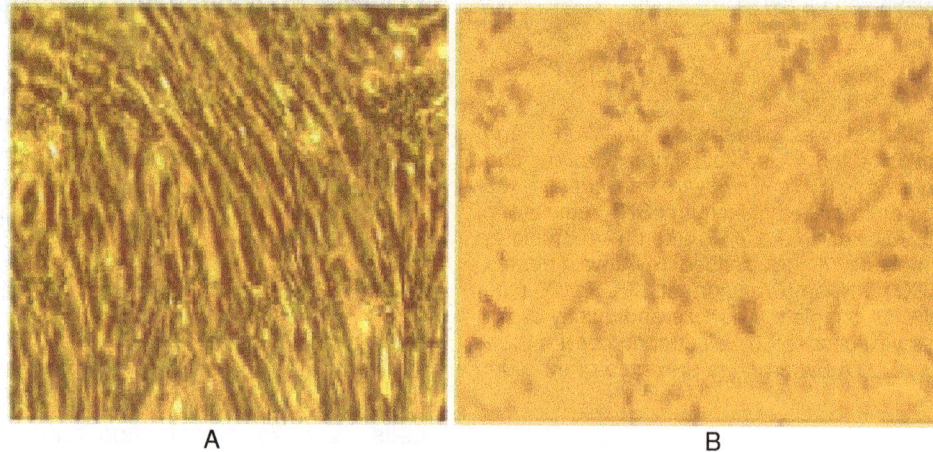

Figure 1. Infectivity of RNA transcript from full length cDNA of Asia 63/72 in BHK 21 cell monolayer cells (A) Mock infected cells (B) Infected with RNA transcript from full length cDNA.

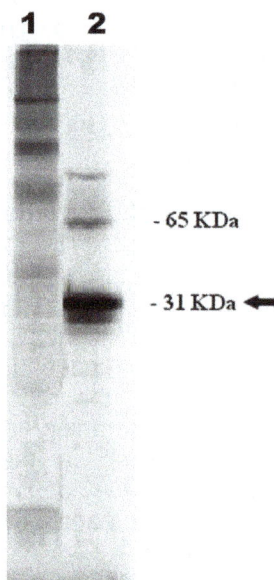

Figure 2. Coupled transcription and translation of Not I linearized genome-length cDNA in rabbit reticulocytes. Lane 1 - Proteins from FMDV serotype Asia-1 genome-length construct. Lane 2 - Proteins from firefly and renilla luciferase positive control.

there are notable differences like broad host range and several unique genetic features. Apart from this, the mechanisms involved in FMDV virulence and pathogenicity are not well known (Mason et al., 2003; Jackson et al., 2003). To elucidate these mechanisms, the

genome-length infective cDNA of FMDV and its recovery of infectious virus in the susceptible cell lines *in vitro* is an important tool. Infectious cDNA clones of O1K (Zibert et al., 1990), A12 (Rieder et al., 1993), OH99 (Liu et al., 2004) and SAT2 (ZIM/7/83) (van Rensburg et al., 2004) have been constructed and characterized. On the Asian continent, FMDV serotype Asia 1 plays a second major role in causing outbreaks after serotype O. The Asia 1 virus displays unique characteristics in its stability, replication kinetics and plaque morphology (Rajdan et al., 1996). In order to study these characteristics we have constructed a genome-length clone of FMDV Asia 1 vaccine strain IND 63/72.

For the synthesis of genome-length cDNA of Asia 1, we have used amplification of multiple fragments, cloning and linking of overlapping fragments strategies. The primers used in this study were initially designed from the published sequences of O1K (Forss et al., 1984; Accession No. X00871), A12 (Robertson et al., 1985; Accession No. M10975) and Asia 1 5'UTR (Accession No. Y17973), L-protease and non-structural proteins (2C, 3C and 3D) (Genbank Accession No. AF207520, AF207524, AF207525 and AF207521). *Nhe*I site which was found as unique in P1-2A of Asia 163/72 was used to link the fragments of L-Protease-P1-2A and 2B. One more *Nhe*I site was introduced into 3D gene without changing the amino acid sequences of gene, and was used for linking the 3ABC to 3D. The conserved Pst 1 was used to link the 3'UTR sequences to the rest of the genome. The unique Xba 1 site was used to link the fragments of small and large 5' UTR with the rest of the genome. The plasmid vectors pBSKS+, pRSETA, pBR322, pcDNA3.1+ and pUC19 were chosen to use the available restriction enzyme sites for the cloning of different overlapping fragments of the IND 63/72 viral genome. The final genome-length construct was made in the pBSKS+ vector, downstream of T7 promoter which was introduced through 5'UTR(S) forward primer. The proof-reading DNA polymerase (pfu DNA polymerase from Fermentas Life Sciences) was used for the synthesis of genome fragments, in order to minimize the PCR mutation. Unique restriction enzyme sites present either in the viral genome or in the vector were introduced in the forward and reverse primers and these amplified fragments were linked to get the genome-length cDNA.

The T7 promoter sequence was introduced at the 5' end to the start of viral genome (5'-UUGAAA-3'), which had two extra G nucleotides (non-viral sequence) added to increase the efficiency of the promoter. It is generally found that the presence of non-viral nucleotides at the 5' end of viral transcripts strongly reduces infectivity (Heaton et al., 1989; Dore et al., 1990). But it had already shown that the extra two G residues did not affect the infectivity of OH99 cDNA (Liu et al., 2004). The 5'UTR of cardioviruses and aphthoviruses contains a homopolymeric Poly (C) tract of unknown function. The mechanisms leading to the size variation of the homopolymer is not clear (Costa et al., 1984). Zibert et al. (1990) considered that poly (C) tract is required for FMDV to replicate in tissue culture. But infectious cDNA clones of FMDV type A12 which contained poly (C) tract of 2 – 35 nucleotides were equally virulent in mice (Rieder et al., 1993). In this study we have constructed the Asia 1 IND 63/72 cDNA containing 12 C residues in the poly (C) tract.

The 3' UTR and poly (A) tail at the 3' end of the transcripts may also play an important role in the infectivity by permitting the hybridization of the 3B-pUpU primer to allow synthesis of the negative sense copy of the viral genome (Barton et al., 1995; Saiz et al., 2001). But the minimum length of the adenosine residue required for infectivity of RNA is unknown. In our genome-length cDNA we have included 19 A residues at the end of 3' UTR. Further, we have included the NotI site to linearize the cDNA before in vitro transcription to minimize the non-viral sequences. When the sequence of the 3' UTR was compared with the other serotype sequences, it was found to be conserved among FMDV serotypes.

The complete genome was characterized at DNA level for the presence of UTR and coding sequences, unique restriction enzymes, open reading frame and at protein level for expression of structural and non-structural proteins. The fragments linked were sequenced before and after linking for confirming the junctions and unique RE sites at the linking region. The 8167 base pairs sequence and the deduced amino acid sequence (2330 aa) were compared with published FMDV sequence which showed 5.3% variation at amino acid level with Chinese strain YNBS/58. The 5' end of the genome upstream from homo polymeric poly (C) tract (S fragment) was 370 nucleotides in length and the remainder of the genome (L-fragment), including the poly (A) tail was 7797 nucleotides. The L-protease fragment contained a single ORF of 6902 nucleotides terminating at a UAA codon 95

bases from the 3' poly (A) sequence.

The in vitro transcribed RNAs were synthesized from genome-length cDNA constructs using T7 RNA polymerase. The purified in vitro transcripts were transfected into BHK21 cell monolayers and the cells were observed for 48 h. There was no specific cytopathic effect of FMDV observed. Initial changes could be observed after two blind passages of transfected cell supernatant and freeze-thawed cell lysates in fresh BHK21 cells; however, the cells recovered showed no lysis indicating that the virus was not stable or cDNA was not infective. However the coupled transcription and translation of the genome-length cDNA showed the production of FMDV specific proteins.

Although obtaining genome-length cDNA clones and/or the corresponding transcripts is a crucial step, it does not necessarily ensure biological activity. Infectivity of transcripts is variable and can in some cases reach 100% as compared to infectivity of wild type virion RNAs. To achieve successful infection, viral transcripts must interact with viral encoded proteins, most particularly with the viral replicase and with host cell components such as the translation machinery. Therefore, the structure of viral transcripts has to mimic that of virion RNA as closely as possible. Several parameters have a dramatic influence on the infectivity of viral transcripts: the heterogeneity of transcript population, the sequence at the 5' and 3' ends and its compatibility with the sequence of the coding region. One may presume the compatibility of 5'UTR with internal sequence may be of primary importance as the 5'UTR used here was amplified at different time point though from single plaque virus. We do not have any data to prove this hypothesis. Our recent studies show that replacement of 5' UTR along with Poly(C) track amplified at a stretch from single viral RNA helps to achieve infectivity (Unpublished data). The heterogeneity of transcript size is responsible for competition between incomplete non-replicable viral copies and genome-length transcripts for interaction with viral and/or host factors involved in the replication process (Hamilton and Baulcombe, 1989). Because of the relatively poor fidelity of the RNA synthesizing enzymes used, in in vitro experiments point mutations are expected, especially with long viral genomes. It was reported that alteration of viral sequences could result from the in vitro transcription step since sequence dependence of T7 and SP6 RNA polymerase fidelity has been observed (Kuhn et al., 1990). Further, the final genome-length cDNA clone can be the result of faithful reverse transcription and amplification of an initial virion RNA which itself may be of mutated version, unable to replicate and which would probably be eliminated in the next round of viral replication. As a consequence, this cDNA clone or the corresponding in vitro transcript would not be infectious. Error prone in vitro transcription coupled with low fidelity of viral RNA dependent RNA polymerases (Domingo et al., 2003) may be responsible for failure in the production of infective virus. The production of non-infectious transcripts has been reported and some showed the possibility of restoring infectivity by exchanging a specific region of the cDNA with a fragment corresponding to the same region but replaced from independent cDNA clone (Boyer and Haenni, 1994).

ACKNOWLEDGEMENT

This work was supported by a grant from National Fund, Indian Council of Agricultural Research (ICAR), India. The authors wish to thank the Director, IVRI, and the Joint Director, Bangalore for providing the facilities.

REFERENCES

Barton DJ, Black EP, Flanegan JB (1995). Complete replication of poliovirus *in vitro*: Pre-initiation RNA replication complexes require soluble cellular factors for the synthesis of VPg-linked RNA. J.Virol., 69: 5516-5527.

Belsham GJ (1993). Distinctive features of foot-and-mouth disease virus, a member virus of the picornavirus family; aspects of virus protein synthesis, protein processing and structure. Prog. Biophys. Mol. Biol., 60: 241.

Boyer JC, Haenni AL (1994). Infectious transcripts and cDNA clones of RNA viruses. Virol., 198: 415-426.

Brooksby JB (1982): Portraits of viruses: foot and mouth disease virus. Intervirol., 18: 1.

Chatarjee NK, Bachrach HL, Polatnick J (1976). Foot-and-mouth disease RNAs presence of 3'polyadenylic acid and absence of amino acid binding ability. Virol., 69: 369.

Costa CMP, Dergmann IC, Scodeller CA, Auge de Mello P, Gomes I, La Torre JL (1984). Heterogeneity of the polyribocytidylic acid tract in aphtho viruses: biochemical and biological studies of viruses carrying polyribocytidylic acid tracts of different lengths. J. Virol., 51: 799-805.

Domingo E, Escarmis C, Baranowski E, Ruiz-Jarabo CM, Carillo E, Nunez JI, Sobrino F (2003). Evolution of foot-and-mouth disease virus. Virus Res. 91: 47-63.

Domingo E, Mateu MG, Martinez MA, Dopazo J, Moya A, Sobrino F (1990). Genetic variability and antigenic diversity of Foot-and-Mouth Disease Virus In: Kurstak, E., Marusyk RG, Murphy FA, Van Ragenmortel MHV (Eds), Appl. Virol. Res., Virus Variability, Epidemiol. Control. Plenum Press, New York. 2: 233-266.

Dore JM, Erny C, Pinck L (1990). Biologically active transcripts of alfalfa mosaic virus RNA3. FEBS Lett., 264: 183-186.

Franki RIB, Fauquet CM, Knudson DL, Brown F (1991). Classification and nomenclature of viruses. Fifth report of the International Committee on Taxonomy of Viruses. Springer, New York. Arch Virol., 2: 324-325.

Fross S, Schaller H (1982). A tandem repeat gene in picornavirus. Nucleic Acids Res., 10: 6441-6450.

Fross S, Strebel K, Beck E, Schaller H (1984). Nucleotide sequence and genome organization of Foot-and-mouth disease virus. Nucleic Acid Res., 12: 6587.

Hamilton WDO, Baulcombe DC (1989), Infectious RNA Produced by *in vitro* Transcription of a Full-length Tobacco Rattle Virus RNA-1 cDNA . J. Gen. Virol. 70: 963-968.

Heaton LA, Carrington JC, Morris TJ (1989). Turnip crinkle virus infection from RNA synthesized in *vitro*. Virol., 170: 214-218.

Jackson T, King AM, Stuart DI, Fry E (2003). Structure and receptor binding. Virus Res. 91: 33-46.

Kitching RP, Knowles NJ, Samuel AR, Donaldson AI (1999). Development of foot-and-mouth disease virus strain characterization – a review. Trop. Anim. Hlth. Prod. 21: 153-166.

Kuhn R, Hong Z, Strauss JH (1990). Mutagenesis in the nontranslated region of Sindbis virus RNA. J. Virol. 64: 1472-1476.

Liu G, Liu Z, Xie Q, Chen Y, Bao H, Chang H, Liu X (2004). Generation of an infectious cDNA clone of an FMDV strain isolated from swine. Virus Res., 104(2): 157-64.

Mason PW, Pacheco JM, Zhao QZ, Knowles NJ (2003). Comparisons of the complete genomes of Asian, African and European isolates of a recent foot-and-mouth disease virus type O pandemic strain (PanAsia). J. Gen. Virol., 84: 1583-1593.

Polatnick J (1980). Isolation of a foot-and-mouth disease polyuridylic acid polymerse and its inhibition by antibody. J. Virol. 33: 774-779.

Rajdan R, Sen AK, Rao BV, Suryanarayana VVS (1996). Stability of foot-and-mouth disease virus, its genome and proteins at 37°C. Acta. Virol. 40(1): 9-14.

Rieder E, Bunch T, Brown F, Mason PW (1993). Genetically engineered foot-and-mouth disease viruses with poly(C) tracts of two nucleotides are virulent in mice. J. Virol., 67: 5139–5145.

Robertson BH, Grubman MJ, Weddel GN, Moore DM, Welsh JD, Yamsura DG, Kleid DG (1985). Nucleotide and amino acid sequence coding for polypeptides of foot and mouth disease virus type A12. J. Virol., 54: 651-660.

Rueckert RR, Wimmer E (1985). Systematic nomenclature of picorna viral proteins. J. Virol., 50: 957-959.

Saiz M, Gomez S, Martinez-Salas F, Sobrino F (2001). Deletion or substitution of the apthovirus 3' NCR abrogates infectivity and virus replication. J. Gen. Virol., 82: 93-101.

Saunders K, King AM, McCahon D, Newman JW, Slade WR, Forss S (1985). Recombination and oligonucleotide analysis of guanidine resistant foot-and-mouth disease virus mutants. J. Virol. 56: 921-929.

Tratschin JD, Hoffman MA, Suryanarayana VVS (1995). Structural and functional analysis of the 5' Untranslated region of Foot and Mouth disease virus Asia-I. Poster presented at IV International Positive strand virus Symposium, Utrecht, The Netherlands.

Vakharia VN, Devaney MA, Moore DM, Dunn JJ, Grubman MJ (1987). Proteolytic processing of foot-and-mouth disease virus polyproteins expressed in a cell-free system from clone-derived transcripts. J. Gen. Virol., 61: 3199-3207.

Van Rensburg HG, Henry TM, Mason PW (2004). Studies of genetically defined chimeras of a European type A virus and a south African Territories type 2 virus reveal growth determinants for foot-and-mouth disease virus. J. Gen. Virology. 85: 61-68.

Zibert A, Maass G, Strebel K, Falk MM, Beck E (1990). Infectious foot and mouth disease virus derived from cloned full-length cDNA. J. Virol. 64: 2467-2473.

11

Antimutagenic activity of aqueous extract of *Momordica charantia*

Meera Sumanth[*] and G. Nagarjuna Chowdary

Department of Pharmacology, Visveswarapura Institute of Pharmaceutical Sciences, 22nd Main, 24th Cross, B. S. K II stage, Bangalore-560070, Karnataka, India.

Toxicological studies have undergone a significant evolution during the past decade, with inclusion and great emphasis on chronic toxicity, carcinogenicity, teratogenicity and mutagenicity. Present study was taken up to evaluate antimutagenicity of aqueous extract of *Momordica charantia* (MC) by bone marrow micronucleus assay (MNT) and chromosomal aberration test (CAT) in mice. Cyclophosphamide (100 mg/kg, i.p) was used as a genotoxic challenge and bone marrow of control and MC treated mice was collected after 24, 48 and 72 h, respectively. In MNT, the bone marrow smears were stained with May-Grunwald's followed by Giemsa stain. Polychromatic and Normochromatic erythrocytes were counted and P/N ratio was calculated. In CAT, colchicine, four mg/kg, i.p, was administered 90 min before sacrifice, bone marrow smears were prepared, stained with Giemsa stain and observed under 100X for different types of chromosomal aberrations. Mitotic index was calculated. The MC has significantly decreased the formation of micronuclei, increased the P/N ratio, inhibited the formation of chromosomal aberrations and increased the mitotic index. Hence, *Momordica charantia* has significant antimutagenic activity.

Key words: Antimutagenic activity, chromosomal aberration test, micronucleus assay, *Momordica charantia*.

INTRODUCTION

All chemicals that produce DNA damage leading to mutation or cancer are described as genotoxic. Toxicological studies have undergone a significant evolution during the past decade, with much greater emphasis being placed on chronic toxicity, carcinogenicity, teratogenicity and mutagenicity. The mutations in somatic cells are not only involved in the carcinogenesis process but also play a role in the pathogenesis of other chronic degenerative diseases, such as atherosclerosis and heart diseases, which are the leading causes of death in the human population (De Flora and Izzotti, 2007). Micronucleus test and chromosomal aberration test are used for studying antimutagenic activity of a drug. One of the best ways to minimize the effect of mutagens and carcinogens is to identify the anticlastogens /antimutagens (substances which suppress or inhibit the process of mutagenesis by acting directly on the mechanism of cell) and desmutagens (substances which some how destroy or inactivate, partially or fully the mutagens, thereby affecting less cell population) in our diets and increasing their use. Nature has bestowed us with medicinal plants. There is a need to explore them for use as antimutagenic and anticarcinogenic food or drug additives.

Momordica charantia, family of Cucurbitaceae, is commonly known as bitter gourd or bitter melon in English and karela in Hindi. Antidiabetic activity of *M.*

---

*Corresponding author. E-mail: meerasumanth@gmail.com.

Abbrevations: **MN,** Micronuclei; **PCE,** polychromatic erythrocytes; **NCE,** normochromatic erythrocytes; **MC,** *Momordica charantia*; **CAT,** chromosomal aberration test.

charantia is well-known. In diabetic patients Karela lowers blood sugar, delays complications such as nephropathy, neuropathy, gastroparesis and cataract, atherosclerosis (Grover and Yadav, 2004; Seham et al., 2006). It also has antiulcer activity (Gurbuz et al., 2000). Diabetes and ulcer are some of the stress induced diseases. Our earlier studies (Meera and Nagarjuna, 2009) indicated antistress and immunopotentiating activity of *M. charantia*. For using Karela or bittergourd as neutraceutical, it is essential to ensure that it is nongenotoxic.

Hence, an attempt is made to carry out *in vivo* mouse bone marrow micronucleus test and chromosomal aberration test to evaluate antimutagenic activity of *M. charantia* by moderating the genetic damage induced by cyclophosphamide.

MATERIALS AND METHODS

Preparation of extract

Fresh karela or fruits of *M. charantia* were procured from local market in July 2008. Fresh unripe fruits were sliced; pulp and seeds were removed and then mechanically squeezed. The juice obtained was dried in hot air oven below 60 C to get dried powder (Grover and Yadav, 2004). The powder obtained was passed through sieve no. 40. The drug solution (10 mg/ml) was made using water as a vehicle for administration to animals.

Experimental animals and treatment

Eight to ten weeks old Swiss albino mice of either sex, weighing 25 - 30 g maintained under standard environmental conditions (25 ± 2°C, relative humidity 45 ± 10%, light and dark cycle of 12 h) and fed with standard pellet diet and water *ad libitum*, were used for the present study. The experimental protocol, which is in accordance with the OECD (Organization of Environmental Carcinogen Detection) guidelines No.470 and WHO guidelines for mutagenicity studies in animals, (www.oecd.org) was approved by the Institutional Animal Ethics Committee before starting the experiments. The animals were divided into eight groups consisting of six animals each. Group one served as normal control, group two, three four was treated with clastogen, Cyclophosphamide 100 mg/kg, i.p. and bone marrow was collected after 24, 48 and 72 h of clastogen administration, respectively. Group five was treated with *M. charantia* (900 mg/kg, p.o.) for seven days. Groups six, seven, eight were treated with *M. charantia* for seven days followed by Cyclophosphamide as a challenge. The dose for mice was calculated based on LD_{50} (91.9 mg/100 gm) values of *M. charantia* (Meera and Nagarjuna, 2009). On seventh day, bone marrow was collected at 24, 48 and 72 h after clastogen administration, respectively.

Bone marrow micronucleus assay (Hayashi et al., 1994)

On seventh day, the animals were anesthetized and the bone marrow was aspirated from femur and tibia into one ml of 5% bovine albumin in phosphate buffered saline (pH 7.2). The cell suspension was centrifuged (1000 rpm for 5 min) and the smears were prepared from the pellet on chemically cleaned glass slides and stained with May-Grunwald's and followed by Giemsa stain. The smears were analyzed under oil immersion using Labomed-

Model Digi 2 microscope (90 - 260 V) for the presence of Micronuclei (MN) in polychromatic erythrocytes (PCE) and normochromatic erythrocytes (NCE). P/N(Polychromatic erythrocyte/Normochromtic erythrocyte) ratio was determined by counting a total of about 500 erythrocytes per animal and 2000 erythrocytes were examined for the presence of micronuclei (Borroto et al., 2003).

Chromosomal aberrations test (Goncalves et al., 2008; Seetharama and Narayana, 2005)

On seventh day, each animal was injected with 0.04% colchicine in a dose of four mg/kg i.p, 90 min prior to death, for mitotic arrest. The bone marrow was aspirated from femur and tibia into suspending medium 0.075 M KCl, centrifuged and supernatant was discarded. The pellet was mixed with fixative (three: one, methanol: acetic acid) and then centrifuged. The preparation was given two changes of fixative and smears were prepared. The slides were flame-dried and stained with 10% Giemsa at pH 6.8 for 15 - 20 min. Smears were screened for different types of chromosomal abnormalities-rings, breaks, exchanges and minute.

Statistical analysis

The results were expressed as Mean ± SEM and analysis was carried out by one-way ANOVA. Post-hock analysis was done by Turkey's multiple comparison tests to estimate the significance of difference between various individual groups. P < 0.001 was considered significant.

RESULTS

Micronucleus assay

The inhibitory effect of *M. charantia* against clastogenicity induced by cyclophosphamide is shown in Tables 1, 2, 3 and Figure 1. In the present study, *M. charantia* showed time dependent inhibitory effect on the frequency of MN in PCE as well as NCE. Decrease in P/N ratio due to Cyclophosphamide was also inhibited by *M. charantia*. Inhibition was found to be more between 24 and 48 h after clastogenic challenge and less there after.

Chromosomal aberration test

As shown from Table 4 and Figure 2b, there was a statistically significant increase in chromosomal aberrations in reponse to cyclophosphamide (100 mg/kg). *M. charantia* significantly inhibited the frequency of various chromosomal aberrations (Figure 2c) and the decrease in mitotic index induced by clastogen.

DISCUSSION AND CONCLUSION

The bone marrow micronucleus test is one of the most suitable genotoxicity tests. Other tests include chromo-somal aberration, peripheral blood micronucleus, and

Table 1. Effect of *M. charantia* after 24 h of clastogenic challenge.

Group	% MNPCE	%MNNCE	P/N
Control	0.46 ± 0.036	0.25 ± 0.030	0.909 ± 0.008
Cyclophosphamide (100 mg/kg)	2.57 ± 0.092[+]	0.88 ± 0.041[+]	0.625 ± 0.009[+]
M. charantia	0.57 ± 0.030	0.34 ± 0.029	0.928 ± 0.010
M. charantia + Cyclophosphamide (100 mg/kg)	1.23 ± 0.566[*]	0.51 ± 0.013[*]	0.818 ± 0.011[*]

n=6, Values are expressed in Mean ± SEM, One way ANOVA followed by Turkey's multiple comparison test. [+]P<0.001 Vs Control, *P<0.001 Vs Cyclophosphamide.

Table 2. Effect of *M. charantia* after 48 h of clastogenic challenge.

Group	% MNPCE	%MNNCE	P/N
Control	0.46 ± 0.036	0.25 ± 0.030	0.909 ± 0.008
Cyclophosphamide (100 mg/kg)	3.71 ± 0.133[+]	1.32 ± 0.048[+]	0.481 ± 0.007[+]
M. charantia	0.57 ± 0.030	0.34 ± 0.029	0.928 ± 0.010
M. charantia + Cyclophosphamide (100 mg/kg)	1.53 ± 0.020[*]	0.64 ± 0.016[*]	0.767 ± 0.014[*]

n=6, Values are expressed in Mean ± SEM, One way ANOVA followed by Tukey's multiple comparison test. [+]P<0.001 Vs Control, [*]P < 0.001 Vs Cyclophosphamide.

Table 3. Effect of *M. charantia* after 72 h of clastogenic challenge.

Group	% MNPCE	%MNNCE	P/N
Control	0.46 ± 0.036	0.25 ± 0.030	0.909 ± 0.008
Cyclophosphamide (100 mg/kg)	4.36 ± 0.049[+]	1.88 ± 0.026[+]	0.399 ± 0.008[+]
M. charantia	0.57 ± 0.030	0.34 ± 0.029	0.928 ± 0.010
M. charantia + Cyclophosphamide (100 mg/kg)	1.82 ± 0.038[*]	0.83 ± 0.019[*]	0.711 ± 0.010[*]

n=6, Values are expressed in Mean ± SEM, One way ANOVA followed by Tukey's multiple comparison test. [+]P<0.001 Vs Control, *P<0.001 Vs Cyclophosphamide.

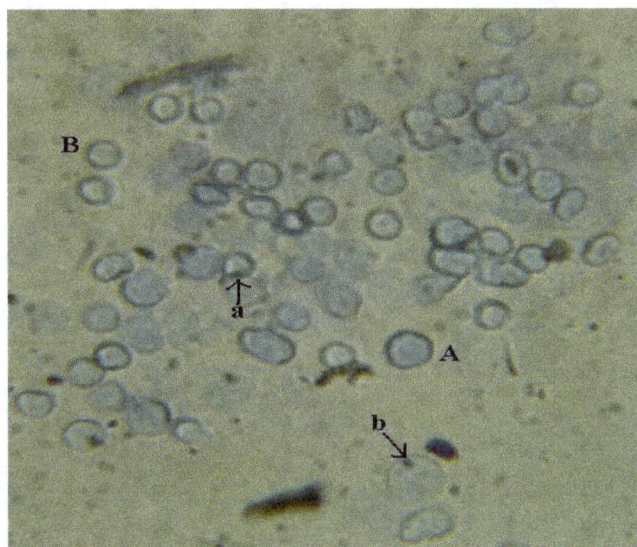

Figure 1. Effect of MC on micronucleus assay.
A-Polychromatic Erythrocyte (PCE), **B**-Normochromatic Erythrocyte (NCE).
a-Micronuclei in PCE, **b**-Micronuclei in NCE (M.G.& G, x100).

Table 4. Effect of M. charantia on chromosomal aberrations and mitotic index.

Group	Treatment	Types of aberrations				Total no. of aberrations	Mitotic index
		Rings	Exchanges	Breaks	Minute		
1.	Normal control	1.16 ± 0.307	0.16 ± 0.166	0 ± 0	1.5 ± 0.223	2.83±0.307	3.51 ± 0.108
2.	Cyclophosphamide (100 mg/kg) p.o., (24 h)	5.16 ± 0.307[+]	1.83 ± 0.307[+]	5 ± 0.365[+]	4.16 ± 0.307[+]	15.83±0.477[+]	1.03 ± 0.022[+]
3.	Cyclophosphamide (100 mg/kg) p.o., (48 h)	5.83 ± 0.307[+]	2.16 ± 0.307[+]	5.66 ± 0.333[+]	5.16 ± 0.307[+]	18.83±1.014[+]	0.93 ± 0.026[+]
4.	Cyclophosphamide (100 mg/kg) p.o., (72 h)	6.16 ± 0.166[+]	2.5 ± 0.223[+]	5.83 ± 0.307[+]	5.66 ± 0.210[+]	20.16±0.654[+]	0.83 ± 0.018[+]
5.	M. charantia (900 mg/kg) p.o.,	1.16 ± 0.307	0.16 ± 1.66	0 ± 0	1.5 ± 0.223	2.83±0.307	3.67 ± 0.144
6.	M. charantia + Cyclophosphamide (100 mg/kg) p.o., (24 h)	2.16 ± 0.307[+]	1.0 ± 0.258	1 ± 0.258[*]	1.66 ± 0.210	5.83±0.477	3.23 ± 0.091[*]
7.	M. charantia + Cyclophosphamide (100 mg/kg) p.o., (48 h)	2.5 ± 0.223[*]	1.5 ± 0.223	1.33 ± 0.210[*]	1.83 ± 0.307[*]	7.16±0.307[*]	3.00 ± 0.081[*]
8.	M. charantia + Cyclophosphamide(100 mg/kg) p.o., (72 h)	2.66 ± 0.210[*]	1.66 ± 0.210	1.66 ± 0.210	2.33 ± 0.333[*]	8.33±0.333[*]	2.74 ± 0.104[*]

sperm morphology tests. The antimutagenic activity of M. charantia was evaluated by measuring their inhibitory effect on cyclophosphamide induced mutagenesis.

Upon administration of cyclophosphamide there was significant rise in % MNPCE, % MNNCE (Figure 1) and it was time dependent, indicating cyclophosphamide induced chromosomal damage in mouse bone marrow cells. These fragmented chromosomes were condensed to form micronuclei which are not included in the main nucleus (Hayashi et al., 1994). Administration of M. charantia alone do not produce any significant variation in % MNPCE and % MNNCE indicating that, it is devoid of any genotoxicity. M. charantia decreased the cyclophosphamide induced formation of micronuclei in PCE and NCE, which may be due to the inhibition of cyclophosphamide induced chromosomal damage. The inhibition was more between 24 and 48 h and less there after. This may be due to the metabolism of the drug after 48 h.

Cyclophosphamide produced a significant, time dependent decrease in the P/N ratio, which may

be due to increase in NCEs which signals a cytotoxic effect. M. charantia significantly inhibited the same, by decreasing the formation of NCE. The inhibition increased linearly at 24 and 48 h after cyclophosphamide. But M. charantia could not bring P/N ratio to normal level. A regimen of M.C treatment/administration for more than seven days may bring P/N ratio to normal level. M. charantia is devoid of any genotoxicity as it does not produce any significant variation in P/N ratio after administration of MC alone.

In chromosomal aberration test, there was a significant, time dependent rise in the total no. of chromosomal aberrations-rings, breaks, exchanges and minute, of cyclophosphamide treated animals (Figure 2b), when compared with normal control animals (Figure 2a). Cyclophosphamide gets metabolized to phosphoramide mustard and acrolein before it can act as a mutagenic agent to promote chromosomal aberrations (Hales, 1982). Chromosomal aberrations are due to lesions in DNA caused by phosphoramide mustard which lead to discontinuities of the DNA helix. M.charantia significantly inhibits the

cyclophosphamide induced chromosomal aberrations (Figure 2c), which may be due to inhibition of cyclophosphamide induced chromosomal damage. The inhibition was more at 24 h and less thereafter. M. charantia is devoid of any genotoxicity as it does not produce any significant variation in total no. of chromosomal aberrations after administration of MC alone.

In our study, we found a significant decrease in mitotic index of cyclophosphamide treated animals, which can be due to the affected cell division in the bone marrow (Goncalves et al., 2008). M. charantia significantly inhibits the disturbances in the cell division of mouse bone marrow.

M. charantia possess antimutagenic activity.

ACKNOWLEDGEMENT

We would like to thank All India Council for Technical Education (AICTE) for providing the stipend during research work to one of the authors-Mr Nagarjuna Chowdary G.

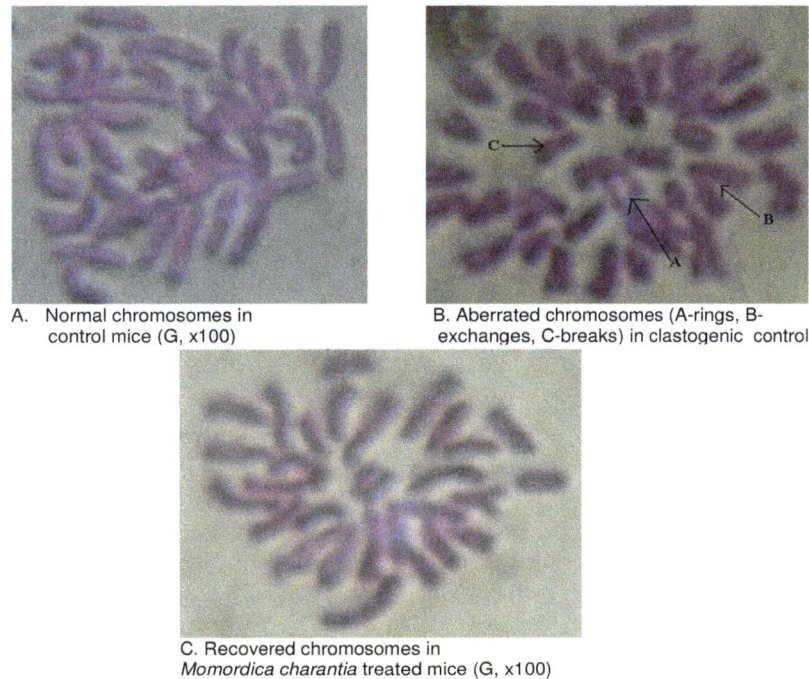

A. Normal chromosomes in
    control mice (G, x100)

B. Aberrated chromosomes (A-rings, B-
   exchanges, C-breaks) in clastogenic control

C. Recovered chromosomes in
   Momordica charantia treated mice (G, x100)

Figure 2. Cytogenetic evaluation of MC.

REFERENCES

Borroto JIG, Creus A, Marcos R, Molla R, Zepatero J (2003). The mutagenic potential of the furylethylene derivative 2-furyl-1-nitroethene in the mouse bone marrow micronucleus test. Toxic Sci., 72: 359-562.

De Flora S, Izzotti A (2007). Mutagenesis and Cardiovascular diseases: molecular mechanisms, risk factors, and protective factors. Mut. Res., 621: 5-17.

Goncalves da Silva RM, Sousa NC, Graf U, Spano MA (2008). Antigenotoxic effects of *Mandevilla velutina* (Gentianales, Apocynaceae) crude extract on cyclophosphamide-induced micronuclei in Swiss mice and urethane-induced somatic mutation and recombination in *Drosophila melanogaster,* Genet. Mol. Biol., 31: 751-758.

Grover JK, Yadav SP (2004). Pharmacological actions and potential uses of *Momordica charantia*: A review. J. Ethnopharmacol., 93: 123-132.

Gurbuz I, Akyuz C, Yesilada E, Sener B (2000). Anti-ulcerogenic effect of *Momordica charantia* L. fruits on various ulcer models in rats. J. Ethnopharmacol., 71: 77-82.

Hales BF (1982). Comparison of the mutagenicity and teratogenicity of Cyclophosphamide and its active metabolites, 4-hydroxycyclophosphamide, phosphoramide mustard and acrolein. Can. Res., 42: 3016-3021.

Hayashi M, Tice RR, Macgregor JT, Aderson D, Blakey DH (1994). *In vivo* rodent erythrocyte micronucleus assay. Mut. Res., 312: 293-304.

Meera S, Nagarjuna CG (2009). Antistress and immunomodulatory activity of aqueous extract of *Momordica charantia.* Pharmacognosy Magazine, 5(19): 69-73.

Seetharama RKP, Narayana K (2005). *In vivo* chromosome damaging effects of an inosine monophosphate dehydrogenase inhibitor: Ribavirin in mice. Ind. J. Pharmacol., 37: 90-95.

Seham AESB, Souad EEG, Osama AES (2006). Some toxicological studies of *Momordica charantia* L. on albino rats in normal and alloxan diabetic rats. . J. Ethnopharmacol., 108: 236-242

www.oecd.org/dataoecd/17/50/1948370.pdf Accessed on 2009 Apr 12th, OECD Guideline no. 423.

12

Molecular characterization of a gene capable of degrading trichloroethylene, an environmental pollutant

Srijata Mitra and Pranab Roy*

Department of Biotechnology, The University of Burdwan Golapbag Burdwan, West Bengal, India.

Trichloroethylene (TCE) is a widely used organic solvent and metal degreasing agent, one of the most frequently detected groundwater contaminants and a potential health hazard. Our novel isolate, *Bacillus cereus* strain 2479 was capable of degrading TCE efficiently. The gene for TCE degradation was PCR amplified from genomic DNA of *B. cereus* 2479. The amplified gene was cloned into expression vector pUC I8 in the *E. coli* host XL1-Blue and expressed under the control of lac promoter and nucleotide sequence was determined. The sequencing results showed that this novel gene (designated as *tce1*, GenBank Accession No: GU183105) contained 342 bp long ORF encoding 114 amino acids with a predicted molecular weight 12.6 kDa and the theoretical pI value of the polypeptide is 5.17. *E. coli* expressing the *tce1* gene overproduces a polypeptide in the presence of the inducer Isopropyl-β-D-thiogalaocto-pyranoside which reacts immunologically to the polyclonal antibody against TCE inducible proteins of the strain 2479. The secondary structure of *Tce1* protein was predicted through internet resources with software, CLC Protein Workbench. The present study suggested that cloned gene product (*Tce1*) was capable of degrading TCE as verified chemically.

Key words: Trichloroethylene, *tce1* gene, molecular cloning, expression, bioinformatics analysis.

INTRODUCTION

Trichloroethylene (TCE) is an Environmental Protection Agency priority pollutant widely used as an industrial metal degreaser and cleaning of cotton, wool and other fabrics in textile industry (Storck, 1987). The contamination of drinking water supply with TCE is increasing in prevalence and concentration (Roberts et al., 1982). Studies on animal models indicate that TCE increases the risk of tumors and lymphoma (Miller and Guengerich, 1983). Exposure to TCE can affect the human central nervous system (CNS). A recent analysis of available epidemiological studies reports TCE exposure to be associated with several kinds of cancers in humans. Biotransformation of TCE to the potent carcinogen vinyl chloride (VC) by consortia of anaerobic bacteria (Vogel and McCarty, 1985) might have role for the aforementioned purpose. Although, TCE acts as potent air, water and soil contaminants, its effect in

drinking water contamination and subsequent health hazard has been reported in great details. For these reasons, there is great interest in implementing processes to remove TCE from drinking water supplies. Bioremediation is one of the environment friendly means of degrading toxic chemicals. Therefore, extensive efforts have been made to study the biodegradation of TCE by bacteria. Many chlorinated organic compounds are known to persist in the environment because of their resistance to microbial attack. For example, TCE was observed to exhibit a half-life of 300 days in one aquifer. However, TCE was found to be co-metabolized by some ammonia-oxidizing bacterial species and by some bacteria able to grow on hydrocarbons such as methane (Oldenhuis et al., 1989) propane (Wackett et al., 1989) and isoprene (Ewers et al., 1990). It was demonstrated that oxygenation reaction by monooxygenase leads to the production of unstable epoxide intermediates (Fox et al., 1990; Little et al., 1988; Miller and Guengerich, 1983). In the case of aromatic inducer substrates for example, phenol or toluene (Harker and Kim, 1990) initial mono or dioxygenases of the degradative pathways may be

---

*Corresponding author. E-mail: pranabroy@rediffmail.com.

responsible for the aerobic TCE degradation.

Several aromatic monooxygenase system, such as a toluene-4-monooxygenase in *Pseudomonas mendocine* KR-1 (Winter et al., 1989; Sheilds et al., 1989) a toluene-ortho-monooxygenase in *Pseudomonas cepacia* G4 (Winter et al., 1989) have been identified to co-oxidize TCE. To our knowledge, only one aromatic dioxygenase system, the toluene dioxygenase in *Pseudomonas putida* F1 has been shown to be involved in TCE degradation. Products generated by toluene dioxygenase from TCE include formic and glyoxylic acids (Nelson et al., 1987). The toluene dioxygenase in *P. putida* F1 was identified as class IIB multicomponent dioxygenase comprising the large subunit of the terminal dioxygenase (*todC1*), which is the most important gene for TCE oxidation (Furukawa et al., 1994). Although no microbial growth on TCE as the sole carbon source has been reported yet, we were the pioneer in reporting a microorganism capable of growing on TCE as the sole carbon source (Dey and Roy, 2009) and we also reported first, *Bacillus cereus* group being used in biodegradation of trichloroethylene (Mitra and Roy, 2010).

In this study, we report the isolation, molecular cloning and characterization of the new gene, designated as *tce1* (GenBank Accession Number: GU183105) involved in trichloroethylene degradation from *B. cereus* strain 2479 belongs to dioxygenase family (Mitra and Roy, 2011a) by phylogenetic analysis. This gene sequence is different from known toluene dioxygenase or monooxygenase genes. Specific antibodies raised against the inducible proteins in *B. cereus* 2479 by TCE reacted with this Isopropyl-β-D-thiogalaoctopyranoside (IPTG) inducible protein synthesized in the recombinant *Escherichia coli* (pSM 101 containing *E. coli*) XL1-Blue.

MATERIALS AND METHODS

Bacterial strain

The strain 2479 was isolated from the soil of industrial belt, situated at Rajbandh (Durgapur, West Bengal, India) where the use of polychlorinated hydrocarbon (including TCE) is quite abundant. Strain 2479 was classified as *B. cereus* on the basis of its morphological and physiological characteristics and ribotyping (Dey and Roy, 2009; Mitra and Roy, 2010).

Chemicals and reagents

TCE (purity by GC analysis > 99%) was obtained from Merck Limited, India. Pyridine was obtained from Qualigen Fine Chemicals, India. Restriction endonuclease, Hind III, T4 DNA ligase, Isopropyl-β-D-thiogalaoctopyranoside, 5-Bromo4-Chloro-3-Indolyl-β-D-galactopyranoside (X-Gal) and Ampicillin were purchased from Sigma-Aldrich Chemical Co, USA. Vector pUC 18 was obtained from Helini, India.

Growth of organism

The isolate was grown in M9 medium (Na$_2$HPO$_4$-6 g; KH$_2$PO$_4$-3 g;

NaCl-0.5 g; NH$_4$Cl-1 g; MgSO$_4$-0.002 M; CaCl$_2$ -0.0001 M; Glucose 0.2%; H$_2$O 1 L pH 7.0). In some experiments, TCE (0.2% v/v) replaced glucose as the sole carbon source. Incubation was done for 2 days with shaking at 31°C.

Amplification of *tce1* from genomic DNA

Genomic DNA of strain 2479 was isolated by Janarthanan and Vincent's method (2007). The quality and concentration of the extracted DNA was checked by 0.8% (wt/vol) agarose gel electrophoresis and measured by UV-VIS spectrophotometer (UV-1700 Pharma Spec, Shimadzu) at 260 and 280 nm respectively. The concentration of DNA was 1.5 mg/ml. The *tce1* gene of strain 2479 was amplified using toluene dioxygenase C1 (*todC1*) gene specific primers from *P. putida* F1 (Romine and Brockman, 1996) synthesized by Sigma-Aldrich Chemical Co, USA. Amplification reaction was performed with reagent supplied by Bangalore Genei, India as follows: Taq DNA polymerase 2 units, magnesium ion conc. of varying concentration of 1, 1 and 2 mM, 10X buffer (containing 100 mM Tris-HCl, pH 9.0; 500 mM KCl; 0.1% Gelatin), 0.2 mM each of the four dNTPs and 100 ng template DNA of strain 2479, 25 pmol each of the primers. The polymerase chain reaction was carried out with initial denaturation at 94°C for 5 min followed by 30 cycles of programmed temperature control: 1 min at 94°C, ½ min at 57°C, 1 min at 72°C with a 5 min final extension at 72°C using strain 2479 DNA as template. The primers were as follows:

Oligo- F: 5'-GCGAGATGAAGCGCTCTTTG-3'.
Oligo- R: 5'-GTATTGATACCTGGGAGGAGG-3'.

The amplified product was analyzed on agarose gel electrophoresis (1.2% wt/vol) and documented using a gel documentation system. PCR amplified product was sequenced using the dideoxynucleotide chain termination method (Chromous Biotech. Pvt. Ltd).

Accession numbers

The partial sequence of *tce1* gene was deposited at EMBL/GenBank/DDBJ database under the accession number GU183105 (454 bp). The 454 bp gene contained 342 bp open reading frame (ORF) encoding 114 amino acids (Accession No. ACZ57347).

Cloning and sequencing of PCR product

PCR amplified product of *tce1* was purified by Spin-Column (Sigma-Aldrich) and cloned into cloning vector pUC 18 at Hind III site. The cloned insert was transformed into *E. coli* XL1-Blue [*recA1 lac endA1 gyrA96 thi hsdR17 supE44 relA1* (F'*proA*$^+$*B*$^+$ *lacI*^{q}ZΔM15Tn10)] and the recombinant *E. coli* was screened on Luria Bertani (LB) agar containing Ampicillin (50 µg/ml), IPTG (50 µM) and X-gal (40 µg/ml) plate for blue-white selection. From white colonies, the plasmid was isolated by alkaline lysis method (Sambrook and Russell, 2001). The isolated recombinant plasmid designated as pSM 101 was amplified using *todC1* gene specific primers as before.

Expression of cloned gene

The recombinant *E. coli* XL1-Blue (pSM 101 containing *E. coli*) was over-expressed with different concentration of IPTG by inducing the lac promoter of the cloning vector. This recombinant strain was grown on LB, Ampicillin (50 µg/ml) medium; after 2 h of incubation at 37°C with shaking, when the absorbance reached 0.5 at 600 nm, different conc. of IPTG (1.0, 1.5, 2.0 and 2.5 mM) were added in four culture flasks and one flask contained no IPTG. Incubation was at 37°C with shaking for 4 h. The cells were centrifuged at 10,000 g for 10 min and lysed by 1% SDS and PMSF (1 mM) and the extract was boiled for 5 min with Laemmli buffer (Laemmli, 1970) (1% SDS, 5% Mercaptoethanol, 0.05% Bromophenol blue in 25 mM Tris-HCl, pH 6.8 and 10% glycerol) and electrophoresed on 12% SDS-PAGE. To visualize the protein bands, the gel was stained by Coomassie blue R-250 (0.2%). Western-blot analysis was done on identical SDS-PAGE (without staining) using antibodies generated as follows:

Preparation of antibodies reacting to Tce1

The strain 2479 was grown separately in two conical flasks; one containing glucose (0.2%) as carbon source and other contained TCE (0.2%) as sole carbon source in minimal M9 medium. The SDS extracted proteins from bacterial cells grown on TCE containing M9 medium were injected into rabbits to obtain polyclonal antibodies. The titre of the antiserum was first determined by Dot-blot in different dilution (1:500; 1:1500; 1:2000) by using the spotted antigens (Mitra and Roy, 2011b). The polyclonal anti-serum was preadsorbed by nitrocellulose membrane (Sigma-Aldrich) on which total cellular proteins from glucose grown cells were immobilized. Thus the antibodies reacting to the common antigens found in both glucose grown and TCE grown cells were removed. The preadsorbed antiserum was found to react specifically with TCE induced protein in Western blot.

Western blot analysis

Following electrophoresis, the recombinant E. coli proteins (without staining) were electrotransferred onto nitrocellulose membrane (Sigma-Aldrich Chemical Co, USA.) for 2 h at 20 V. The membrane was blocked with 3% skimmed milk for 1 h at room temperature and probed for 1 h with 1:1000 diluted preadsorbed antiserum (raised against TCE inducible proteins). After washing with the buffer A (10 mM Tris-HCl pH-8.0; 0.9% normal saline; 0.2% Tween 20), the membrane was labeled with secondary antibody (alkaline phosphatase -conjugated goat anti-rabbit IgG, dilution 1:30,000) for 1 h at room temperature. After thoroughly washed with buffer A, the membrane was visualized with alkaline phosphatase substrates (5-bromo-4-chloro-3-indolyphosphate and nitroblue tetrazolium) (BCIP/NBT) in color development buffer-B (100 mM Tris-HCl with 5 mM Mg^{2+} at pH 9.5) and allowed to dry.

Measurement of TCE degradation by Fujiwara test

Fujiwara test performed to estimate the concentration of the polychlorinated hydrocarbon in the medium (Moss and Rylance, 1966). In the said test TCE was treated with pyridine in an alkaline environment. The observance of the red aqueous phase was determined at 470 nm by spectrophotometer. The absorbance of the red aqueous phase is proportional to the concentration of free TCE. If the test was performed without inoculation of the bacterium in the medium, the red upper layer was formed and it indicated that the medium contains the compound TCE. The same test was done with inoculation of the bacteria (capable of degrading trichloroethylene) and the absence of red color indicated that the medium contains no free TCE. Out of four TCE containing M9 media, three were inoculated each with recombinant E. coli (pSM 101 containing E. coli), non recombinant E. coli, B. cereus 2479 (positive control) and last one was not inoculated (negative control). These were incubated for 48 h at 37°C and Fujiwara test was performed. Optical density (OD$_{470}$) was measured by UV-VIS

spectrophotometer (UV-1700 Pharma Spec, Shimadzu) at the beginning (just after inoculation) and after 12, 24, 36 and 48 h. The Fujiwara test was done repeatedly.

Bioinformatics analysis

The nucleotide sequence, deduced amino acid sequence and ORF encoded by tce1 gene were analyzed and sequence comparison was conducted through database searches using the Gene Runner (version 3.01) software and BLAST program at NCBI http://www.ncbi.nlm.nih.gov) respectively. The molecular masses and the theoretical pI values of the polypeptide were predicted using the ProtParam tool (http://www.expasy.org/tools/protparam.html). Extensive structural analysis of the primary sequence of Tce1 protein was performed at NPS@ (network protein sequence analysis: web server http://npsa-pbil.ibcp.fr) using novel software such as DPM, DSC, GOR4, PHD, PREDATOR, SIMPA96 and SOPM. The information regarding 'secondary structure prediction' thus obtained was supplemented by using PSIPRED software at the web server: http://bioinf4.cs.ucl.ac.uk. Physico-chemical properties were computed using the software package CLC-Workbench (Version 5.0). Kyte-Doolittle method (Kyte and Doolittle, 1982) method was used for obtaining the hydropathy profile of the protein using default parameters.

RESULTS

Cloning and characterization of Tce1

The ~600 bp PCR amplified product was obtained from B. cereus 2479 by using todC1 gene specific primers (Figure 1A) and sequencing was performed with the PCR product. This gene is a new one and designated as tce1. To confirm the presence of insert, amplified product was inserted into plasmid, pUC 18 and then transformed into E. coli. XL1-Blue. Then recombinant plasmid pSM 101 was isolated and amplified by using the same two primers (OligoF and OligoR). In this PCR reaction the same ~600 bp amplified product was obtained as shown in Figure 1B.

Over-expression of cloned tce1 gene

In SDS-PAGE, one ~14.3 kDa recombinant protein was over-produced at 2 and 2.5 mM IPTG and less intense bands with 1 and 1.5 mM IPTG concentration but there was no such band with the negative control (without IPTG) (Figure 2A). In Western Blot studies, the over-expressed recombinant E. coli protein (14.3 kDa) specifically reacted with preadsorbed antiserum against TCE inducible protein (Figure 2B).

TCE degradation by pSM 101 containing E. coli

Fujiwara test was done to examine that the recombinant E. coli XL1-Blue (pSM 101 containing E. coli) could degrade TCE. Figure 4 shows OD value of red upper

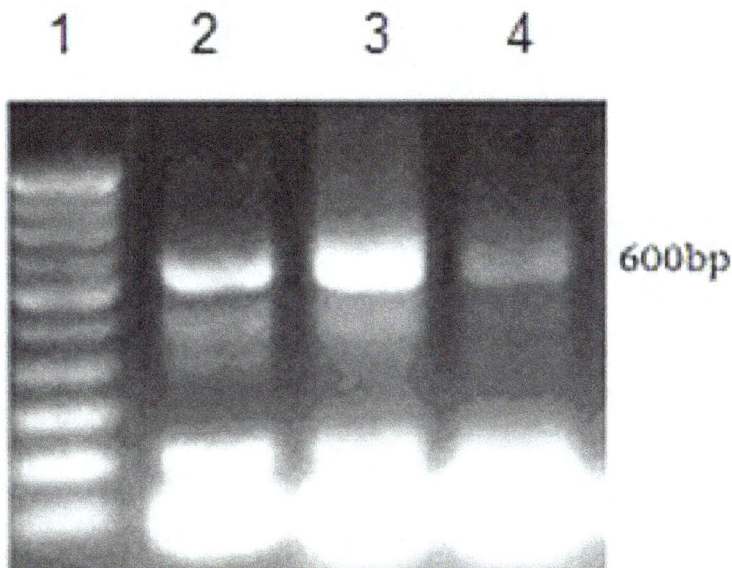

Figure 1A. 1.2% Agarose gel electrophoresis of polymerase chain reaction of *Bacillus cereus* 2479 genomic DNA using *todC1* gene specific primers. Presence of distinct bands of size 600 bp was observed with different magnesium ion concentration in the PCR mix: 2, 1.5 and 1 mM Mg^{++} in lane 2, 3, 4 respectively and lane 1 shows molecular weight marker, 100 bp ladder.

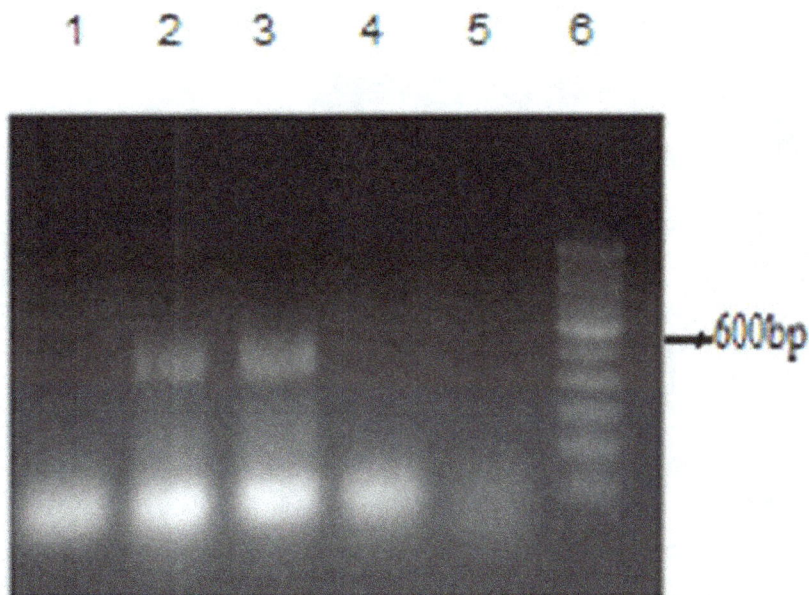

Figure 1B. PCR amplification with different templates: lane 1: genomic DNA of *E. coli*; lanes 2, 3: pSM 101; lane 4: pUC 18; lane 5: negative control (without any template) lane 6: shows 100 bp ladder.

layer gradually decreases with time (h) in case of recombinant *E. coli* and *B. cereus* 2479 during Fujiwara test. After 48 h OD reached 0.11 and 0.10 in recombinant *E. coli* and *B. cereus* 2479 respectively. It indicates that the medium contained no more free TCE. The plot shows no declination of OD value in the uninoculated medium as well as the non recombinant *E. coli* (without pSM 101) medium. It conferred that medium contained TCE.

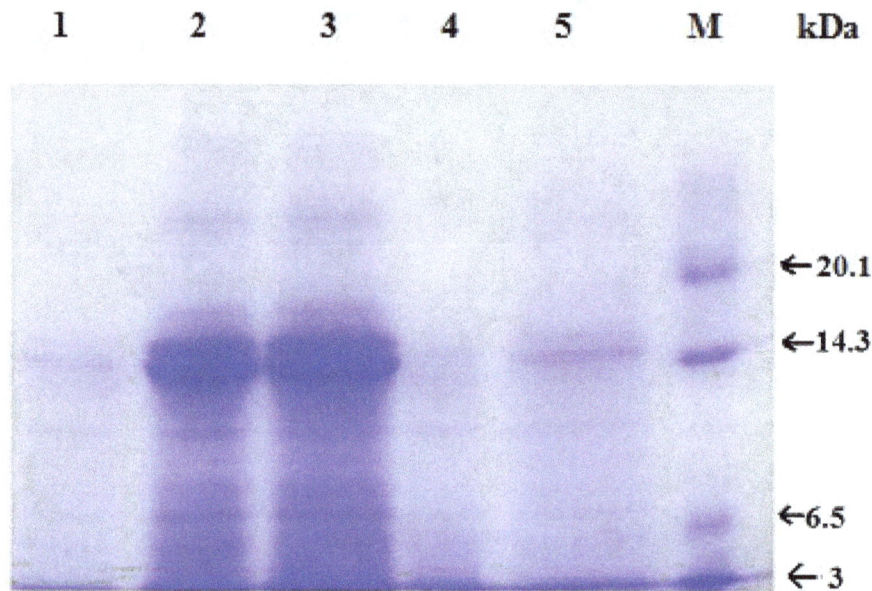

Figure 2A. SDS-PAGE profile of *E. coli* cells grown in LB, presence of IPTG: lanes 1, 2 and 3 cells grown at 1.0, 2 and 2.5 mM IPTG conc. respectively; lane 4 cells grown without any IPTG; lane 5 cells grown at 1.5 mM IPTG conc. lane 6 shows molecular weight marker.

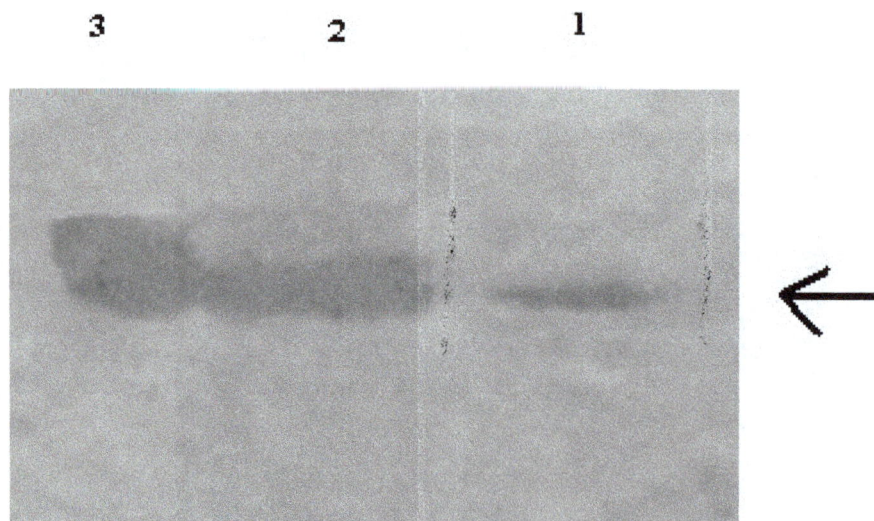

Figure 2B. In Western Blot: lanes 1, 2 and 3 shows specific reaction of anti-serum against TCE inducible protein with proteins from recombinant *E. coli* cells grown in presence of IPTG concentration at the rate of 1.0, 2, 2.5 mM respectively.

Nucleotide and amino acid sequence analysis

The partial sequence of *tce1* gene was deposited at EMBL/GenBank/DDBJ database under the accession No GU183105 (454 bp). The 454 bp sequence contained 342 bp open reading frame (ORF) encoding 114 amino acids (Accession No. ACZ57347) (Figure 5). The amino acid sequence was submitted to SWISS-PROT database with practical ProtParam procedures (http://www.expasy.ch/tools/protparam.html). Some physicochemical properties of deduced *Tce1* protein were as follows: molecular weight 12.6 kDa, molecular formula $C_{559}H_{886}N_{152}O_{165}S_8$, total number of atoms 1770, theoretical isoelectric point 5.17, aliphatic index 101.93,

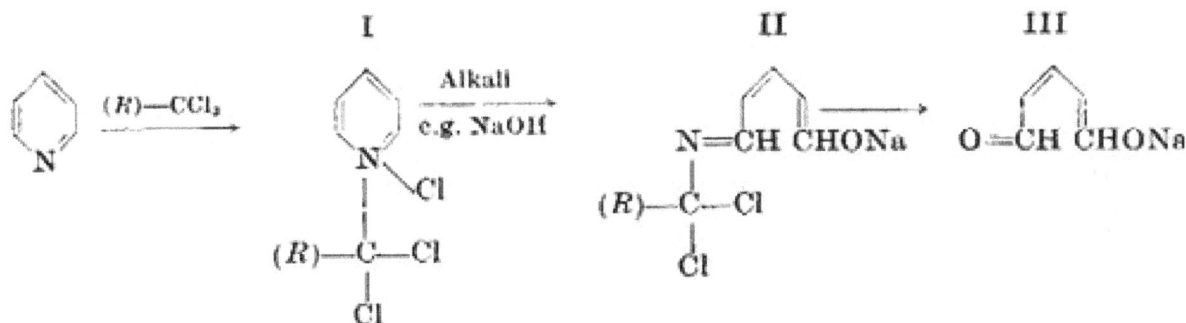

Figure 3. Reaction mechanism of Fujiwara test.

instability index 28.60, grand average of hydropathicity (GRAVY) 0.152.

Hydrophathy profile of *Tce1* protein

The hydropthy profile of TCE protein is presented in Figure 6 wherein the amino acids point score an averaged of nine residues individual scale is plotted against the position of amino acids in the sequence from N-terminal to C-terminal region. The plot shows that the protein possesses both hydrophobic and hydrophilic regions of which the later is dominating over the former. Hydrophobicity and hydrophilicity apparently show region specificity in that N-terminal portion is more hydrophobic and C-terminal portion is more hydrophilic. Hydrophilicity is more prominent than hydrophobicity as the score of the former reaches as high as -2.5 whereas the later is only +1.5 (Figure 6). The profile further identifies a region of about 25 residues long (from residue position 80 to 105) with very high hydrophilicity.

Secondary structure analysis of *Tce1* protein

Primary structure of proteins possesses multilayered information that helps to form functional structure of the proteins. Identification of local structures such as secondary structures and their sequences help to understand the fold topology of proteins. We used authentic web-based procedures (materials and methods section) to deduce the secondary structural characteristics of the protein whose result is presented in Figure 7. As seen in the figure that the protein is almost composed of right handed helix which constitutes about 85% of total secondary structures. Only about 5% of total secondary structures is constituted by beta-type (including turns) and rest 10% are of disorder type (random coil). Helical region are populated both in the N-terminal and C-terminal region of the sequence where as the middle portion of the sequence of *Tce1* protein possesses coil, turn and beta-structures.

DISCUSSION

Trichloroethylene a suspected carcinogen is the ground water contaminant (Richmond et al., 2001; Ma et al., 2002). It can readily be degraded by indigenous soil microbial population in presence or absence of toluene or phenol. One possible mechanism is toluene dioxygenase (TDO; EC 1.14.12) catalyzes the first reaction in the degradation of trichloroethylene or toluene by *P. putida* F1 (Yeh et al., 1977; Nelson et al., 1988). Nelson et al. (1987) and Wackett and Gibson (1988) showed that mutants of *P. putida* FI in which toluene dioxygenase activity was absent failed to degrade TCE because *todC1* is the crucial gene for initial oxidation of TCE. The *todC1* gene product being capable of degrading TCE as shown in *P. putida* F1 (Romine and Brockman, 1996) was chosen for synthesizing the primers used in this study to amplify the novel gene (*tce1*) from *B. cereus* 2479. Toluene monooxygenase (*tmoA*) gene of *P. putida* F1 specific primers were also tried for amplification with our isolated DNA but no specific product was obtained. So we chose to concentrate on *todC1* gene for further research. The sequence of *tce1* showed homology to other known toluene dioxygenase gene on the basis of phylogenetic analysis. This is the first instance when *B. cereus* containing *todC1* gene can degrade TCE efficiently (Mitra and Roy, 2011). In this study, the theoretical molecular weight of the recombinant protein was 12.6 kDa. The molecular weight of the induced protein from pSM 101 containing *E. coli* was found to be 14.3 kDa. This discrepancy may be due to the expression of *tce1* gene sequence that was partial instead of the full-length coding region. The recombinant *E. coli* could degrade TCE efficiently was analyzed by Fujiwara test. When any 1:1:1-Trichloro compound is heated in presence of pyridine and aqueous caustic alkali, a red color is imparted to the upper layer. The reaction is extremely sensitive. The nature of the red material has not been elucidated, but the mechanism has been proposed (Moss and Rylance, 1966) (Figure 3). From the observations in Fujiwara test, it can be concluded that the compound, TCE was metabolized by the bacterium *B.*

Result of Fujiwara test

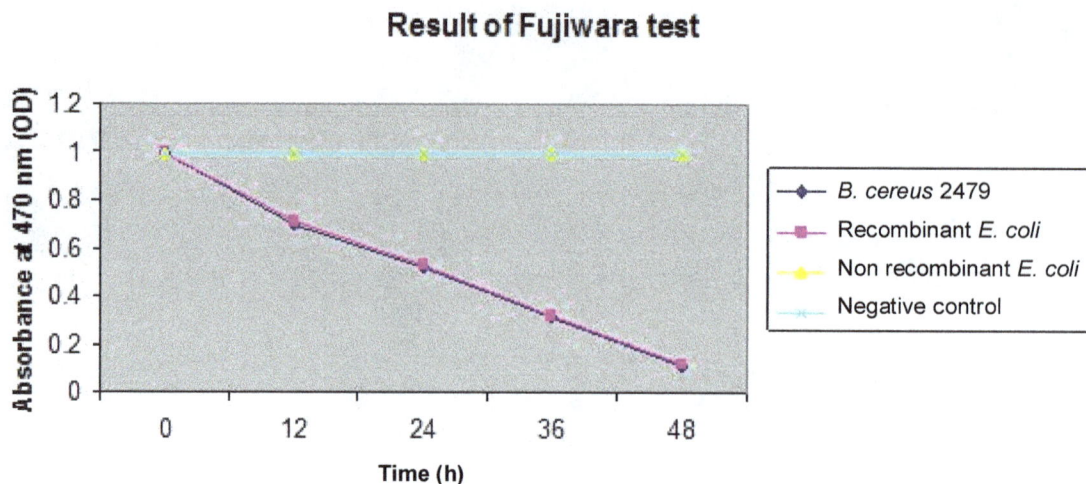

Figure 4. Optical density of *B. cereus*, recombinant *E.coli*, non recombinant *E.coli* and without any bacteria in the TCE containing M9 media recorded in different time (h) intervals, measured as OD $_{470}$ (nm).

```
CCG AAT ACT CGA CGC CAT TAT GTT TCG AGT CAC AAG TAA TGG CCC GCA AAT TTC

TTT TGG CTA TTG CGA CCT GGC ATC TGC GCC GGC GAA CTT TGA ACT AGG TCA AAT

TTA ATG GCA AAG TTA CTC GCT TTA GGT GAC TCC CAT CTC GAA GCC CTT AAA CTT
     M   A   K   L   L   A   L   G   D   S   H   L   E   A   L   K   L

GCA GCT GAC TTG AAT CTT CTA GCT GTT GAT GAG GTT AGG TTT TGC ATC GTG CCT
 A   A   D   L   N   L   L   A   V   D   E   V   R   F   C   I   V   P

GGA GCT ACT GCA GTC GGG ATG CGC AAC CCT AAC TCG ATT ACC AAT GCG CTG ACC
 G   A   T   A   V   G   M   R   N   P   N   S   I   T   N   A   L   T

CTG TTT CGC ACG GCG GCC TCC AGC ATG CAA GAT GCG ACG CAT ATT CTA GTA CAT
 L   F   R   T   A   A   S   S   M   Q   D   A   T   H   I   L   V   H

CTT GGC GAA GTG GAC TGT GGT TTC GTA ATG TGG TGG AGA CAG CAA AAA TAT GGC
 L   G   E   V   D   C   G   F   V   M   W   W   R   Q   Q   K   Y   G

GAG CCG ATA GAA CAT CAA ATG CGT GAA TCG TTG GCC GCC TAC AGC GAC TTC ATT
 E   P   I   E   H   Q   M   R   E   S   L   A   A   Y   S   D   F   I

TTA GAA TTA CAA TCG ATG AAT T
 L   E   L   Q   S   M   N
```

Figure 5. Nueclotide sequence and deduced amino acid sequence of *tce1* gene (Accession no. GU183105). The deduced amino acid sequences are shown in one letter code.

cereus strain 2479 and recombinant *E. coli*. No degradation was found in the control and non recombinant *E. coli* containing media (Figure 4). The structure and biochemical function of *Tce1* protein require further investigation.

Hydrophobic force is the dominant among all weak forces for protein folding. Hydropathy profile determination based on amino acids hydropathy scale would help in understanding region of protein forming hydrophobic domain, membrane spanning region,

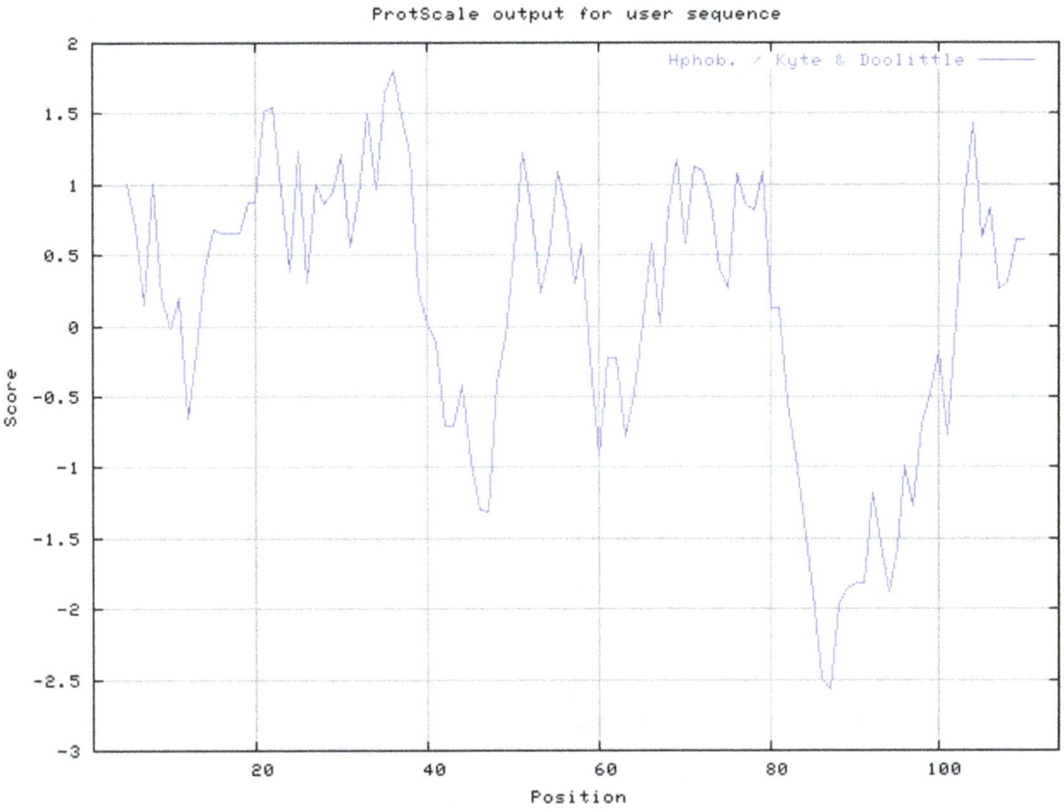

Figure 6. Hydrophobicity analysis of *Tce1* protein. using the scale Hphob. / Kyte and Doolittle, the horizontal axis indicates the location of amino acids and the vertical axis indicates hydrophobic percentile; hydrophobicity appears when percentile is higher than 0, while hydrophilicity appears when percentile is lower than 0.

```
MAKLLALGDSHLEALKLAADLNLLAVDEVRFCIVPGATAVGMRNPNSITNALTLFRTAASSMQDATHILV
ccbhhhhhhhhhhhhhhhhhhhhhhhhhhhhhhhhhhhhheeccbhhhhthhtcccccbhhhhhhhhhhhhhhhhhhhhhh
HLGEVDCGFVMWWRQQKYGEPIEHQMRESLAAYSDFILELQSMN
hhhhhhhhhhehhhhhhhhchhhhhhhhhhhhhhhhhhhhhhhhcc
Sequence length :    114
DPM :
    Alpha helix      (Hh) :    97 is   85.09%
    3₁₀ helix        (Gg) :     0 is    0.00%
    Pi helix         (Ii) :     0 is    0.00%
    Beta bridge      (Bb) :     0 is    0.00%
    Extended strand  (Ee) :     3 is    2.63%
    Beta turn        (Tt) :     2 is    1.75%
    Bend region      (Ss) :     0 is    0.00%
    Random coil      (Cc) :    12 is   10.53%
    Other states          :     0 is    0.00%
```

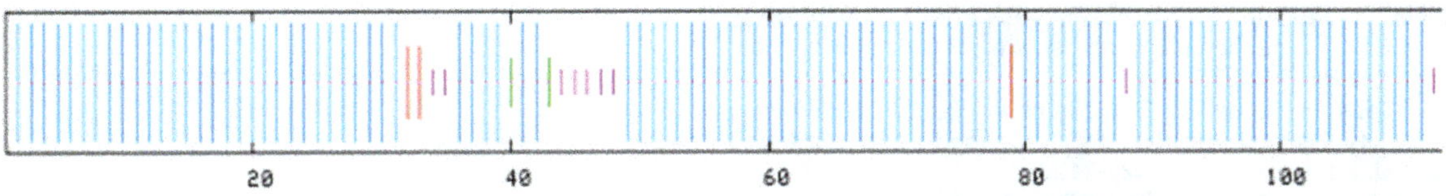

Figure 7. Secondary structure prediction of *Tce1* protein. Blue (longest): alpha helix; purple (shortest): random *coil*; red (medium): extended strand.

potential antigenic site and region that exposed on the protein's surface. Kyte and Doolittle method is the most popular one for determination of hydropathy profile which was used for the present study. Hydropathy profile analysis (Figure 6) shows presence of both hydrophobic and hydrophilic regions indication the possibility of balance interplay of all kind of weak forces for the formation of folded protein. Sequence region from 80 to 105 seems to form the surface of the protein and it also possesses strong hydrophilicity. Prediction of local order structure and their sequence from the primary sequence information help to understand folding topology, folding classes and other structural insights. Our analysis of secondary structural properties of the protein using authentic web server provides insightful observation in that the protein predominately composed of alpha-type structure with very little content of other structures indicating the possibility of formation of alpha-type domain. In this context it is widely known that the protein belong in the family dioxygenase whose hydroxylase component was shown to be oligomeric protein constituted either by alpha-type or alpha and beta-type structures (Butler and Mason, 1997). Thus it seems reasonable to think that our protein seems to possess former type hydroxylase properties.

ACKNOWLEDGMENTS

The authors gratefully acknowledge Dr Ashis Kumar Mondal for generously supporting the program. The authors are indebted to Dr. Amal Kumar Bandyopadhyay for providing his valuable suggestions in analysis of protein as well as helping in manuscript preparation. The authors would like to thank Sri Kaushik Dey of Durgapur College of Commerce and Science for providing the organism 2479 and Sri Sushil Kumar Sinha for providing technical assistance during computer work.

REFERENCES

Butler CS, Mason JR (1997). Structure-function analysis of the bacterial aromatic ring-hydroxylating dioxygenases. Adv. Microb. Physiol., 38: 47–84.

Dey K, Roy P (2009). Degradation of trichloroethylene by Bacillus sp.: Isolation strategy, strain characteristics, and cell immobilization. Curr. Microbiol., 59: 256-260.

Ewers JD, Schroder DF, Knackmus HJ (1990). Selection of trichloroethylene (TCE) degrading bacteria that resist inactivation by TCE. Arch. Microbiol., 154: 410-413.

Fox BG, Broneman JG, Wackett LP, (1990). Haloalkene oxidation by the soluble methane monooxygenese from Methylosinus trichosporiumOB3b. mechanistic and environmental implications. Biochemistry, 29: 6419-6427.

Furukawa K, Hirose J, Hayashida S, Nakambura K., (1994). Efficient degradation of trichloroethylene by a hybrid aromatic ring dioxygenase. J. Bacteriol., 176: 2121-2123.

Harker AR, Kim Y (1990). Trichloroethylene degradation by two independent aromatic-degradation pathways in Alcaligenes eutrophus JMP134. Appl. Environ. Microbiol., 56: 1179-1181.

Janardhanan S, Vincent S (2007). Practical Biotechnology, Methods and Protocols. University Press (India) Private Ltd.

Kyte J, Doolittle RA (1982). Simple method for displaying the hydropathic character of a protein. J. Mol. Biol., 157: 105-132.

Laemmli UK (1970). Cleavage of structural proteins during the assembly of the head of bacteriophageT4. Nature, 227: 680-685.

Little C, Palumbo D, Herbes AV, Lidstrom SE, Tyndall MERL, Gilmer PJ, (1988). Trichloroethylene Biodegradation by a methane oxidizing bacteria. Appl. Environ. Microbiol., 54: 951-956.

Ma HW, KY Wu, CD Ton (2002). Setting information priorities for remediation decisions at a contaminated- groundwater site. Chemospere, 1: 75-81.

Miller RE, Guengerich FP (1983). Oxidation of trichloroethylene by liver microsomal cytochrome P-450.evidence of chlorine migration in a transition state not involving trichloroethylene oxide. Biochemistry. 21: 1090-1097.

Mitra S, Roy P (2010) Molecular identification by 16S rDNA sequence of a novel bacterium capable of degrading Trichloroethylene. J. Biol. Sci., 10: 637-642.

Mitra S, Roy P (2011a) Molecular Phylogeny of a Novel Trichloroethylene Degrading Gene of Bacillus cereus 2479. J. Biol. Sci., 11: 58-63.

Mitra S, Roy P (2011b). Protein Profile of the Bacterium Capable of Degrading Trichloroethylene. Res. J. Microbiol., 6:503-509.

Moss MS, Rylance HJ (1966). The fujiwara reaction: Some observation on the mechanism. Nature, 210: 945-946.

Nelson MJK, Montgmery SO, Mahagley WR, Pritchard PH (1987). Biodegradation of trichloroethylene and involvement of an aromatic biodegradation pathway. Appl. Environ. Microbiol., 53: 949-954.

Nelson MJ K, Montgmery SO, Pritchard PH (1988). Trichloroethylene metabolism by microorganisms that degrade aromatic compounds. Appl. Environ. Microbiol., 54: 604-606.

Oldenhuis R, Vink RLJM, Janssen DB, Withelt B (1989). Degradation of chlorinated aliphatic hydrocarbons by Methylosinus Trichosporium OB3b expressing soluble methane monooxygenase. Appl. Environ. Microbiol., 55 : 2819-2826.

Richmond SA, Lindstrom JE, Braddrock JF (2001). Assessment of natural attenuation of chlorinated aliphatic groundwater. Environ. Sci. Technol., 20: 4038-4045.

Roberts PV, Schreinger JE, Hopkins GC (1982). Field-study of organic-water quality changes during groundwater recharge in the Palo-Alto Baylands. Water Res., 16: 1025-1035.

Sambrook J, Russell DW (2001). Molecular Cloning: A Laboratory Manual. Cold Spring Harbor Laboratory Press, New York.

Sheilds MS, Montgmery SO, Chapmen PJ, Cuskey SM, Pritehard PH (1989). Novel pathway of toluene catabolism in the trichloroethylene degradation bacteriumG4. Appl. Environ. Microbiol., 55: 1624-1629.

Storck W (1987). Chlorinated solvent use hurt by federal rules. Chem. Eng. News, 65: 11.

Vogel TM, McCarty P (1985). Biotransformation of tetrachloroethylene to trichloroethylene, dichloroethylene, vinyl chloride, and carbon dioxide under methanogenic conditions. Appl. Environ. Microbiol., 49: 1080-1083.

Wackett LP, Brusseau GA, Householder SR, Hanson RS, (1989). Survey of microbial Oxygenase. trichloroethylene degradation by Propane-Oxidizing bacteria. Appl. Environ. Microbiol., 55: 2960-2964.

Winter RB, Yen M, Enley BD (1989). Efficient degradation of trichloroethylene by recombinant E. coli. Biotechnology, 7: 282-285.

Yeh WK, Gibson DT, Liu TN (1977). Toluene dioxygenase: a multicomponent enzyme system. Biochem. Biophys. Res. Commun., 78: 401–410.

13

Genetic polymorphism of growth hormone locus and its association with bodyweight in Grati dairy cows

Sucik Maylinda

Faculty of Animal Husbandry, Brawijaya University, Malang – Indonesia. E-mail: sucikmaylinda@yahoo.com

The aim of this research was to study genetic polymorphism in growth hormone (GH) locus and its association with body weight of 43 Grati dairy cows ranging from 2 to 4 years old. Polymorphism of GH locus was ascribed using a PCR-RFLP method involving restricted enzyme Msp1 on 2% agarose gel. All data were subjeced to statistical analysis based on one way classification model using a statistical software package of Genstat 200 version 2. The results showed that frequencies of normal allel (Msp1$^+$) and mutant allel (Msp$^-$) were 0.34 and 0.66 respectively with 0.22 polymorphism was found in this locus. There was significant relationship between Msp1$^+$/Msp1$^+$, Msp1$^+$/Msp1$^-$ and Msp1$^-$/Msp1$^-$ genotypes and body weight (P < 0.05) but Msp1$^+$/Msp1$^+$ and Msp1$^+$/Msp1$^-$ genotypes had a stronger correlation to the higher body weight than Msp1$^-$/Msp1$^-$ genotype.

Key words: Grati cows, polymorphism, growth hormone, body weight.

INTRODUCTION

Grati cattle (GC) is regarded as indigenous breed that was a result of crossing between Friesian Holstein (FH) and Ongole crossbred cattle in Pasuruan regency of East Java since 19th century (Huitema, 1986). They have been adapted to relatively harsh environment especially to hot and humid climate and low-quality feed to produce milk and power to plough a farm land prior to planting. The cow yields milk between 2,500 to 4,500 kg/lactation which is far below the potential of FH dairy cows. Nevertheless, GC plays a significant role in the income generation for smallholder dairy farms in the region. Recently, selection for better performance of such important indigenous breed has received more attention especially with the advancement of genetically molecular biotechnology. Cheong et al. (2006) and (Kish, 2008) stated that growth hormone (GH) plays a vital role in post-natal growth and general metabolism including for lactation. Thus it is not surprising if GH has been the most intensive object of studies in ruminant animals to associate mutation of GH with the productive traits (Lara et al., 2002; Zhou et al., 2005; Khatami et al., 2005; Ferraz et al., 2006; Pawar et al., 2007; Mouzavisadeh et al., 2009). Yardibi et al. (2009) reported a significant effect of GH hormone gene polymorphism on milk yield of Eastern and South Anatolian Red breed cows. More recently, Komisarek et al. (2011) carried out a study on 209 dairy cows in Poland and found a significant decrease in milk production upon GHR-F279Y and shorter calving interval upon GHL127V polymorphism, respectively. Growth is a complex trait, which is controlled by multiple genes.

During the peak lactation milk is often synthesised at the expense of tissue catabolism, especially when the intake of feed energy is inadequate and a decline in body weight results. Under field conditions smallholder dairy farmers usually use body weight for selection criterion due to the lack of proper milk recording. This study aims at investigation of GH locus polymorphism in Grati cows and its association with body weight that potentially can be used as a marker assisted selection.

MATERIALS AND METHODS

Animals

43 Grati dairy cows that were kept at a dairy stall belonging to Suka Makmur Agricultural Farming Cooperative at Grati district, Pasuruan regency of East Java were obeserved during 2004 to 2005. The cows aged between 27 and 30 months old and varied between 1 and 3 lactation periods. All cows received a diet consisting of elephant grass (*Pennisetum purpureum*) and local concentrates.

Table 1. Band pattern of the fragment after being digested by Msp1 enzyme.

Length of DNA of particular band (bp)	Identified as allele	Genotype
224 105	Normal allele (Msp1⁺) *)	Msp1+ /Msp1+
329	Mutant allele (Msp1⁻) **)	Msp1+ /Msp1-
224 105 329	(Msp1⁺) and (Msp1⁻)	Msp1- /Msp1-

Cut by enzyme ; **) uncut by enzyme.

The bodyweight was estimated from the chest girth (CG) prior to blood collection using a Schoorl formula:

$BW_{kg} = (CG_{cm} + 22)^2/100$ (Sudono et al., 2003)

The reason for estimating body weight by chest girth measurement was poorly due to the unavailability of weighing scale in the site of study, and based on the many reports that chest girth is the best predictor for body weight (ILRI, 2011)

Genotyping for GH

Approximately 10 ml of blood was taken from jugular vein of each cow using venojet spuit and mixed with EDTA 10% in the Falcon tube and stored at -25°C. Following the genomic DNA isolation, the cows were genotyped for GH locus using PCR-RFLP (polymerase chain reaction-restriction fragment length polymorphism) and 2% agarose gel electrophoresis (Sulandari and Zein, 2003). Amplification of fragment of 329 bp at intron 3' (Dybus, 2002) was done with PCR using primer F: 5'– CCCACGGGCAAGAATGAGGC - 3' dan R:5' – TGAGGAACTGCAGGGGCCCA – 3'. To digest this fragment, a protocol of RFLP (restricted fragment length polymorphism) with restriction enzyme Msp1 was used to recognize the particular site of CC↓GG. The protocol for PCR analysis is stated in Table 1. PCR protocols to amp lily that fragment were by predenaturation at 94°C for 5 min for 1 cycle, denatauration at 94°C for 40 s, annealing at 60°C for 40 s and elongation at 72°C for 30 s for 30 cycles (Dybus, 2002).

Alleles identification

Following the end of PCR and RFLP processes, the products were then subsequently electrophorated using 2% agarose gel to identify polymorphism of alleles based on the length of the band (Table 1).

Statistical analysis

One-way analysis of variance was carried out to compare polymorphism level using a statistical software Genstat 2000 version 2. Polymorphism level was calculated using formula of Budak et al. (2003) that is:

$PIC_i = 1 = \sum p^2_{ij}$

where PIC_i is polymorphic information content at i-th locus, p_{ij} is the frequency of j-th allele and i-th locus.

RESULTS AND DISCUSSION

Growth hormone gene polymorphism

The number cows that met the criteria for this study was only 43. This mainly due to the large variation in farm size and husbandry practise between smallholder dairy farms in the village under study. Figure 1 indicates the number of dairy cows showing polymorphism at GH locus bearing Msp1⁺/Msp1⁺, Msp1⁺/Msp1⁻ and Msp1⁻/Msp1⁻ genotype were 7, 15 and 21, respectively (Table 2). It was found that the frequency of Msp1⁺ and Msp1⁻ alleles was 0.34 and 0.66. The level of polymorphism at that locus was 0.22 indicating that Grati dairy cows are polymorphic as has been generally accepted that the minimum value of polymorphism is 1% (Hyperdictionary, 2000; Dorak, 2006). This finding suggests that in Grati dairy cows there was high variability in the GH locus and offers opportunity to use GH genotype for selection criteria. From the previous study of Jakaria et al. (2007) on 132 indigenous cattle breed (Pesisir cattle) in West Sumatra, the MspI-GH gene showed significant relationship with bodyweight as estimated by chest girth measurement. Based on the Chi Square test (Table 2), it was found that alleles frequency were under genetic equilibrium.

Similar findings were reported in dairy cattle by Pawar et al. (2007) that polymorphism occurred in exon 5 of GH1 gene. Allelic frequency in genetic equilibrium means allele frequency will never change resulting in alleles and genotypic frequency stability.

Relationship of GH locus polymorphism with body weight

Table 3 shows the significant association (p < 0.05) between genotypes and body weight. However, the Msp1⁻/Msp1⁻ genotype gave less influence on body weight than those of Msp1⁺/Msp1⁺ and Msp1⁺/Msp1⁻ genotypes. Another locus in growth factor such as somatomedine or IGF1 gene has also been reported plays an important role in growth performance of beef

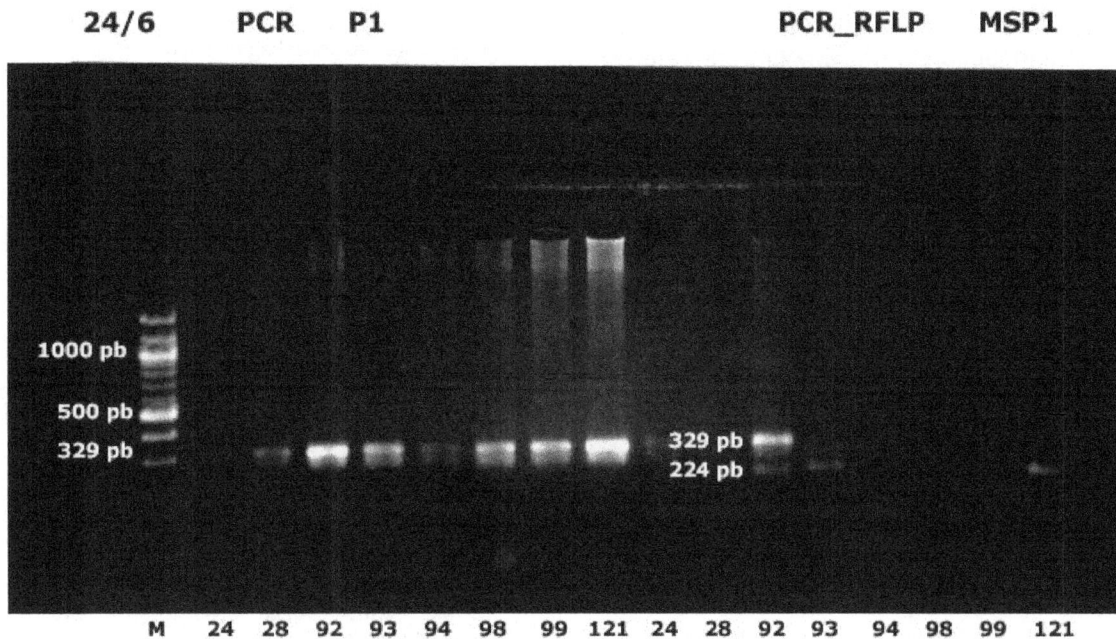

Figure 1. Eelectrophoresis result of PCR-RFLP products.

Table 2. Frequency of alleles Msp1+ and Msp1 at GH1 locus.

Allele frequency	Genotypic frequency
$(Msp1^+) = 0.34$ *)	Msp1+ /Msp1+ = 7
$(Msp1^-) = 0.66$ *)	Msp1+ /Msp1- = 15
	Msp1- /Msp1- = 21

*) The values denote that ppopulation were in genetic equilibrium based on the Chi square test ($p < 0.05$).

Table 3. PCR-RFLP product and mean of body weight of dairy cows in each genotype.

Genotype	N (**)	Mean of body weight (kg)
Msp1+ /Msp1+	7	448.8 ± 48.60^b
Msp1+ /Msp1-	15	447.7 ± 45.77^b
Msp1- /Msp1-	21	408.8 ± 50.83^a
Total	43	428.9 ± 51.58

*) Body weight was estimated using Schoorl formula, **) number of dairy cow. Values bearing unlike superscript in the same column differ significantly ($p < 0.05$).

cattle (Reyna et al., 2010). This notion supports the previous findings in other species of animals that GH gene can be promoted as candidate gene to improve animal performance such as body weight and carcass weight in poultry (Thakur et al., 2006) and Angus cattle (Zhao et al., 2004). As has been known that the growth of animals is under the hormonal control of GH, growth hormone receptor (GHR) and insulin-like growth factor I (IGF-I). Polymorphism that occurs in the regulatory region (for example promoter region) and coding region (exons) of the genes responsible for those three hormones will influence the expression of the genes and the function of protein during the translation process (Gee et al., 1999). There is strong indication that in all animals the level of blood GH reflects the GH genotype. Current knowledge in dairy biology indicates that genetically superior animals differ from lesser animals mainly in their regulation of nutrient utilization and that growth hormone (GH) release

(Bauman, 1992). The report of Sørensen et al. (2002) demonstrated that induction of growth hormone releasing hormone (GHRH) in Danish Jersey calves affected the GH Leu/Leu genotype to increase the level of GH in blood as compared with their counterparts with GH Val/Val genotype. Such an effect was only intermidiate when applied to GH Leu/Val genotype.

Conclusion

It was concluded that dairy cow population at Grati is polymorphic at locus GH with 22% polimorphism level. $Msp1^+/Msp1^+$ and $Msp1^+/Msp1^-$ genotypes can be used as the candidate genes in dairy cow selection to improve body weight.

ACKNOWLEDGEMENTS

The author wishes to express sincere gratitude to Prof. Muladno of Bogor Institute of Agriculture for his tangible help during the study and most particularly his skillful assistance in DNA analysis. I am deeply indebted to the manager and staff of Suka Makmur, Agricultural Farming Cooperative for warm and friendly welcome during the study that enabled me to accomplish data collection as schedulled. Thanks are due to the Dean of the Faculty of Animal Husbandry, Brawijaya University for providing valuable information to find the research grant and the Rector of Brawijaya University for his valuable support to publish this result. Last but not least, my special thank goes to Prof. Hendrawan Soetanto for his comment and willingnesss to discuss on the manuscript untill he was ready to submit it for publication.

REFERENCES

Bauman DE (1992). Bovine somatotropin: review of an emerging animal technology. J. Dairy Sci., 75: 3432-3451.

Budak H, Pedraza F, Cregan PB, Baenziger PS, Dweikat I (2003). Development and utilization of SSRs to estimate the degree of genetic relationships in a collection of Pearl Millet germplasm. Crop Sci. Ame., 43: 2284-2290.

Dorak MT (2006). Basic Population Genetics. www.dorak.info/genetics/popgen.html

Dybus A (2002). Associations of growth hormone (GH) and prolactin (PRL) genes polymorphism with milk production traits in Polish Black-and-White cattle. Anim. Sci. Papers Reports, 20(4): 203-212.

Huitema H (1986).Peternakan di daerah tropis:Arti ekonomi dan kemampuannya.Penerbit Yayasan Obor Indonesia dan P.T. Gramedia,Jakarta.

Hyperdictionary (2000). Meaning of Polymorphism. Copy Right © 2000-2003 Webnox Corp.

ILRI (2011). Cattle linear measurements. In: Research on farm and livestock productivity in the central Ethiopian highlands. FAO Corporate Document Repository.

Jakaria J, Duryadi D, Noor RR, Tappa B (2007). The Relationship of MspI Growth Hormone Gene Polymorphism and Body Weight and Body Measurements of West Sumatera Pesisir Cattle. J. Ind. Trop. Anim. Agric., 32(1): 33-40.

Kish S (2008). Exploring How Growth Hormones are Released in Animals. National Institute of Food and Agriculture. USDA.

Komisarek J, Michalak A, Walendowska A (2011).The effects of polymorphisms in DGAT1, GH and GHR genes on reproduction and production traits in Jersey cows. Anim. Sci. Papers Reports, 29(1): 29-36.

Pawar RS, Tajane KR, Joshi CG, Rank DN, Bramkshtri BP (2007). Growth hormone gene polymorphism and its association with lactation yield in dairy cattle. Indian J. Anim. Sci., 77(9): 884-888.

Reyna XFDLR, Montoya HM, Castrellon VV, Rincon AMS, Bracamonte MP, Vera WA (2010). Polymorphisms in the IGF1 gene and their effect on growth traits in Mexican beef cattle. Genet. Mol. Res., 9(2): 875-883.

Sørensen P, Grochowska R, Holm L, Henryon M, Løvendahl P (2002). Polymorphism in the Bovine Growth Hormone Gene Affects Endocrine Release in Dairy Calves. J. Dairy Sci., 85: 1887-1893.

Sudono A, Rosdiana F, Setiawan BS (2003). Intensive Dairy Cattle Keeping. 1st Ed. Agromedia Publisher.

Sulandari S, Zein MSA (2003). Protocols in DNA Laboratory, Center of Biology Research, The Indonesian Institute of Sciences. Pp. 23–45.

Thakur MS, Parmar SNS, Tolenkhomba TC, Srivastava PN, Joshi CG, Rank DN, Solanki JV, Pillai PVA (2006). Growth Hormone gene polymorphism in Kandaknath breed of poultry. Ind. J. Biotechnol. 5: 189-194.

Yardibi H, Hosturk HT, Paya I, Kaygisiz F, Ciftioglu G, Mengi A, Oztabak K (2009). Associations of Growth Hormone Gene Polymorphisms with Milk Production Traits in South Anatolian and East Anatolian Red Cattle. J. Anim. Vet. Adv., 5: 1040-1044.

Zhao Q, Davis ME, Hines HH (2004). Associations of polymorphisms in the Pit-1 gene with growth and carcass traits in Angus beef cattle. J. Anim. Sci., 82: 2229-2233.

14

Authentication of fusion genes in chronic myeloid leukemia

Aamir Rana[1,2], Shahzad Bhatti[1], Ghulam Muhammad Ali[2], Shoukat Ali[4], Nazia Rehman[3], Sabir Hussain Shah[3] and Ammad Ahmad Farooqi[1]

[1]Institute of Molecular Biology and Biotechnology, University of Lahore, Pakistan.
[2]National Institute for Genomics and Advanced Biotechnology, NARC, Islamabad, Pakistan.
[3]Pakistan Agricultural Research Council (PARC), Institute of Advanced Studies in Agriculture, Islamabad, Pakistan.
[4]Plant Biotechnology Program, NARC, Islamabad, Pakistan

Chronic myeloid leukemia is a multifactorial molecular anomaly that confounds the standardization of therapy to date. It is triggered by a broad spectrum of "fused oncoproteins" which are entailed in the disease refractoriness. In Pakistan the molecular diagnosis for leukemia is still in its infancy, as the diagnosis does not efficiently encompass a wide range of the fusion transcripts which are generated as a result of exclusive genomic rearrangements. Two point mutations C944T and T932C of ABL gene were detected which cause complete/partial imatinib resistance with limelight NUP98-LEDGF fusion transcript. It will be helpful in understanding primary resistance of molecularly targeted cancer therapies.

Key words: Allele-specific oligonucleotide polymerase chain reaction (ASO-PCR), leukemia, myeloid, chronic, imatinib.

INTRODUCTION

Hematological malignancies are the most prevalent molecular anomalies around the world. Fusion transcripts have been widely characterized in various diseases. These fusion transcripts are generated by unfaithful repair of DNA that results in illegitimate genomic rearrangements (Ammad et al., 2011). Imatinib mesylate (Gleevec) is the frontline drug for the clinical management of chronic myeloid leukemia (CML) (O'Dwyer et al., 2003). It is a matter of great concern that BCR-ABL kinase domain (KD) is the hotspot for point mutations. These mutations induce steric hindrance and drastically abrogate Imatinib mesylate binding by impairing critical interactions between ABL and Imatinib mesylate (Azam et al., 2003). Lens epithelium derived growth factor (LEDGF) is a transcriptional co-activator that contains coiled-coil domains that mediate self-association of the NUP-X chimaeric proteins and are documented in leukemic patients (Hussey et al., 2001; Hussey and Dobrovic, 2002; Grand et al., 2005; Morerio et al., 2005). These fusion transcript profiles are still unexplored in the leukemic patients of Pakistan. We tried to check out the prevalence of fusion transcripts with focus on NUP98-LEDGF, this is first study conducted and documented.

MATERIALS AND METHODS

Forty blood samples of patients with CML, which were receiving imatinib treatment from January 2009, to November 2010 were obtained from different hospitals of Lahore. DNA/RNA extraction and PCR amplification were done by using materials and method previously used by Catherine et al. (2002). Blood samples were subjected to genomic DNA extraction method.

*Corresponding author. E-mail: aamirrana_grg@yahoo.com.

Abbreviations: ASO-PCR, allele-specific oligonucleotide-PCR; **CML,** chronic myeloid leukemia; **LEDGF,** lens epithelium derived growth factor; **NUP-98,** Nucleoporin-98.

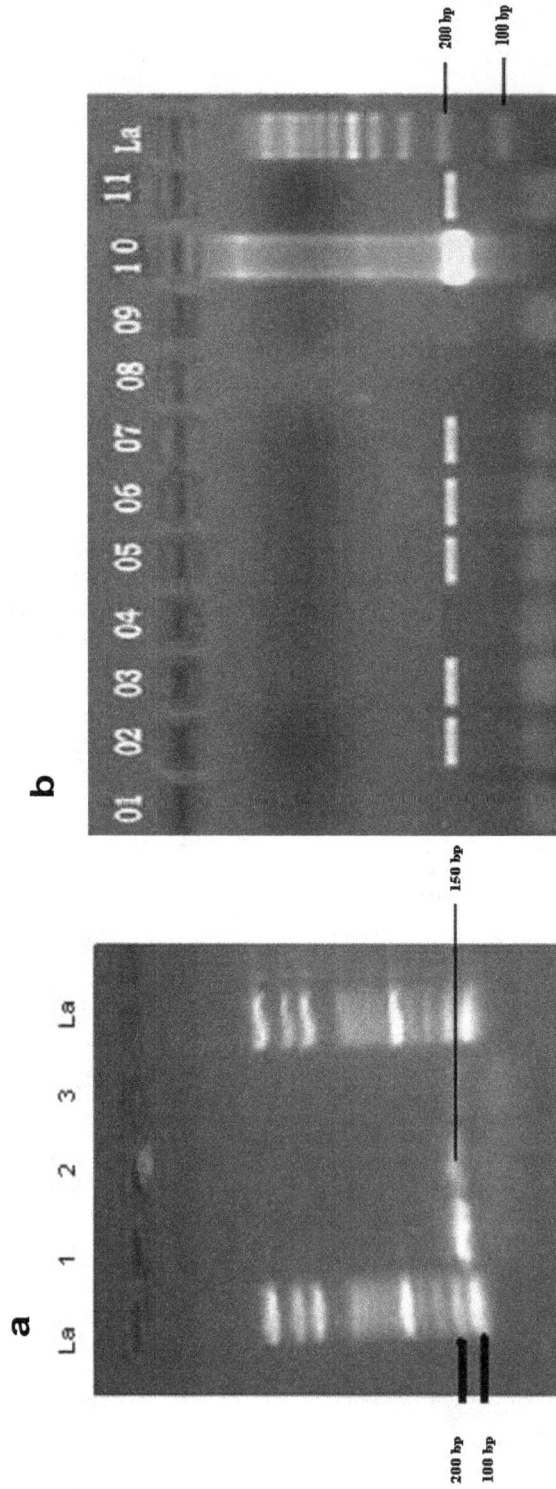

Figure 1. Analysis of C944T and T932C mutations by ASO-PCR. (a) Lane 1 and 5: DNA ladder of 1 kb, Lane 2, 3 and 4: patients positive with C944T mutation. (b) Lane 1: Negative control, Lane 2, 3, 5, 6, 7, 10 and 11: Patients positive with T932C mutation. The rest of the patients were negative.

PCR amplification

PCR mix preparation: A 20 µl PCR reaction was performed containing 1 µl of DNA (50 ng/µl), 2 µl 10X PCR buffer (Fermentas, USA), 25 pM of both forward and reverse primers, 300 M each of dATP, dGTP, dCTP and dTTP, 1.5 mM MgCl$_2$, 1.5 U Taq polymerase and 12 µl PCR water. To characterize fusion genes in the given samples, the extracted DNA from the blood was subjected to ASO-PCR. For the detection of two point mutations, cytosine to thymine at the position 944 of ABL gene and thymine to cytosine at the position 932 of ABL gene, mutation specific primers were used, and it was previously reported by Catherine et al. (2002). For the amplification of NUP98-LEDGF, forward primer to NUP98 exon 5 (1F: 5'-AGTACTAGCAGTGGAGGACTCTT-3') and a reverse primer to LEDGF exon 9 (1R: 5'-CTCTTCATCCTTCTTAGGCTGCT-3') were used and it

was previously reported by Grand et al. (2005).

RESULTS AND DISCUSSION

ABL is a protein kinase that is involved in the phosphorylation of various downstream molecules but myristoylation in the hydrophobic region of the protein makes it silent by auto inhibition. Contrarily, if there is a fusion of BCR and ABL, it de-represses the activity of ABL. The activation of ABL is restored as soon as there is a fusion of BCR with ABL and this fusion protein is hyper-activated to disturb the spatial-temporal behavior of the signaling (Mian et al., 2009). To treat this pathology, imatinib is used to extinguish and

dampen the effect of BCR-ABL. According to contemporary findings there is a paradigm shift from sensitivity towards resistance. In the patients with CML, C944T mutation was detected (in 3 patients). Cytosine to thymine mutation (mutation 1) at the position 944 of ABL gene was confirmed by appearance of the amplicons at 150 bp (Figure 1a). Thymine to cytosine mutation (mutation 2) at position 932 of the ABL gene was detected in (seven) imatinib resistant CML patients by using mutation specific primers. On 2% agarose gel, 174 bp specific bands were observed in patients positive for thymine to cytosine mutation at 932 of ABL gene (Figure 1b).

In present study we screened two mutations in CML patients which are in concordance with the

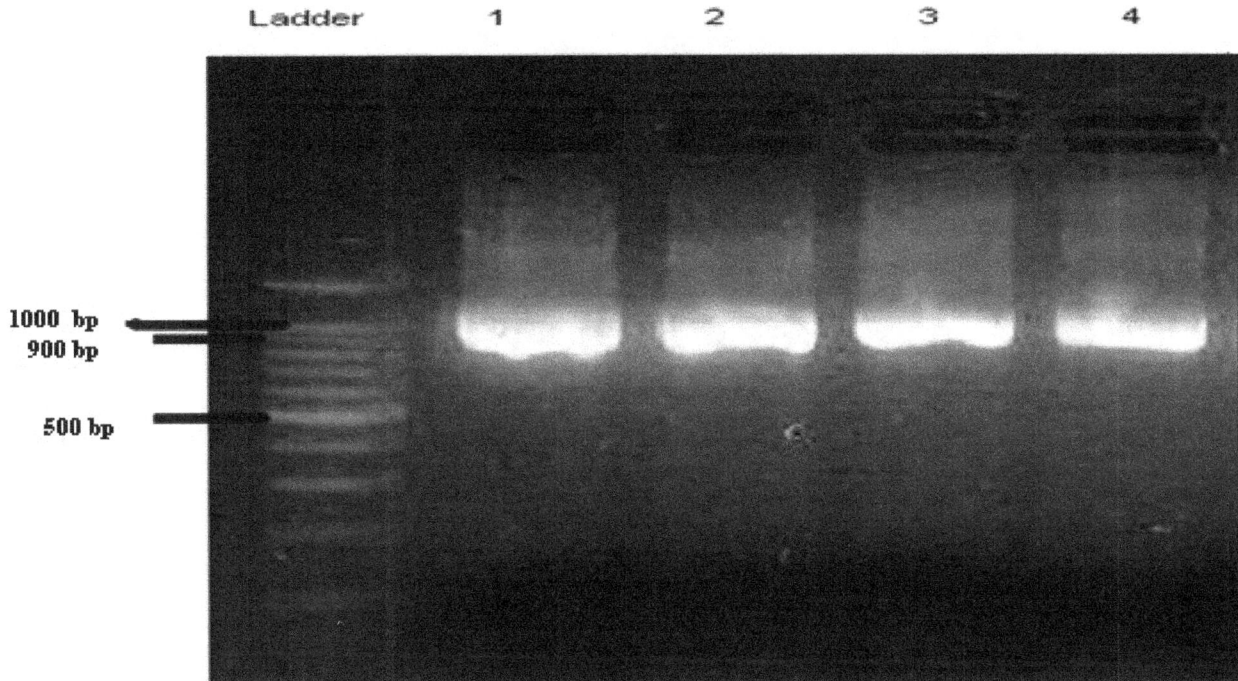

Figure 2. Amplification of NUP98-LEDGF. The bands show amplification of NUP98-LEDGF at 971 bp in 1, 2, 3 and 4.1 kb Ladder is used.

results of Catherine et al. (2002). We did not find a double mutation in a single patient which is in discordance with the results reported by Zafar et al. (2004). Samples were either positive for C944T or T932C. We have characterized and documented yet another fusion transcript NUP98-LEDGF in CML patients from Lahore (Figure 2). In the previous studies only case study was conducted to elucidate the molecular aspects of the fused genes. However, we have identified this chimeric transcript in (four) CML patients. It will be a great utility in clinical management of drug treatment and holds a potential for providing a reliable tool in diagnosis.

ACKNOWLEDGEMENTS

The corresponding author sincerely express thanks the family and research supervisor Ammad Ahmad Farooqi for participating in the study. The entire work was supported by National Institute for Genomics and Advanced Biotechnology, NARC Islamabad, Pakistan.

REFERENCES

Ammad AF, Qaisar M, Aamir R, Taskeen MM, Maryam I, Syed AN, Zia UR, Shahzad B (2011). SMURF and NEDD4 interference offers therapeutic potential in chaperoning genome integrity. J. Exp. Integr. Med., 1: 1-8.

Azam M, Latek RR, Daley GQ (2003). Mechanisms of autoinhibition and STI-571/imatinib resistance revealed by mutagenesis of BCR-ABL. Cell., 112: 737-740.

Catherine R, Cornu V, Nathalie D, Jean L, Thierry F, Preudhomme C (2002). Mutations in the ABL kinase domain pre-exist the onset of imatinib treatment. Seminars in Hematol., 100: 80-85.

Grand FH, Koduru P, Cross NC, Allen SL (2005). NUP98-LEDGF fusion and t(9;11) in transformed chronic myeloid leukemia. Leuk Res., 29: 1469-1472.

Hussey DJ, Dobrovic A (2002). Recurrent coiled-coil motifs in NUP98 fusion partners provide a clue to leukemogenesis. Blood, 99: 1097-1097.

Hussey DJ, Moore S, Nicola M, Dobrovic A (2001). Fusion of the NUP98 gene with the LEDGF/p52 gene defines a recurrent acute myeloid leukemia translocation. BioMed Central Genet., 2: 20-29.

Mian A, Latek RR, Daley Q (2009). Mechanism of autoinhibition and STI-571 resistance revealed by mutagenesis of BCR-ABL. Cell. 112: 831-843.

Morerio C, Acquila M, Rosanda C, Rapella A, Tassano E, Micalizzi C, Panarello C (2005). t(9;11)(p22;p15) with NUP98-LEDGF fusion gene in pediatric acute myeloid leukemia. Leuk. Res., 29: 467-470.

O'Dwyer ME, Mauro MJ, Druker BJ (2003). STI571 as a targeted therapy for chronic myeloid leukemia. Cancer Invest., 21:429-438.

Zafar I, Rubina T, Javed AQ (2004). Two different point mutations in ABL gene ATP-binding domain conferring primary imatinib resistance in a chronic myeloid leukemia (CML) patient: A case report. Biol. Proced. Online. 6: 144-148.

15

Genomic DNA extraction protocols from a Moroccan medicinal and aromatic plant *Artemisia herba-alba* Asso for RAPD-PCR studies

Ourid Ibtissam[1,2], Ghanmi Mohamed[1], EL Ghadraoui Lahsen[2], Kerdouh Benaissa[1] and Bakkali Yakhlef Salah Eddine[1]*

[1]Centre de Recherche Forestière, Rabat, Marocco.
[2]Faculté des Sciences et Techniques, Fès, Marocco.

This is the first report on development of protocol for high purity genomic DNA isolation from the Moroccan *Artemisia herba-alba* schrub leaves and optimization of conditions for RAPD-PCR analysis. Two DNA extraction protocols were specifically developed: QIAgen DNA Kit and protocol developed by Ouenzar et al. (1998). DNA yield and purity were monitored by gel electrophoresis and by determining absorbance at UV (A_{260}/A_{280}). Both ratios were between 1.7 and 2.0, indicating that the presence of contaminating metabolites was minimal. The Ouenzar and collaborators protocol gave higher yield but was more time consuming compared to QIAgen Kit. However, both techniques gave DNA of good quality that is amenable to RAPD-PCR reactions. Additionally, restriction digestion and PCR analyses of the obtained DNA showed its compatibility with downstream applications. Randomly Amplified Polymorphic DNA profiling from the isolated DNA was optimized to produce scorable and clear amplicons. The presented protocols allow easy and high quality DNA isolation for genetic diversity studies within *A. herba-alba*.

Key words: *Artemisia herba- alba*, genomic DNA extraction, PCR-RAPD.

INTRODUCTION

Artemisia herba-alba is a medicinal and aromatic dwarf shrub that grows wild in arid areas of the Mediterranean basin, extending into northern Himalayas (Vernin et al., 1995). This species has a vegetative growth in autumn (large leaves) and then at the end of winter to spring (small leaves). It is traditionally known for its essential oils. It has a very pronounced purgative effect and playing a major role in the control of intestinal worms (Idris et al., 1982). Extracts from *A. herba-alba* have antidiabetic effect (Al-Waili, 1986; Al-Khazraji et al., 1993; Al-Shamaony et al., 1994; Jouad et al., 2001) and strong antibacterial activities (Hatimi et al., 2001; Neerman, 2003). It also shows an allelopathic role against some other plants (Escudero et al., 2000). Flavonoids from this plant have a neurological action (Medhat Salah and Jäger, 2005).

The *A. herba alba* is also a big economic interest in Morocco. Its content in essential oil varies from 1 to 1.5% of dry material. Its essence is intended for the industry of the beauty care and the perfume shop. Two countries, Morocco and Tunisia, were divided in the international market to cater for this oil, but the big part returns to Morocco which holds 90% of the world market. More so, the *A. herba-alba* is recommended for the protection of the pastoral potential and the restoration of the degraded ecosystems.

In view of its medical value, mainly as natural antioxidant plant, *in situ* and *ex situ* conservation strategies is needed through forestry policies and legislations to protect this species.

DNA molecular techniques were commonly used for studying the genetic relationship among accessions and analysis of genetic diversity between and within the populations. Random amplified polymorphic DNA (RAPD)

---

*Corresponding author : bakkali_yse@yahoo.fr

markers have been widely used in the reconstruction of phylogenetic relationships for many organisms (Williams et al., 1990; Devos and Gale, 1992; Echt et al., 1992; Bakkali et al., 2010). However a commonly experienced problem with RAPD analysis, is its poor reproducibility (Devos and Gale, 1992). It is therefore essential to optimize the PCR to obtain reproducible and interpretable results.

Furthermore, isolation of high quality of DNA for use in many molecular markers is one of the most important and time-consuming steps (Zidani et al., 2005). Various protocols for DNA extraction have been successfully applied to many plant species (Doyle and Doyle, 1987; Guillemaut and Marechal-Douard, 1992; Ziegenhagen et al., 1993).

In the present work, we report two total genomic DNA isolation protocols: the DNeasy Plant Mini Kit and the protocol of Ouenzar et al. (1998). We also optimized RAPD marker technique that would be used to understand the species genetic diversity and population structure.

MATERIALS AND METHODS

Plant material

The A. herba-alba leaves were collected in Oujda region, Morocco, quickly frozen in liquid nitrogen, and then placed in -80°C until genomic DNA extraction.

DNA extraction protocols

Two different genomic DNA extraction methods were tested to recover genomic DNA from A. herba-alba. The first procedure assayed to extract genomic DNA from A. herba-alba was the DNeasy Plant Mini Kit (QIAGEN). Leaves of A. herba-alba were ground to a fine powder in CryoMill MM 400. Fifty milligrams of the powder was transferred to a pre-chilled microcentrifuge tube containg 400 µl of lysis buffer AP1 and 4 µl of RNase A (stock solution 100 mg/ml). The suspension was vortexed vigorously until a complete emulsion is formed and then incubated for 10 min at 65°C. Clumped tissues were dispersed using a disposable micropestle to allow higher yields of DNA. 130 µl of buffer AP2 was added to the lysate then the mixture was incubated for 5 min on ice. After centrifugation for 5 min at maximum speed and room temperature, the supernatant was transferred into mini columns filtration "QIAshredder spin column" placed in 2 ml collection tubes using a wide bore pipette and centrifuged for 2 min at maximum speed. The flow-through fraction was transferred into clean eppendorf tube. Occasionally, a light colored cell-debris pellet may appear and must not be disturbed. 0.5 volume of buffer AP3 and 1 volume of ethanol (96 to 100%) were added to the cleared lysate and mixed by pipetting. 650 µl of the mixture was transferred to DNeasy mini spin column and centrifuged for 1 min at 8000 rpm. This step was with the remaining sample and the flow-through and collection tubes were discarded. The DNeasy column was placed in a new collection tube and 500 µl of AW buffer (washing buffer) was added onto the column. After centrifugation for 1 min at 8000 rpm, the supernatant was discarded. An additional 500 µl of AW buffer was added to column and then centrifuged for 2 min at maximum speed. An additional washing step with 500 µl ethanol (96 to 100%) was applied to avoid coloration of the final eluted DNA. The column

was placed in a new collection tube and span for 5 min under a vacuum in order to completely dry the column membrane. This additional step was introduced because it is crucial that no residual ethanol is carried over during elution of DNA from the column. Finally, the DNeasy column was transferred to a 1.5 ml microcentrifuge tube and 100 µl of preheated (65°C) buffer AE (elution buffer) was added directly onto the DNeasy column membrane. After incubation for 5 min at room temperature, the DNA was eluted in buffer AE by centrifugation for 1 min at 8000 rpm.

The second procedure used to extract genomic DNA was adapted from Ouenzar et al. (1998) with some modifications. 100 mg of leaves powder were ground in 5 ml of lysis buffer containing 50 mM Tris-HCl (pH 8), 5 mM EDTA (pH 8), 0.05% BSA, 1% PEG$_{6000}$ and 0.5% beta-mercaptoethanol. The suspension was transferred in tubes containing 60 µl of SDS (20%) and 40 µl of Sodium Acetate (3 M, pH 8), and incubated at 65°C for 30 min. The DNA was extracted once with chloroform-isoamyl alcohol (24/1) and once with phenol/chloroform/isoamyl alcohol (25/24/1), followed by an additional extraction with chloroform-isoamyl alcohol (24/1). Every extraction was preceded by a centrifugation at 10000 × g for 10 min. The final aqueous supernatant was recovered in fresh tubes and mixed with an equal volume (v/v) of cold Isopropanol then was incubated at -20°C for one hour. Finally, the mixture was centrifuged at 10000 × g for 20 min. The resulting pellet was washed twice with cold ethanol (70%), air-dried and dissolved in TE buffer (10 mM Tris, 1 mM EDTA, pH 7.4) then treated with proteinase-K (20 µg/ml) at 37°C for 30 min. Proteinase-K was removed by one extraction with phenol/chloroform/isoamyl alcohol. After centrifugation, nucleic acids were precipitated by adding two volume of absolute cold ethanol and 1/10 of NaCl (5 M), followed by incubation over night at -20°C. The DNA was pelleted by centrifugation at 10 000 × g for 20 min, washed with 70% cold ethanol, air-dried, and then dissolved in 200 µl of TE buffer. The extracted DNA was further treated with 2 µl of RNAase (10 µg/ml) at 37°C for 30 min.

DNA quality was examined following electrophoresis on agarose gel. Spectrophotometric analysis was performed at 260 and 280 nm using a Nanodrop ND-1000 spectrophotometer (NanoDrop Technologies, Wilmington, DE, USA). Absorbance at 260 nm and the A_{260}/A_{280} ratio provided an estimate of quantity and purity of extracted DNA, respectively.

Restriction analysis and electrophoresis

One microgram of the extracted DNA was analysed by digestion with AluI, CfoI, EcoRI and MspI restriction enzymes according to manufacturer's specifications (Promega, Madison, WI, USA). Effective digestion of DNA by these enzymes was regarded as indicator of absence of polysaccharides (Do and Adams, 1991). The digested genomic DNA was separated on a 3% TBE agarose gel at 80, stained with ethidium bromide, and then viewed under UV illumination and photographed using.

Optimization of RAPD-PCR reaction

For the optimization of RAPD reaction using the DNeasy Plant Mini Kit (QIAGEN) extracted DNA, five decamer primers, OPA-01 (CAGGCCCTTC), OPA-02 (TGCCGAGCTC), OPA-03 (AGTCAGCCAC), OPA-04 (AATCGGGCTG) and OPA-05 (AGGGGTCTTG) from Operon Technologies Inc. (Promega, Madison, WI, USA), were used. The PCR conditions were optimized by varying the quantity of the DNA template, the concentration of MgCl$_2$, the units of Taq polymerase, dNTPs and primer concentrations and the annealing temperature (Tm). The initial protocol tested for RAPD reaction was the one recommended

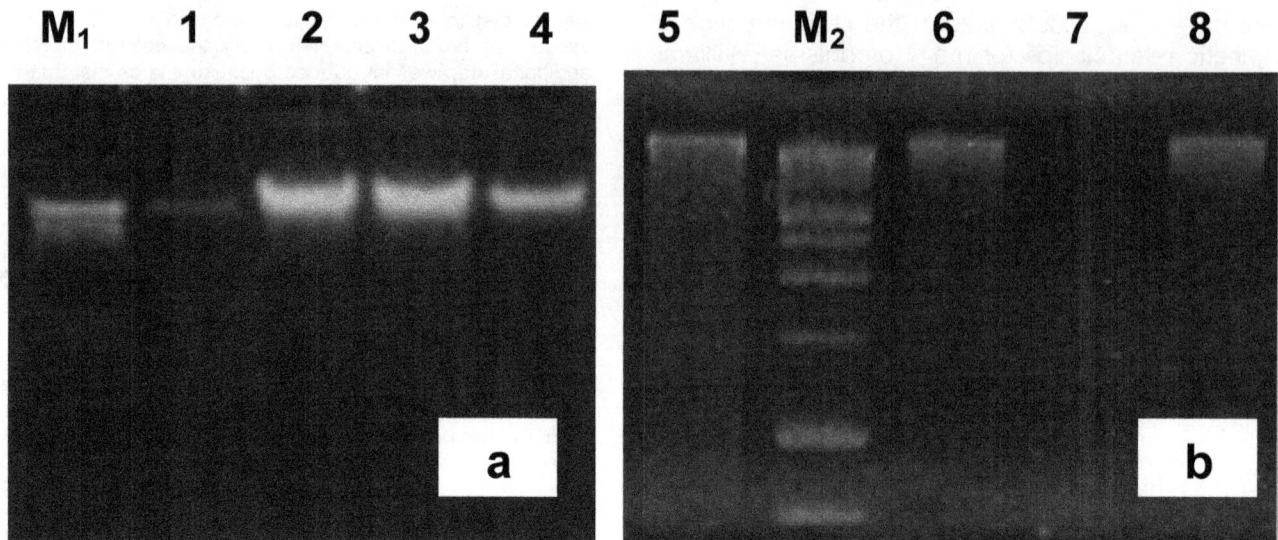

Figure 1. (a) Agarose gel analyses of DNA isolated from *A. herba-alba* leaves obtained by the two different procedures; lanes 1 and 4 represent DNA obtained by the DNeasy Plant Mini Kit; lane 2 and 3 indicate DNA samples isolated by the method described in Ouenzar et al. (1998). (b) Agarose gel analyses of *Artemisia* DNA digested with restriction enzymes. Lane 5, 6, 7, and 8 represent restriction digestion with *Alu*I, *Cfo*I, *Eco*RI and *Msp*I, respectively. M_1: λDNA/Hind III marker. M_2: 10 kb marker.

by Vural and Dageri (2009). Each 20 µl reaction volume contained about 50 ng of template DNA, 4 µl 1x buffer, 3 mM $MgCl_2$, 0.2 mM dNTPs, 0.5 µM Operon primer and 0.2 units of *Taq* polymerase. The thermal cycles used were: 1 cycle of 3 min at 94°C, followed by 30 cycles of 45 s. at 94°C, 1 min at 37°C, extension was carried out at 73°C for 1 min and final extension at 72°C for 7min and a hold temperature of 10°C at the end. The parameters tested in RAPD-PCR reactions were as follows: 7 different concentrations of DNA template (5, 10, 20, 40, 70, 100, 200 ng), 8 different concentrations of $MgCl_2$ (1.5, 2, 2.5, 3, 3.5, 4, 4.5, 5 mM), 9 different concentrations of *Taq* polymerase (0.2, 0.4, 0.6, 0.8, 1, 1.2, 1.4, 1.6, 1.8, 2 unit), 4 different concentrations of primers and 3 different annealing temperatures (33, 37 and 40°C). Amplification products were separated by electrophoresis in 2% agarose gels. It was performed in 1xTBE (trisborate-EDTA) running buffer at 100 volts and then the revelation of RAPD profiles was done by ethidium bromide staining under ultra-violet light. Molecular weights of the amplified products were estimated using 100 bp and 10 kb DNA ladder.

RESULTS

Isolation of good quality DNA from *A. herba-alba* proved difficult because leaves are very rich in polyphenolic compounds. In the present study, two different genomic DNA extractions were performed in order to obtain polysaccharide free genomic DNA from *Artemisia* shrub leaves, namely, the method described in Ouenzar et al. (1998) and the QIAgen DNeasy Plant Mini Kit (Valencia, CA, USA). Both procedures were used with some modifications. The steps and reagents of the first procedure are routinely used for plant genomic DNA extraction. For the second procedure, we improved some modifications according to Bakkali et al. (2010). The

concentrations and purity of DNA obtained from those procedures were determined with spectroscopy and gel-electrophoresis. The modified Ouenzar et al. (1998) extraction procedure produced a relatively high yield product (2 µg/50 mg sample tissue). While the average DNA yield by the DNeasy Plant Mini Kit procedure is around 0.8 to 1 µg per 50 mg sample tissue. Our procedures were capable of extracting, high-quality genomic DNA as indicated by the absorbance ratio $(A_{260})/(A_{280})$ of approximately 2.0.

In addition, the suitability of the genomic DNA obtained was also checked in restriction digestion and PCR-RAPD reactions. Digestion reactions with four different restriction enzymes, *Alu*I, *Cfo*I, *Eco*RI and *Msp*I, suggest that the extracted DNA was of low polysaccharide contaminants (Figure 1a and 1b).

In this study, we optimized RAPD-PCR reaction conditions for *Artemisia* shrub. A summary of the parameters tested and the corresponding results is presented in Table 1. The effects of the template DNA quantity (Figure 2), $MgCl_2$ concentration (Figure 3), primer and dNTP concentrations, *Taq* polymerase units (Figure 4) and the annealing temperature (Figure 5) on RAPD-PCR reaction were investigated. Optimum conditions were chosen that give more discriminatory band profiles (Figure 6).

DISCUSSION

Quantity and purity of extracted genomic DNA plays crucial role for analysis of molecular diversity and

Table 1. Summary of the parameters of RAPD-PCR tested and the optimum conditions selected.

PCR parameter	Range tested	Optimum conditions
DNA concentrations (ng)	5-10-20-30-40-50-70-100-200	5 ng
Magnesium chloride (mM)	1.5-2.5-3-3.5-4-4.5-5	1.5 mM
Primer concentration (µM)	0.1-0.3-0.5-0.7	0.3 µM
Taq polymerase (units)	0.2-0.4-0.6-0.8-1- 1.2-1.4-1.6-1.8-2	0.4 unit
dNTPs concentration (mM)	0.05-0.1-0.15-0.2-0.25	0.1mM
Temperatures of annealing (°C)	33, 37 and 40	37°C

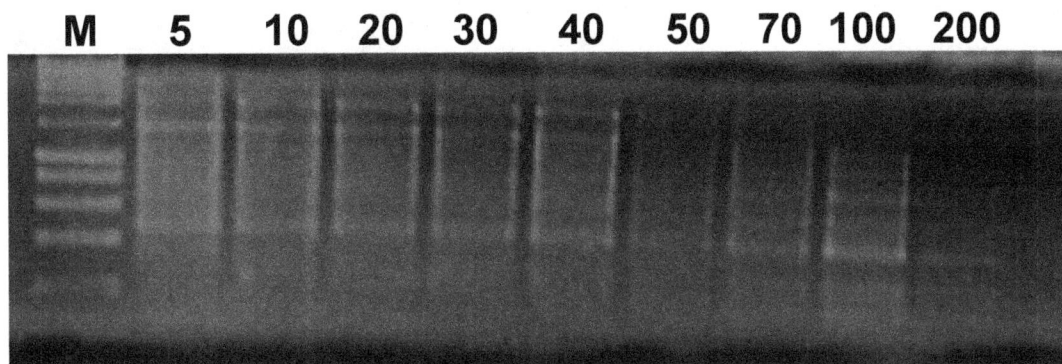

Figure 2. RAPD-PCR profile of nine DNA concentrations ranging from 5 to 200 ng obtained by the Operon primer OPA-05 in 1.8% agarose gel. M: 10 kb DNA ladder.

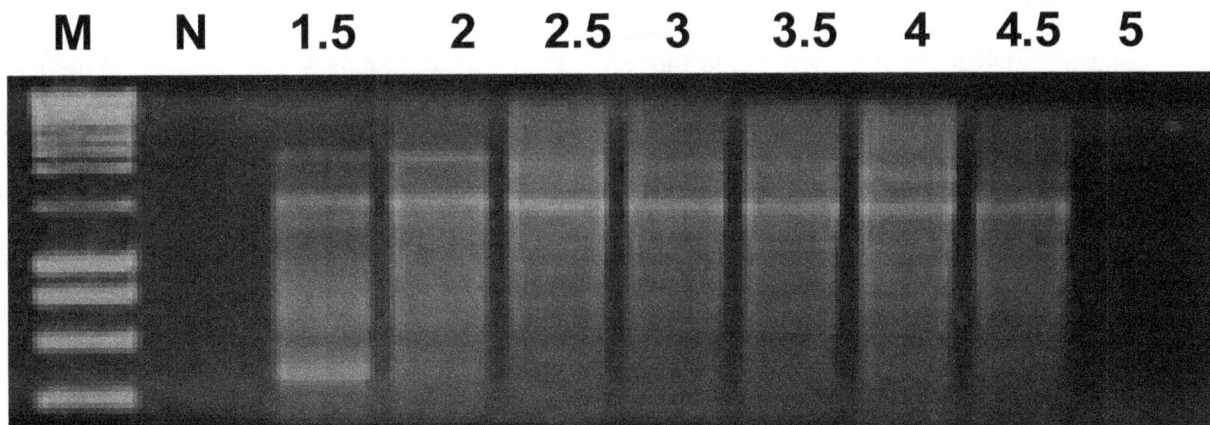

Figure 3. Optimization of RAPD-PCR parameters with eight MgCl$_2$ concentrations using the Operon primer OPA-05 and 5 ng of *A. herba-alba* genomic DNA. M: 10 kb DNA ladder. N: PCR negative control (no template DNA).

optimization of different parameters for PCR (Weeden et al., 1992; Staub et al., 1996). There are difficult to get plant DNA free from contaminating proteins and polysaccharides. These compounds can link covalently to DNA which becomes useless in enzymatic based reactions. It is for that reason that our first efforts concerned the development of an extraction protocol capable of giving a DNA which can be used as support

for the enzymatic reactions. To do it, two basic techniques were used. The technique of DNA extraction by the QIAGEN Kit, it is effective, simple and fast. The best adapted to the plant material of *A. herba-alba*. Our modified Kit methods are those proposed by Bakkali et al. (2010). It could be suitable for increasing and standardizing the quality and quantity of genomic DNA extracted. The modified Ouenzar et al. (1998) extraction

M   N   0.2 0.4 0.6 0.8   1   1.2   1.4   1.6   1.8   2

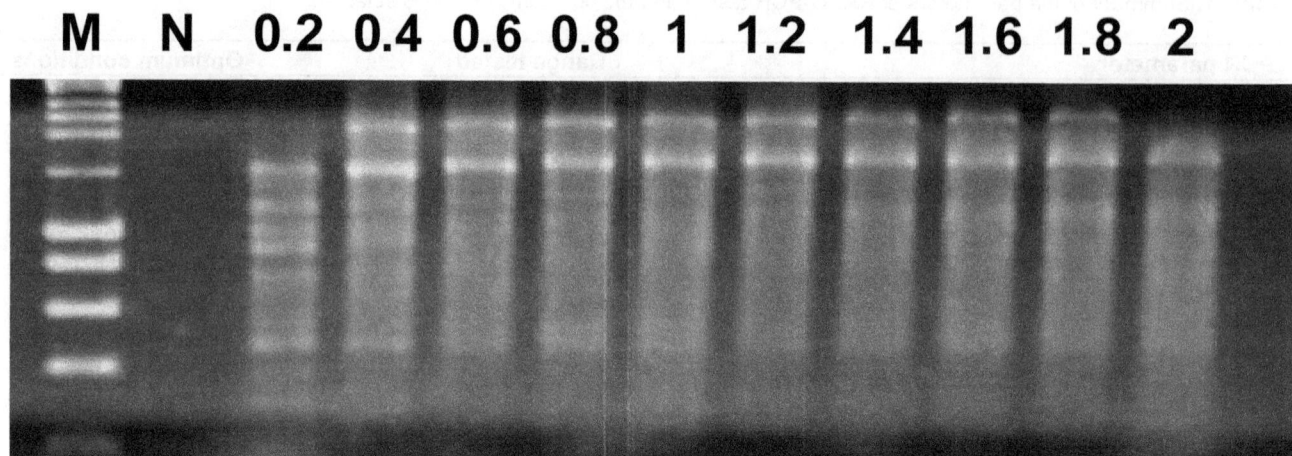

Figure 4. RAPD-PCR profile of *A. herba-alba* DNA (5 ng) using ten different amounts of *Taq* DNA polymerase using the Operon primer OPA-05. M: 10 kb DNA ladder. N: PCR negative control (no template DNA).

protocol allows the obtaining of translucent, good quality and intense genomic DNA. This procedure is a maxi-scale preparation which is very time-consuming. Its adaptation to the *A. herba-alba* leaf would probably be due, in part, to its lysis buffer containing the Polyethylene Glycol (PEG$_{6000}$) and on the other hand, to a long processes of deproteinisation. The PEG is considered by some authors as a chemical agent having a positive and effective impact on the dissociation of the tannin-protein complex responsible for the anchoring of several elements or cellular constituents of certain plants (Jone and Mangan, 1977; Priolio et al., 2000).

RAPD is one of the molecular marker that had been used in genetic study. RAPD technique had been used in genetic study for Moroccan forest tree (Konaté, 2007; Bakkali et al., 2010) and also had used for diversity analysis for medicinal and aromatic plant (Padmalatha and Prasad, 2006). To obtain reproducible and interpretable results with RAPD analysis, it is essential to optimize some PCR parameters.

In this study, the PCR-RAPD protocol was optimized by introducing several modifications to the original (Vural and Dageri, 2009) protocol in some PCR-components such as template DNA, primer, magnesium chloride, *Taq* polymerase, dNTPs, as well as in amplification cycles especially annealing temperature.

Amount of template DNA strongly influences the outcome of the reaction. More than 30 ng / 25 µl give the premium amplification (Henegariu et al., 1997). In the present study, identical RAPD profiles were found when DNA amounts varied from 5 to 200 ng. DNA concentration of 5 ng / 20 µl was found optimum. Optimization of MgCl$_2$ is an important factor for precise amplification; 1.5 mM of MgCl$_2$ was found optimum in 20 µl final volume. Moreover, concentration of dNTPs in reaction mixture is also strongly correlated to the Mg ions

concentration due to the interaction between mononucleotides and the Mg^{2+}. 0.1 mM found optimum in this study.

In PCR, 2 to 2.5 units of Taq Polymerase are normally used in 100 µl final volume. Higher *Taq* Polymerase concentration (above 4 units/100 µl) can generate non-specific products and may reduce the yield of the desired product (Saiki, 1988). However, in the present study, 0.4 unit/20 µl reaction was used to amplify the DNA of *A. herba-alba*. Annealing temperature is one of the most important parameters that need adjustment in the PCR. The normal range of annealing temperature is 35 to 55°C. The annealing temperature 37°C was found optimum to amplify with primer pairs OPA-01 to OPA-05. The oligonucleotide primers OPA-01 to OPA-05 generate relatively more diverse and reproducible genomic fingerprints for a number of plant species and were used for the optimization of RAPD parameters (Doulis et al., 2000; Padmalatha and Prasad 2006; Bakkali et al., 2010). We used the same primers to optimize the primer concentration for RAPD-PCR analysis of *A. herba-alba*. Results showed that 0.3 µM of the primer generated distinct and reproducible DNA profile.

In this study, the two different extraction methods gave a DNA of quality suitable for RAPD-PCR analysis. The protocol described in Ouenzar et al. (1998), used with some modifications, gave higher DNA yield compared to QIAgen. These modifications were made in both the lysis buffer (5mM EDTA, 0.05% BSA and 0.5% beta-mercaptoethanol) as the various steps in the implementation of the Protocol. The standard reaction developed included: 5 ng of DNA template, 1.5 mM MgCl$_2$, 0.3 mM primer, 1 mM dNTPs (for each) and with 0.4 U of *Taq* polymerase per 20 µl PCR reaction. The protocols described here and the optimized RAPD parameters constitute a strong beginning for future

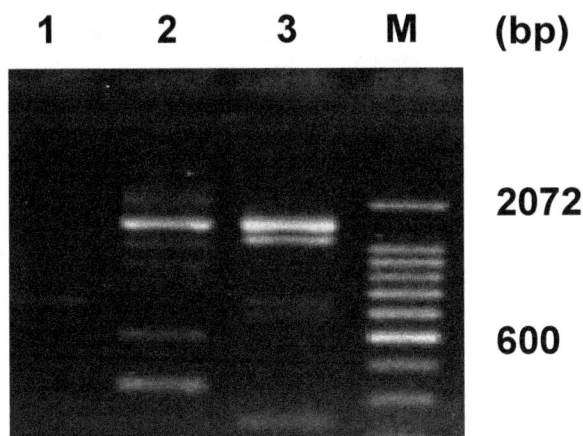

Figure 5. RAPD-PCR profile of *A. herba-alba* DNA (5 ng) in 1.8% agarose gel with three different annealing temperatures: (1) 33°C, (2) 37°C and (3) 40°C. M: 100 bp DNA ladder. N: PCR negative control (no template DNA).

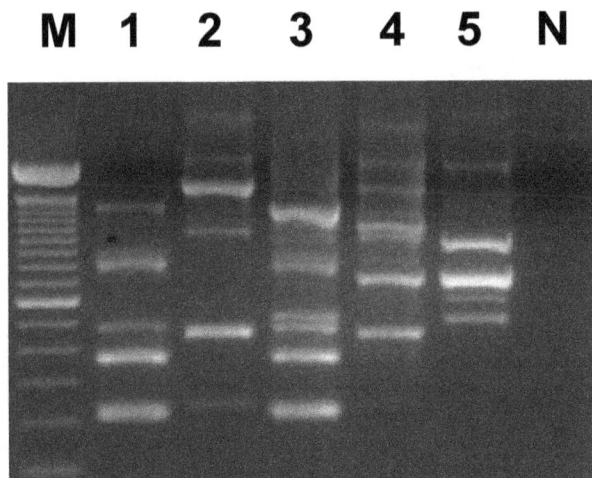

Figure 6. A representative agarose gel fractionation of RAPD amplification products from A. herba-alba genomic DNA (5 ng) derived from a single Artemisia shrub using the Operon primers: OPA-01 (1), OPA-02 (2), OPA-03 (3), OPA-04 (4), OPA-05 (5). Amplification products were fractionated in a 1.8% agarose gel. M: 100 bp DNA ladder. N: PCR negative control (no template DNA).

molecular characterization and genetic improvements studies in *A. herba-alba*.

ACKNOWLEDGEMENT

The authors gratefully acknowledge the financial support provided by the RS 11/2011 project funded by CNRST-Morocco (Centre National pour La Recherche Scientifique et Technique, Maroc).

REFERENCES

Al-khazraji SM, Al-Shamaony LA, Twaij HA (1993). Hypoglycemic effect of *Artemisia herba-alba*. I. Effect of different parts and influence of the solvent on hypoglycemic activity. J. Ethnopharmacol., 40(3): 163-166.

Al-Shamaony L, Al-Khazraji SM, Twaij HA (1994). Hypoglycemic effect of *Artemisia herba-alba*. II. Effect of a valuable extract on some blood parameters in diabetic animals. J. Ethnopharmacol., 43 (3): 167-171.

Al-Waili NS (1986). Treatment of diabetes mellitus by *Artemisia herbaalba* extracts: preliminary study. Clin. Exp. Pharmacol. Physiol., 13(7): 569-573.

Bakkali YS, Guenoun I, Kerdouh B, Hamamouch N, Abourouh B (2010). Efficient DNA Isolation from Moroccan Arar tree [*Tetraclinis articulata* (Vahl) Masters] leaves and Optimization of the RAPD-PCR Molecular Technique. Acta Botanica Malacitana, 35: 77-86.

Devos KM, Gale MD (1992). The Random Amplified Polymorphic DNA markers in wheat. Theorical Applied Genetics, 84: 567-572.

Do N, Adams RP (1991). A simple technique for removing plant polysaccharide contaminants from DNA. *BioTechniques*, 10: 163-166.

Doulis AG, Harfouche AL, Aravanopoulos FA (2000). Rapid high quality DNA isolation from Cypress (*Cupressus sempervirens*) needles and optimization of the RAPD marker technique. Plant Mol. Biol. Reptr. 17: 1-14.

Doyle JJ. Doyle JL (1987). A rapid isolation procedure for small quantities of fresh leaf tissue. Phytochem. Bul., 19: 11-15.

Echt CS, Erdahl LA, Mccoy TJ (1992). Genetic segregation of random amplified polymorphic DNA in diploid cultivated alfalfa. Genome, 35: 84-87.

Escudero A, Albert MJ, Pita JM, Pérez-Garcia F (2000). Inhibitory effects of *Artemisia herba-alba* on the germination of the gypsophyte *Helianthemum saquamatum*. Plant Ecol., 148: 71-80.

Guillemaut P, Marechal-Drouard L (1992). Isolation of plant DNA: A fast, inexpensive, and reliable method. Plant Mol. Biol. Reptr., 10: 60-65.

Hatimi S, Boudouma M, Bichichi M, Chaib N, Guessous Idrissi N (2001). Evaluation *in vitro* de l'activité antileishmanienne d'*Artemisia herba-alba* Asso. Rencontres franco-africaines de pédiatrie No14, Paris, FRANCE (07/10/2000) 2001, 94(1): 29-31, 57-70.

Henegariu O, Heerema NA, Dlouhy SR, Vance GH, Vogt PH (1997). Multiplex PCR: Critical parameters and step-by-step protocol. Biotechniques, 23: 504-511.

Idris UE, Adam SE, Tartour G (1982). An anthelmintic efficacy of *Artemisia herba-alba* against *Haemonchus contortus* infection in goat. Natl. Inst. Anim. Health, 22(3): 138-143.

Jone WT, Mangan JL (1977). Complexes of the condensed of sainfoin (Onobrychis viccifolia Scop.) with fraction 1 leaf protein and with submaxillary mucoptotein, and their reversal by polyethylene glycol and pH. J. Sci. Food. Agric., 28: 126-136.

Jouad H, Haloui M, Rhiouani H, El Hilaly J, Eddouks M (2001). Ethnobotanical survey of medicinal plants used for the treatment of diabetes, cardiac and renal diseases in the North centre region of Morocco (Fez-Boulemane). J. Ethnopharmacol., 77: 175-182.

Konaté I (2007). Diversité phénotypique et moléculaire du caroubier (*Ceratonia siliqua* L.) et des bactéries endophytes qui lui sont associées. Thèse de doctorat National: Faculté des Sciences de Rabat (Maroc) : 189p.

Medhat SS, Jäger AK (2005). Screening of traditionally used Lebanese herbs for neurological activities. J. Ethnopharmacol., 97: 145-149.

Neerman MF (2003). Sesquiterpene lactones: a diverse class of compounds found in essential oils possessing antibacterial and antifungal properties. Int. J. Aroma, 13: 114-120.

Ouenzar B, Hartmann C, Rode A, Benslimane A (1998). Date palm DANN minipreparation without liquid nitrogen. Plant Mole. Biol. Reporter, 16: 263-269.

Padmalatha K, Prasad MNV (2006). Optimization of DNA isolation and PCR protocol for RAPD analysis of selected medicinal and aromatic plants of conservation concern from Peninsular India. Afr. J. Biotechnol., 5(3): 230-234

Priolio A, Waghorn GC, Lanza M, Biondi L, Pennisi P (2000). Polyethylene glycol as a means for reducing the impact of condensed tannins in carob pulp: Effects on lamb growth performance and meat quality. J. Anim. Sci., 78: 810-816.

Saiki RK, Gelfand DH, Stoffel S, Sharf SJ, Higuchi R, Horn GT, Mullis KB, Erlich HA (1988). Primer-directed enzymatic amplification of DNA with a thermostable DNA polymerase. Science, 239: 487-491.

Staub JJ, Bacher J, Poeter K (1996). Sources of potential errors in the application of random amplified polymorphic DNAs in cucumber. Hort. Sci., 31: 262 – 266.

Vernin G, Merad O, Vernin GMF, Zamkotsian RM, Parkanyi C (1995). GC–MS analysis of *Artemisia herba-alba* Asso essential oils from Algeria. In: Charalambous G (Ed.), Food Flavors: Generation, Analysis and Process Influence. Elsevier Science BV, Amsterdam, pp. 147-205.

Vural HC, Dageri A (2009). Optimization of DNA isolation for RAPD-PCR analysis of selected (*Echinaceae purpurea* L. Moench) medicinal plants of conservation concern from Turkey. J. Med. Plants Res., 3(1): 016-019.

Weeden NF, Timmerman M, Hermmat M, Kneen BE, Lodhi MA (1992). Applications of RAPD Technology to Plant Breeding, Symposium Proceeding, Minneapolis. pp. 12 –17.

Williams JGK, Kubelik AR, Livak KJ, Rafalski JA, Tingey SV (1990). DNA polymorphisms amplified by arbitrary primers are useful as genetic markers. Nucleic Ac. Res., 18: 6531-6535.

Zidani S, Ferchichi A, Chaieb M (2005). Genomic DNA extraction method from earl millet (*Pennisetum glaucum*) leaves. Afr. J. Biotechnol., 4(8): 862-866.

Ziegenhagen B, Guillemaut P, Scholz F (1993). A procedure for mini-preparations of genomic DNA from needles of silver fir (*Abies alba* Mill.). Plant Mol. Biol. Reptr., 11: 117-121.

16

Molecular cloning, expression, sequence analysis and *in silico* comparative mapping of trehalose 6-phosphate gene from Egyptian durum wheat

Ayman A. Diab[1]*, Ahmed M. K. Nada[2] and Ahmed Ashoub[3]

[1]Molecular Markers and Genome Mapping Department, Agricultural Genetic Engineering Research Institute (AGERI), Agricultural Research Center (ARC), Giza, 12619-Egypt.
[2]Plant molecular Biology department, Agricultural Genetic Engineering Research Institute (AGERI), Agricultural Research Center (ARC), Giza, 12619-Egypt.
[3]Nucliec acid and protein structure department Center (ARC), Agricultural Genetic Engineering Research Institute (AGERI), Agricultural Research Center (ARC), Giza, 12619-Egypt.

Trehalose is a non-reducing disaccharide which consists of two glucose units that functions as a compatible solute to stabilize the membrane structures under heat and desiccation stress. Trehalose-6-phosphate synthase (TPS) and trehalose-6- phosphate phosphatase (TPP) are the key enzymes for trehalose biosynthesize in the plant kingdom. On the basis of bioinformatics prediction, fragment containing an open reading frame of 945 bp was cloned from durum wheat. Sequence comparison and analysis of conserved domains revealed the presence of a *TPP* domain. Full length of the gene was isolated using gene race technology. Semi-quantitative RT-PCR and real time quantitative PCR indicated that the expression of this gene is up-regulated in response to drought stress. The biochemical assay of the trehalase activity showed that the enzyme's activity decreased under the dehydration stress. The obtained phylogenic tree showed that the isolated TPP protein forms a distinct clad close to the *Oryza sativa* trehalose-6- phosphate phosphatase. *In silico* and comparative mapping indicated that the isolated *TPP* gene is localized on rice chromosome 8, durum wheat chromosome 20, bread wheat chromosome 3B, oat linkage group E, sorghum chromosome 4 and barley 5H.

Key words: Abiotic stress tolerance, trehalose-6- phosphate phosphatase (TPP*)*, durum wheat, trehalose, real time PCR, cloning, full length gene, drought stress.

INTRODUCTION

The global food situation is currently being redefined by many driving forces like globalization, urbanization, energy prices, and climate change. According to the report of the food and agriculture organization of the United Nations (FAO) 2010, the number of undernourished people around the world in 2010 has declined but remains abnormal and unacceptable. The renewed global attention is being given to the role of agriculture and food in development policy. One of the required actions that is suggested by the Egyptian Cabinet, Information and Decision Support Center (IDSC) to solve the food problem in Egypt was to focus on the agricultural research to enhance the capability of crop plants to withstand different abiotic stresses, such as salt, drought,

*Corresponding authors. E-mail: aymanalidiab@gmail.com.

cold and Heat shock which will lead to higher yields by either increasing the crop set and/or by extending crop cultivation in the areas previously denied due to abiotic stresses.

Understanding the gene networks that represent the biological system of plants under abiotic stress and their defense mechanism makes it necessary to characterize the candidate genes that are responsible for the physiological response to the stress. Trehalose is an important building block to build up sugars that create cellular signaling and communication. It is included in the building of a number of cell wall glycolipids. Trehalose is a disaccharide sugar widely distributed in bacteria, fungi, insects, plants and invertebrate animals. In microbes and yeast, trehalose is produced from glucose where trehalose-6-phosphate synthase (TPS) and trehalose-6-phosphate phosphatase (TPP) function together as a large complex to synthesize trehalose. Moreover, TPS and TPP serve as sugar storage, metabolic regulator and protect living organisms against abiotic stress (Wiemken, 1990; Strom and Kaasen, 1993). Before 1997, it was thought that trehalose is present only in a few desiccation-tolerant plants. However, its role in plants was not yet fully elucidated. Under osmotic stress, trehalose was shown to accumulate at high levels in resurrection plants such as *Selaginalla lepidophylla* (Wingler, 2001) and as a sugar reserve or stress protectant in *Arabidopsis thaliana* (Goddijn and Smeekens, 1998; Vogel et al., 2002; Schluepmann et al., 2003). Interestingly, TPP and TPS genes are broadly found in the genomes of higher plants and form large gene families (Leyman et al., 2001; Schluepmann et al., 2004). Mellor (1992) considered trehalose as a symbiotic determinant between higher plants and microorganisms. However, there is no direct evidence supporting this hypothesis so far. It was found that the precursor of trehalose, trehalose-6-photophate, (T-6-P) is the key regulator in the glycolytic pathway (Blazquez et al., 1998). It targets the initial step of glycolysis to reduce the entrance of glucose into glycolysis. The same role of trehalose in the sugar metabolism was invistigated (Vogel et al., 1998; Paul, 2001; Wingler, 2001; Eastmond and Graham, 2003). From the results of genetic and reverse genetic analysis, trehalose was found to have an essential role in carbohydrate metabolism and development of higher plants. In Arabidopsis, loss of *AtTPS1* function is an embryo-lethal phenotype (Eastmond et al., 2002; Schluepmann et al., 2003; Gomez et al., 2006); and a mutation in the maize TPP gene caused abnormalities in the inflorescence architecture (Satoh-Nagasawa et al., 2006). Trehalose content increased in rice as a result of the over expression of fused bacterial *TPS* and *TPP* proteins (named *TPSP*). The bi-functional *TPSP* protein enhanced the rice tolerance to abiotic stresses (Garg et al., 2002; Jang et al., 2003). Over expression of *TPSP* had a direct effect on the photosystem II damage under abiotic stress

(Garg et al., 2002; Jang et al., 2003). El-Bashiti et al. (2005) reported the possible role of trehalose as osmoprotectant compound in wheat species under salt and drought stress conditions. The accumulation of trehalose in wheat under abiotic stresses was found to be tissue and species specific.

Martı́nez-Barajas and his colleagues (2011) analyzed *T6P* content and *SnRK1* activities in wheat (*Triticum aestivum*) grain. The data shows a correlation between *T6P* and sucrose overall that belies a clear effect of developmental stage and tissue type on *T6P* content, consistent with tissue-specific regulation of *SnRK1* by *T6P* in wheat grain. Homologs of SNF1-related protein kinase1 (*SnRK1*) marker genes designated in Arabidopsis (Baena-González et al., 2007) was used to prove that regulation of *SnRK1* by T6P could operate *in vivo*, using Wheat Estimated Transcript Server (WhETS; Mitchell et al., 2007).

In long term, the overexpression of trehalose biosynthetic genes in wheat may seem to be promising for improvement of abiotic stress tolerant transgenic wheat.

This work aimed at the isolation, cloning and characterization of functional trehalose-6-phospate phosphatase (*TPP*) gene from durum wheat under dehydration stress to investigate the trehalose 6 phosphatase (*TPP*) gene ability for drought tolerant in Durum wheat in order to examine the magnitude of the *TPP* gene response to drought stress.

MATERIALS AND METHODS

Plant materials, growth conditions and stress treatments

Durum wheat plants, (variety Sohag 3) presumably holding genes of resistance to drought were subjected to dehydration stress. Seeds of durum wheat (*Triticum turgidum. L.* var. durum wheat) were sterilized in 10% sodium hypochloride for 30 min and then rinsed with ddH$_2$O for 1 min. Seeds were planted in soil composed of sand and clay (1:1) for three weeks and watered daily under controlled conditions (28°C day/25°C night, 12 h photoperiod, ~500 mol m^{-2} s^{-1} photon flux density and 83% relative humidity). Drought treatment was applied as described by Ozturk et al. (2002) where, seedlings were removed from soil, washed carefully and placed on paper towels under the same growing conditions. Leaves were harvested after 2, 4, and 6 h of drought treatment, frozen in liquid nitrogen and stored at -80°C. Control seedlings were planted and grown concurrently in the same conditions without any drought regime (well-watered) then leaves were harvested at the same time and frozen in liquid nitrogen and stored at - 80°C. For the estimation of water loss, leaves were weighted at the same time intervals as that used in the dehydration experiments (zero, 2, 4 and 6 h). The ratio of the leaves weight in comparison to the control was used as indication of water loss.

Total RNA isolation

Total RNA was extracted according to Chomczynski (1993) where, 100 mg of the control and drought treated leaves (0, 2, 4 and 6 h) were ground in liquid nitrogen. 1 ml of TriPure reagent (Cat. No. 1 667 165, Roche) was added to the fine leave powder and shacked gently. The mixture was left for 5 min at room temperature before adding 0.2 ml of chloroform. The mixture was left at room temperature for 10 min then centrifuged for 15 min at 4°C. Half ml isopropanol was added to the aqueous phase and incubated at room temperature for 10 min. The samples were centrifuged at 12,000 x g for 10 min at 4°C. The RNA pellets were re-suspended in 75% ethanol then centrifuged at 7500 x g for 5 min at 4°C. The RNA pellets were dried and re-suspended in diethylpyrocarbonate (DEPC)-treated RNase-free water and stored at-80°C.

Reverse transcription PCR (RT-PCR)-based cDNA cloning

A pair of primers, (5'-ATGGATTTGAGCAATAGCTC-3' and 5'-ACACTGAGTGCTTCTTCCAT-3') were synthesized and used to perform RT-PCR amplification using ImProm-IITM reverse transcription system (Cat. No. A3800, Promega). According to Liang and Pardee (1995), a cDNA of the TPP gene was generated using a RT-PCR based approach. The PCR cycle condition consists of three segments. The first one was a pre-denaturation for 4 min at 94°C. The second variable segment was consists of 40 cycles each one was 1 min at 94°C, 1.5 min at 55°C and 2 min at 72°C; the last segment was an extension for 10 min at 72°C. The amplified cDNA fragment was purified and cloned for sequencing.

Cloning of PCR product

The PCR products were cloned in pGEM-T Easy plasmid (Promega, USA) and transferred into Escherichia coli DH5α. The white colonies were picked and screened for the presence of the cloned gene of interest through digestion with EcoRI (Sambrook et al., 1989). The pGEM®-T plasmid having TPP cDNA was selected by PCR using T7 and SP6 primers that amplify the 945 bp fragment having the TPP gene. In this reaction, 1.25 units Taq DNA polymerase, 20 pmol primers and 200 mM dNTPS were added to 15 ng of plasmid in a buffer containing 10 mM KCl, 10 mM (NH4)2SO4, 20 mM Tris -HCl-pH 9 and 0.1% Triton® X-100. The PCR cycle condition consists of three segments. The first was a pre-denaturation for 4 min at 94°C. The second variable segment was of 40 cycles each one was 1 minutes at 94°C, 1 min at 55°C and 2 minutes at 72°C; the last segment was an extension for 7 minutes at 72°C.

DNA sequencing and Bioinformatics analysis

The TPP clone was sequenced according to Sanger et al. (1977) using a Big Dye Terminator Cycle sequencing FS Ready Reaction Kit (Applied Biosystems, Foster City, CA) and an ABI PRISM 310 DNA sequencer (Applied Biosystems). A homology search was performed using BLASTX against the NCBI protein database (http://www.ncbi.nlm.nih.gov). Sequences of the trehalose phosphate phosphatase genes that showed similarity to the TPP gene were obtained from the NCBI non-redundant and dbEST data sets using BLASTX or BLASTP ver. 2.0.10 (Altschul et al., 1997). The full amino acid sequences of the proteins were aligned using CLUSTAL- X ver. 1.8 (Thompson, et al., 1997) and subjected to phylogenetic analysis. Phylogenic trees were constructed using the neighbor-joining (NJ) method (Saitou, and Nei, 1987) with parsimony and heuristic search criteria and 1000 bootstrap replications to assess branching confidence.

Rapid amplification of cDNA PCR (RACE-PCR)

Rapid amplification of cDNA PCR was done to obtain the full length of TPP gene. According to Frohmann (1994), preparation of cDNA and anchor primers was conducted using Roche kit cat No.

1734792 (Clontech Lab, Inc). PCR was performed by using SP2 primer from advantage cDNA PCR kit (Clontech) and the primer 5'-CCTCCAGCACTTCGTTTACGAG-3' designed according to the gene sequence. PCR products were migrated by electrophoresis on 2% (W/V) agarose gel. The glass-milk (BIO 101) was used to recover and purify the DNA fragment, which was then ligated to pGEM®-T easy vector and finally transferred into Escherichia coli DH5α (invetrogen, cat.No.18265-017). the cloned full length gene was sequenced using ABI PRISM big dye terminator cycle sequencing ready reaction kit (PE Applied Biosystem, USA).

Expression patterns using semi- quantitative RT-PCR

Template cDNA was prepared using Super-Script II (Invitrogen) with 1 mg total RNA. 1 µl of cDNA reaction mixture was diluted with 9 µl DEPC treated water, then, 1 µl of diluted mixture was used to perform Semi-quantitative RT-PCR reaction as follows: 1.0 µl dNTPS (10 mM), 2.5 µl MgCl2 (25 mM), 5.0 µl 10X buffer, 5.0 µl Forward primer (10 pmol/µl), 5.0 µl Reverse primer (10 pmol/µl) , 1.0 µl Template cDNA (25 ng/µl) , 0.5 µl Taq (5 U/µl), up to 50 µl dd H2O. The amplification was carried out in Hybrid PCR Express system programmed with specific primers for TPP and 18S (as a control to normalize the amount of cDNA present in each sample) genes as follows: 5 min at 95°C, followed by 35 cycles at 95°C for 45 s, 55°C for 60 s, 72°C for 2 minutes, 72°C for 5 minutes. For each sample, 10 µl of the amplification reaction was size-fractionated on a 2% (w/v) agarose gel and stained with ethidium bromide. Bands were detected on UV-transilluminator and photographed by a Gel Documentation system 2000 Bio-Rad to ensure that amplifications were in the linear range, for each template and primer pair. A Gene Ruler™ 1 kb DNA ladder was used as a standard.

Real-time PCR data analysis

Primers of TPP and 18S used for semi-quantitative RT-PCR were used in real time PCR analysis. The most commonly used method for relative quantification is the $2^{-\Delta\Delta Ct}$ method. Derivation and examples of this method have been described by Livak and Schmittgen (2001). The relative difference in gene expression using the $2^{-\Delta\Delta Ct}$ method was calculated as follows:

Relative fold change in gene expression = $2^{-\Delta\Delta Ct}$,
Where, $\Delta\Delta C_t$ = ΔC_t treated - ΔC_t untreated and ΔC_t = (C_t target gene - C_t reference gene).

Trehalase enzyme assay

Samples weights about 100 mg from drought treated leaves at 0 hr (control), 2, 4, 6 hrs, were ground in liquid nitrogen. The powder was suspended in ice-cold suspension solution containing 0.1 M citrate (Na+), pH 3.7, 1 mM PMSF, 2 mM EDTA and insoluble polyvinylpyrrolidone (10 mg/g dried weight)). 2 ml of extraction buffer was added to each 1g dry weight of sample. The homogenate was filtered through two layers of cheesecloth and centrifuged at 31,500 rpm (48,000 g) for 30 min at 4°C in Sorval Combi Plus with T-880 type rotor. The supernatant was used for the enzyme assays. Adapted from Vandercammen et al. (1989). The protein concentration was determined according to Bradford (1976) using bovine serum albumin (BSA) as standard.

Trehalase enzyme activity was measured using glucose oxidase-peroxidase kit (Bicon) according to Müller et al. (1992). The reaction mixture was composed of 10 mM trehalose, 50 mM MES (K+), pH 6.3 and 0.2 mg crude extract in a final volume of 1 ml.

Table 1. Leaves' weight of durum wheat under dehydration shock treatment.

Time of dehydration shock (h)	Loss in water content of leaves (%)
Control 0	0
2	15.79
4	26.36
6	34.5

The reaction was incubated at 37°C for 30 min and then started by the addition of trehalose to the reaction mixture, which was preincubated at 37°C for 10 min. 100 µl of samples were taken from the reaction mixture and immediately put in thermostat at 100°C for 3 min to stop the reaction. Precipitates were removed by centrifugation at 8700 rpm for 10 min. For the analysis, 10 µl of the supernatant was mixed with µl of glucose oxidase - peroxidase kit solution, mixed by vortex and then the mixtures were incubated at 37°C for 15 min. The absorbance of the samples was measured at 470 nm in Schimadzu UV-1201 spectrophotometer against blank solution.The increase in the absorbance against time was assumed to be equal to the amount of glucose formed. One unit of trehalase activity is defined as the amount of enzyme that catalyzes the hydrolysis of 1 mmole of trehalose/ minutes at 37°C at pH 6.3.

In silico and comparative mapping of *TPP*

For *in silico* mapping, the isolated sequence was compared to rice and oat sequences using BLAST (with an e-value threshold of 1e-1000). The matches were used to identify markers from the genetic linkage map (http://www.tigr.org). The results obtained from this stage were used to construct a comparative map between durum and bread wheat, rice, sorghum, barley and oat to identify the tentative chromosomal location of the gene understudy using comparative mapping strategy (Diab et al., 2007; Abou Ali et al., 2009).

RESULTS AND DISCUSSION

Physiological parameters and dehydration stress-specific transcript profiles

As shown in Table 1, leaves' weight was gradually decreased with dehydration time compared with the control (zero time dehydration). This weight losses indicates the decline in the water content of experimented leaves by 15.79, 26.36 and 34.5% for 2, 4, 6 h dehydration respectively. These results are in agreement with Ozturk et al. (2002) who found that the water content of barley declined by 10% within the initial 4 h, and then more rapidly by 30% (6 h) and 36% at 10 h of stress. Xue et al. (2008) reported that the change in relative leaf water content (LWC) of different genotypes indicated their different susceptibility to water scarcity. The dehydration treatment provides reliable, fast and easy way to detect genes responsible to abiotic stress response in physiological term (Talame et al., 2006).

Molecular cloning of *TPP* fragment

The RT-PCR reaction produced by *TPP*1 gene fragment

with a length of ≈ 1000 bp is shown in Figure 1. The amplified *TPP* cDNA fragment was ligated into the pGEM-T easy vector (3015 bp) and transformed in *Escherichia coli* competent cells. The cloned *TPP*1 fragment was screened using T7 and SP6 primerd. Positive colonies having the insert displayed a band about ≈ 1300 bp (*TPP* fragment with a length of ≈ 1000 bp linked to the region between Sp6 and T7 in native pGEM-T easy plasmid ≈ 300 bp) (Figure 2).

Sequence analysis of trehalose-6-phosphate phosphatase (*TPP*) fragment

The isolated fragment was sequenced using ABI PRISM. Figure 3 shows the sequence obtained for the *TPP* fragment. Sequencing of the isolated fragment revealed that the length of Trehalose-6-phosphate phosphatase (*TPP*) fragment was 945 bp. The obtained sequence was subjected to the BLASTx analysis which proves that the sequence has different degrees of similarity with other *TPP* genes. The *TPP* fragment showed similarity to the *TPP* genes from *O. sativa*, (EU559275.1) 88%, *A. thaliana* (AY093147.1) 66% and *Z. mays* (NM_001158750.1) 66%.

Isolation, cloning and characterization of the full-length *TPP* gene

first strand of cDNA was synthesized according to Frohmann (1994) from total RNA using a gene specific cDNA primer SP2, (5'GGACGAACCTCTAAAACCATTC3'). The terminal transferase was used to add a homopolymeric A-tail to the 3' end of the cDNA. Since eukaryotic coding sequences and 5'untranslated RNA regions tend to be biased toward G/C residues, the use of a poly (A)-tail decreases the likelihood of inappropriate truncation by the Oligo dT-anchor primer. Additionally, poly(A)-tail was used due to the weaker A/T binding than G/C binding, therefore longer stretches of A residues were required before the Oligo dT-anchor primer will bind to an internal site and truncate the amplification product. As shown in Figure 4, the full length of *TPP* gene that was obtained by RACE PCR was ≈1500 bp. PCR product was ligated into PGEM-T Easy Vector and then transformed into *E. coli*. The recombinant plasmid was digested by EcoR1 res-

bp

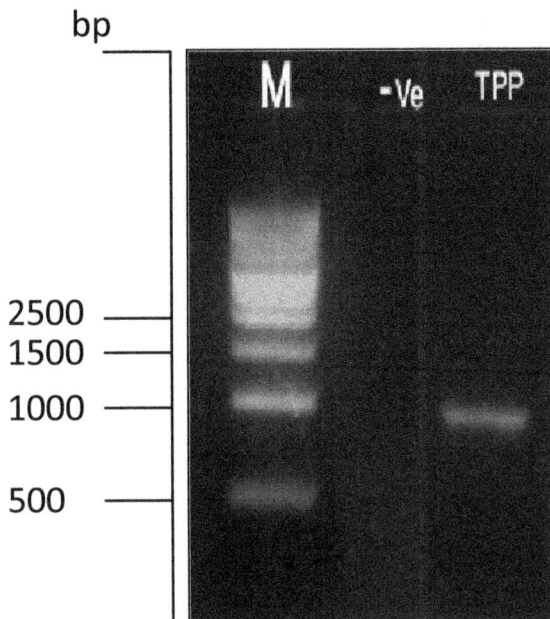

Figure 1. Agarose gel showing 1 kb marker (M), negative contr l (-ve), and *TPP* candidate band for *TPP* fragment.

bp

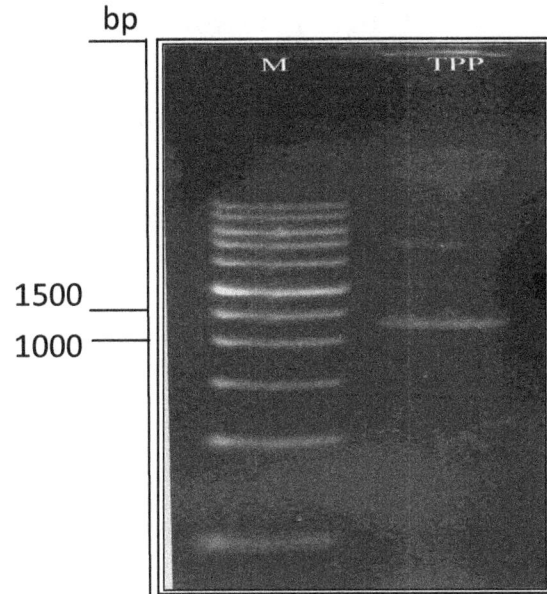

Figure 2. Agarose gel shows (M) 1 kb marker and amplified (*TPP*1) fragment using Sp6 and T7 primers.

triction enzyme to release the cloned gene. Two bands were obtained as a result of the digestion reaction of the recombinant plasmid. One was around 3000 bp representing the vector (3015 bp) and the other was around 1500 bp representing the insert (Figure 5).

The *TPP* gene(s) were isolated before in several studies. Shima et al. (2007) isolated Os*TPP*1 and Os*TPP*2 representing the two major trehalose-6-phosphate phosphatase genes expressed in rice, and they found that the rice genome contains nine *TPP* genes. The Os*TPP*2 gene encodes a 42.6 kDa protein (382 amino acid residues). The same results were obtained by Pramanik and Imai (2005). They found nine putative *TPP* genes in the rice genome sequence. In *Arabidopsis*, 11 *TPS* and 10 *TPP* genes have been identified (Leyman et al., 2001; Eastmond and Graham, 2003). While Alexandrov et al. (2009) isolated 1883 bp trehalose-phosphate phosphatase from *Z. mays* and Ge et al. (2008) isolated 1478 bp *TPP*1 gene from *O. sativa*.

The isolated full length gene was sequenced using ABI PRISM (310 Genetic Analyzer); the sequence data is shown in Figure 6. This sequence was utilized to run a homology search using blast tool provided by NCBI (http://blast.ncbi.nlm.nih.gov/Blast.cgi) (Altschul et al., 1997). The results of the homology research revealed that the isolated gene displayed different degrees of similarities to other *TPP* genes. The isolated durum wheat *TPP* showed similarity with the *O. sativa*, (AB120515.1) *TPP* by 93%, *A. thaliana*, AY059840.1 by 68%, *Z. mays* (NM_001158750.1) by 76%.

According to the open reading frame of the isolated gene, the length of the protein that was expressed from the isolated gene was of 481 amino acid residues. The deduced amino acid sequence of the isolated *TPP* gene had a molecular weight of 53546.61 Daltons, and its isoelectric point was 10.37. The amino acid analysis results of *TPP* protein revealed that it contains 71 strongly basic amino acids, 40 strongly acidic amino acids, 172 hydrophobic amino acids, and 127 polar ami-no acids. Amino acids sequence of *the TPP* was analyzed using BLASTP. The results show that these sequences have high homologous sequence with different *TPP* amino acids sequences. The Durum wheat *TPP* had a similarity of 93% to *O. sativa* (AB120515.1), to *Nicotiana tabacum* (BAI99253.1) by 56%, *A. thaliana* (NP_193990.1) by 56%, *Z. mays* (NP_001152222.1) by 53% and *Arabidopsis lyratanigra* (XP_002887719.1) by 42%.

Alignment of the predicted amino acid sequence of *TPP* with proteins from other species identified several conserved regions are as shown in Figure 7. Pramanik and Imai (2005) reported that the alignment of the Os*TPP*1protien sequence with other *TPP* gene products from *Saccharomyces cerevisae* (Sc*TPS*2), *E. coli* (Ec*OtsB*) and *Arabidopsis* (At*TPPA* and At*TPPB*), revealed that *TPP* sequences are moderately conserved with exception in the N-terminal region. The two distinct phosphatase boxes that are unique features of phosphatases are highly conserved (Vogel et al., 1998). Eastmond et al. (2002) investigated the *TPP* and *TPS* multigene family in plant sequences and suggested that trehalose biosynthesis is highly regulated by environmental changes in plants. Van Dijck et al. (2002) reported that, all *TPS* proteins in plants contain a conserved N-terminal extension that not found in fungal or bacterial *TPS* proteins.

TTTGTTAAACTAAAGGAGCTTAAGTAT AACGGAAACCATGACACAGCCACCGTATCA
AATTTCT CAGAGGCATAA CAACATGACT AATAATAGTAAA CAAGCCAATATC AAT
CGACCTA AACGCAATTTTCTAA CAGGGGACGG CAACATCATT CGACCCTAGC TAT
GGATTGCC AACGCTGCAAAACAGTCGCCTGTCGGAAACTA CAGTCATCTC CTGTTT
GTTC ATTATCGCAACGTTGCAGCGAGTG CTTGGAAACTGGTCTTCCATCTCACCAAG
G AGTTCAGGAACGCT CTCCATCACT TCAGACGGGT CCCTGAGCGA GTAGAAGGCC
TCGGATTCCT TCGGCGCCTG ACGAGACCAGT ATCCCGTATC CGCAGTTCCT CTCCC
GAAGCACCTTGACTACG CGTCTTCGTC GGTGCGGTCG TCGCCGATGT AGATAGGGA
T CACTTTCTCG GACTCGCTCA GCGCCAGCGA CTGAAGCAGG AATCGACGGC CTTC
CCCTTG TCCCAGTCAAGAAGAGATCACTGGACG AACCTCTAAA ACCATTCGTC CGT
TGGGAGATCAC TTTGAGACGG GGGAAGTCCT CCAGCACTTC GTTTGACGAGC CGT
GCGACCA GCTCCCAGTCAAGAAGA CTTCTCGTCC ACGTTGCGGTAATGTACAGA CA
CGCAGAAC TTGTTGTAGCCT CAACGCTTGC GCCTTCGATT CGGCTCGTGACTTCCA
AGAG GACCTTGGAA ACCTCATCGA TCATAGGCAG AAAATCGCGA GCAGTTGGAA
GAGGTTGGCT TCTTTGCCCT TTTCAGCATT GCGTTCATAA TGTGCTGAAG GATGGT
TCATTA TGTCCATCGCC ATTGTAGACTACCA GCATAGCAGA GTTCCTTCAG ATTTA
CGAAT TCAAGAACCTTATTCGGGGA CCTTCCATTG TCCGTGTCAA AAAGAAAAAA
AGTAAAAAA

Figure 3. Nucleotide sequence of (*TPP*) fragment representing 945 bp.

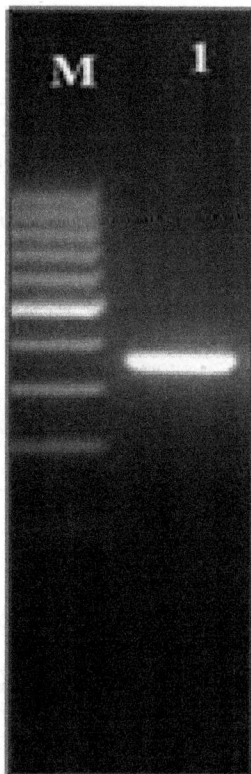

Figure 4. Agarose gel electrophoresis of gene-specific fragment (*TPP*1) gene ≈1500 bp isolated by RACE-PCR.

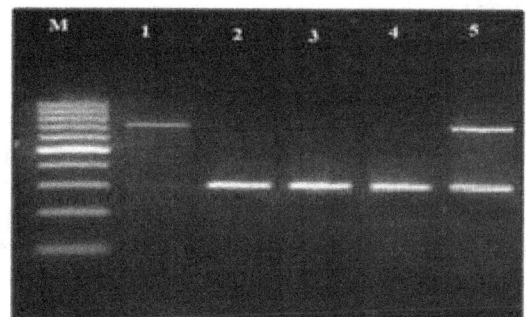

Figure 5. Agarose gel representing digestion of candidate colonies with *EcoRI* enzyme, 1kb marker (M), pGEM-T vector (1), (*TPP*) gene (2, 3, 4) and positive colony plasmid (5) DNA digestion.

To determine the evolutionary relatedness of *TPP* protein to Trehalose 6 phosphate phosphatase proteins isolated from other species, the neighbor joining method (NJ) was used to generate a phylogenic tree based on amino acid sequence homology. The tree showed that *TPP* protein forms a distinct clad on phylogenetic trees derived from various *TPP* sequences (Figure 8). Bootstrap analysis placed the durum wheat *(Triticum durum)* sequence close to *O. sativa* with a high degree of confidence, demonstrating that the two species descent from common ancestor.

Protein sequence analysis homology modeling

The results of BLAST search against PDB program exhibited a high level of sequence similarity to the crystallized structure for modeling *TPP* protien (Figure 9). This protein structure is very important to study the mode of action of disaccharides, like trehalose that appear to be one of the most effective stabilizers for dried enzymes

(A)

```
ATGGATTTGA GCAATAGCTC ACCTGTCATC ACCGATCCGG TGTCGATCAG CCA
GCAGCTGTTG CGCGCCC TGCCTTCAAA TCTGATGCAG ATTTCAGTCA TGCGCGG
TGG CTACTCCAGCTCTCGCATGG ACGTTGGTGT CAGTAGGCTC ATAATCGAGG
AAGACCTTGTCAATGGACTGCTTGATGCGA TGAAATCCTG CTCACCTCGC AGGA
GGCTGA GTGTAGCAAT TGGCGAGGACAATTCATCTG AAGAAGATGACGCTGCTT
AC AGCGCTTGGA TGGCAAAATG TCCTTCTGCATTGGCTTCCT TCAAGCGAAT TG
AAGCGAGT GCACAAGGGA AGGAGATTGC TGTGTTTCTAGACTATGACG GCACA
CTGTC GCGTATTGTG GATGATCCTG ACAAAGCAGT GATGTCTCCCGTGATGAGT
G CTGCTGTGAG AGATGTTGCG AAGTACTTCC CCACTGCAAT TGTCAGCGGA  A
GGTCCCGCA ATAAGGTGTTTGAATTTGTTAAACTAAAGGAGCTTAAGTAT AACG
GAAACCATGACACAGCCACCGTATCAATTTCT CAGAGGCATAA CAACATGACT A
ATAATAGTAAA CAAGCCAATATC AATCGACCTA AACGCAATTTTCTAA CAGGG
GACGG CAACATCATT CGACCCTAGC TATGGATTGCC AACGCTGCAAAACAGTC
GCCTGTCGGAAACTA CAGTCATCTC CTGTTTGTTC ATTATCGCAACGTTGCAGCG
AGTG CTTGGAAACTGGTCTTCCATCTCACCAAGG AGTTCAGGAACGCT CTCCAT
CACT TCAGACGGGT CCCTGAGCGA GTAGAAGGCCTCGGATTCCT TCGGCGCCTG
 ACGAGACCAGT ATCCCGTATC CGCAGTTCCT CTCCCGAAGCACCTTGACTACG C
GTCTTCGTC GGTGCGGTCG TCGCCGATGT AGATAGGGAT CACTTTCTCG GACTC
GCTCA GCGCCAGCGA CTGAAGCAGG AATCGACGGC CTTCCCCTTG TCCCAGTC
AAGAAGAGATCACTGGACG AACCTCTAAA ACCATTCGTC CGTTGGGAGATCAC
TTTGAGACGG GGGAAGTCCT  CCAGCACTTC GTTTGACGAGC CGTGCGACCA GC
TCCCAGTCAAGAAGA CTTCTCGTCC ACGTTGCGGTAATGTACAGA CACGCAGAA
C TTGTTGTAGCCT CAACGCTTGC GCCTTCGATT CGGCTCGTGACTTCCAAGAG G
ACCTTGGAA ACCTCATCGA TCATAGGCAG AAAATCGCGA GCAGTTGGAA GAG
GTTGGCT TCTTTGCCCT TTTCAGCATT GCGTTCATAA TGTGCTGAAG GATGGTT
CATTA TGTCCATCGCC ATTGTAGACTACCA GCATAGCAGA GTTCCTTCAG ATTT
ACGAAT TCAAGAACCTTATTCGGGGA CCTTCCATTG TCCGTGTCAA AAAGAAAA
AA AGTAAAAAA
```

(B)

```
MDLSNSSPVITDPVSISQQLLRALPSNLMQISVMRGGYSSSRMDVGVSRLIIEEDLVNG
LLDAMKSCSPRRRLSVAIGEDNSSEEDDAAYSAWMAKCPSALASFKRIEASAQGKEI
AVFLDYDGTLSRIVDDPDKAVMSVMSAAVRDVAKYFPTAIVSGRSRNKVFEFVKLKE
LKYNGNHDTATVSISQRHNNMTNNSKQANINRPKRNFLTGDGNIIRPLWIANAAKQS
PVGNYSHLLFVHYRNVAASAWKLVFHLTKEFRNALHHFRRVPERVEGLGFLRRLTRP
VSRIRSSSPEAPLRVFVGAVVADVDRDHFLGLAQRQRLKQESTAFPLSQSRRDHWTN
LNHSSVGRSLDGGSPPALRLTSRATSSQSRRLLVHVAVMYRHAELVVASTLAPSIRLV
TSKRTLETSSIIGRKSRAVGRGWLLCPFQHCVHNVLKDGSLCPSPLTTSIAEFLQIYEFK
NLIRGPSIVRVKKKKSKK.
```

Figure 6. (A) Nucleotide sequence of (*TPP1*) gene representing 1458 bp as obtained from the ABI PRISM 310 DNA sequencer. (B) The amino acid sequence (481 aa).

and cell membranes *in vitro* and *in vivo*. However, the interaction of trehalose with biological membranes has been studied more than its interactions with other proteins.

Carpenter (1993) has reported that trehalose might interact with dry protein by hydrogen bonding to the polar amino acid residues in the protein. On the other hand, the interaction between trehalose and biological membranes indicates that trehalose can replace H_2O molecules around the polar head groups of the phospholipid in the dry state (Gaber et al., 1986). This hypothesis has been studied by Potts (1994) where trehalose binds to dry phospholipid vesicles. During desiccation, the interaction of trehalose with the biological membrane decreases the melting temperature (Tm) of the membrane to keep its

liquid crystalline phase (Crowe et al., 1993). This molding could be used to predict the interaction between the trehalose and other protein candidates in the biological membranes for more understanding of the trehalose mode of action for protecting plants against abiotic stress conditions.

Expression analysis

The results obtained in this work indicate that the expression of the *TPP* gene was up-regulated under dehydration stress compared to control (Figure 10). The highest expression level of *TPP* gene under dehydration stress was at 4 h. This up-regulation is important for the synthesis and accumulation of trehalose where, trehalose

Figure 7. Alignment of the predicted amino acid sequence of *TPP* and those from other plants including *Arabidopsis lyrata* (*ArLTPP*), *Zea mays* (*ZmTPP*), *Arabidopsis thaliana* (*ArTPP*), *Nicotiana tabacum*(*NtTPP*), *Oryza sativa* (*OsTPP*).

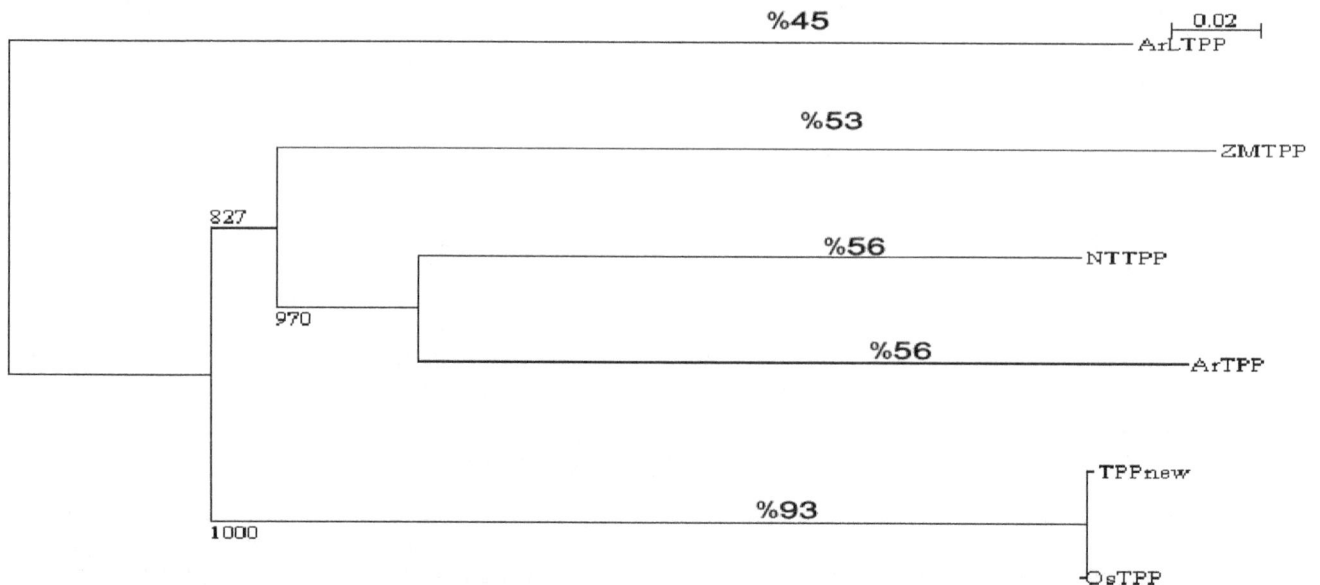

Figure 8. Phylogenetic tree based on sequences of *TPP1* with other *TPP*'s isolated from other plants. The phylogenetic tree was constructed by the neighbor-joining including *Arabidopsis lyrata* (*ArLTPP*), *Zea mays* (*ZmTPP*), *Arabidopsis thaliana* (*ArTPP*), *Nicotiana tabacum* (*NtTPP*), *Oryza sativa* (*OsTPP*).

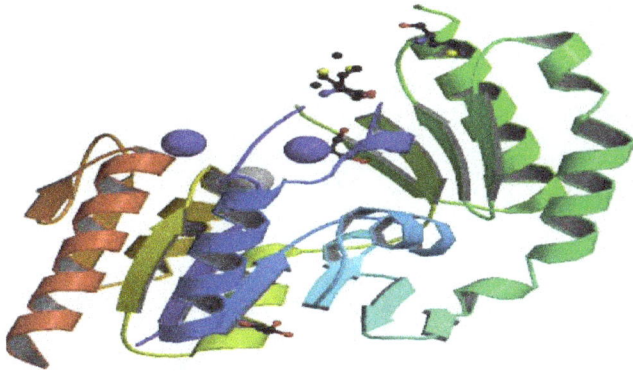

Figure 9. 3D Model of *TPP* protein structure.

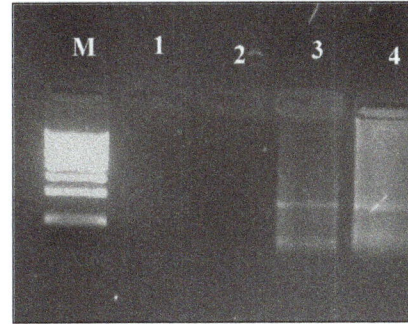

Figure 10. Agarose gel represents the *TPP* expression patterns of leaves under dehydration stress treatment after 0 (1), 2 (2), 4 h (3), 18S gene (4). M is 1Kb lader DNA marker.

is accumulated in large quantities under abiotic stresses (Elbein et al., 2003 and Wolf et al., 2003).

Reserve transcription combined with the polymerase chain reaction (RT-PCR) has proven to be a powerful method to quantify gene expression according to Murphy et al. (1990). Real-time PCR technology has been adapted to perform quantitative RT-PCR (Heid et al., 1996). The results of the real time PCR of the *TPP* gene showed that the expression level of *TPP* was slightly increased (up-regulated) after 4 h of dehydration treatment in leaves of Durum wheat compared with the control (0 h) and the relative fold change calculated by $\Delta\Delta$CT method, respectively which in agreement with the semi-quantitative PCR results (Figures 11 and 12). The plant's response to dehydration is accompanied by the activation of a group of genes, which are responsible for regulatory proteins that further regulate the transduction of the stress signal and modulate gene expression (Shinozaki and Yamaguchi, 2006).

Higo et al. (2006) studied the expression of trehalose gene synthesis (*mts* and *mth*, encoding maltooligosyl trehalose synthase and hydrolase) and trehalose hydrolysis (*treH*) in *Anabaena sp*. The genes (mts and mth) were up-regulated markedly upon dehydration. Gene disruption of *mth* resulted in a decrease in the trehalose level and in tolerance during dehydration stress. In contrast, gene disruption of *treH* resulted in an increase in both the amount of trehalose and tolerance. Trehalose did not stabilize proteins and membranes directly during dehydration; the expression of the two genes, one of which encodes a cofactor of a chaperone DnaK, correlated with trehalose content, a chaperone system induced by trehalose is important for the dehydration tolerance of *Anabaena* sp.

Cumino et al. (2002) found that many other genes, including *spsA*, encoding sucrose-6-phosphate synthase were up-regulated constantly during dehydration stress. Many genes related to hotosynthesis and ribosomal protein was down-regulated in the early dehydration phase, whereas genes for nitrogen fixation and photosynthesis I was down-regulated in the late dehydration phase.

Determination of trehalase activity under dehydration stress

As shown in Table 2, the activity of trehalase enzyme under dehydration stress was decreased from 1.004 to 0.781 after 2 h and to 0.427 after 4 h. The activity was then elevated after 6 h of dehydration treatment compared to the control (Figure 13). The elevation of trehalse activity after 6 h might be due to internal regulation mechanism in the system biology of the plant to prevent the uncontrolled increase of the trehalose which is important to prevent detrimental effects of trehalose accumulation on the regulation of carbon metabolism (Brodmann, 2002).

These results are in agreement with Brodmann (2002) who showed that trehalase activity normally keeps cellular trehalose concentrations low in order to prevent detrimental effects of trehalose accumulation on the regulation of carbon metabolism. The role of trehalase may be of particular importance in interactions of plants with trehalose-producing microorganisms. In support of this hypothesis, expression of the Arabidopsis trehalase gene and trehalose activity were found to be strongly induced by infection of Arabidopsis plants with the trehalose-producing pathogen *Plasmodiophora brassicae*. Penna (2003) found that trehalose was thought to protect biomolecules from environmental stress, as suggested by its reversible water-absorption capacity to protect biological molecules from desiccation-induced damage. The low levels of trehalose in transgenic plants can be explained by specific trehalase activity, which degrades trehalose; hence, it might be possible to increase trehalose accumulation by down regulating trehalase activity. El-Bashiti et al. (2005) found that trehalase activity in different wheat cultivar was increased under control conditions in both root and shoot of Bolal cultivar compared with salt and drought stress treatments. However, under drought conditions, there was no significant change in trehalase activity of shoot tissues. Trehalase is ubiquitous in higher plants and

Amplification Plot

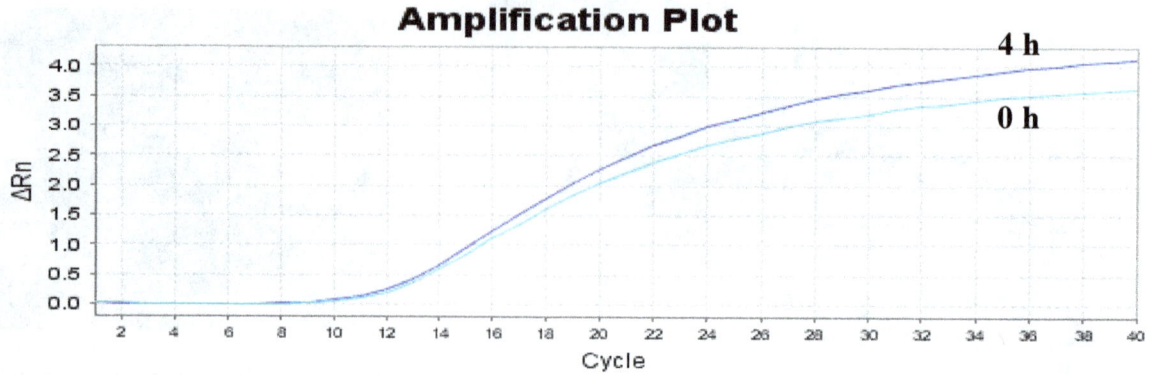

Figure 11. Variation of fluorescence amplification plot of *TPP1* gene for the dehydration shock treatments (0 and 4 h)

Amplification Plot

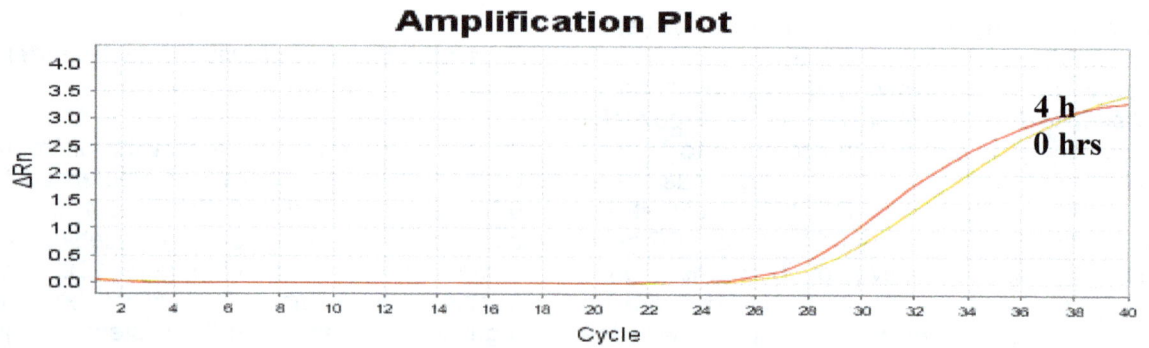

Figure 12. Normalization of fluorescence amplification plot of 18S gene (reference gene) for the three dehydration shock treatments (0, 4 h).

Enzyme activity

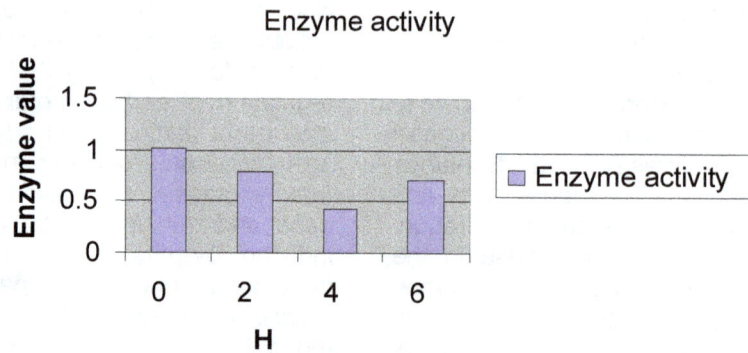

Figure 13. The trehalase activity after dehydration shock treatments.

Table 2. Trehalase activity under dehydration shock treatments.

Hour	Enzyme activity
0	1.004
2	0.7813
4	0.4273
6	0.7170

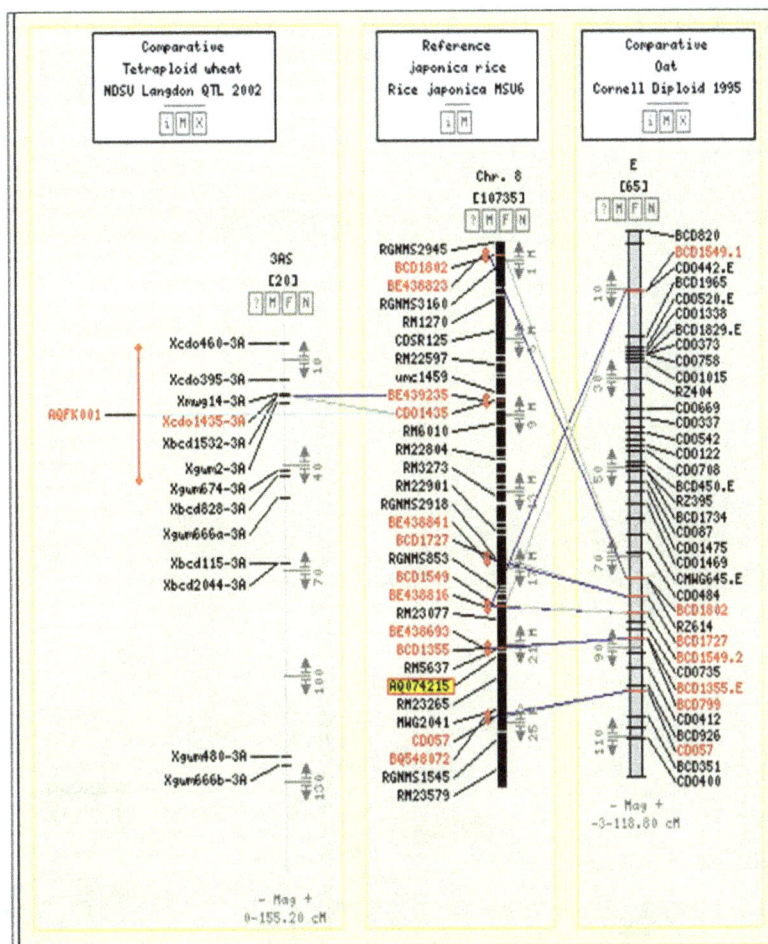

Figure 14. Comparative map showing the locus AQ074215 on Rice chromosome 8 that is closely linked to other locus on Oat and Tetraploid wheat.

single-copy trehalase genes have been identified and functionally characterized from soybean (*Glycine max*) and Arabidopsis (Aeschbacher et al., 1999; Müller et al., 2001). It is likely that trehalase is the sole route of trehalose breakdown in plants (Müller et al., 2001). Katoh et al. (2004) concluded that although *TPS* catalyses the transfer of glucose from UDP-glucose to glucose 6-phosphate to produce trehalose 6-phosphate and *UDP*, and *TPP* catalyses the dephosphorylation of trehalose 6-phosphate to trehalose. The low level of accumulation of trehalose may be attributed to the unique gene structure for trehalose metabolism. They investigated the expression of genes for trehalose synthesis, *mth* (maltooligosyl trehalose hydrolase) and *mts* (maltooligosyl trehalose synthase), as well as that for trehalose degradation, *treH* (trehalase), exhibited marked increase upon dehydration. So trehalose did not accumulate so much.

In silico and comparative mapping

Comparative maps can be used to study genome evolu-

tion; how the genome has been rearranged through time, and to make inferences about gene organization (Liang et al., 2008). *In-silico* mapping indicated the matching of the *TPP* gene sequence with the sequence of rice *TPP* on chromosome 8 linked to the locus (AQ074215).

Comparative mapping showed that the rice AQ074215 locus on chromosome 8 was also mapped on barley chromosome 5H. The marker (lwgsc) on bread wheat chromosome 3B was found to be closely liked to the rice locus (AQ074215) on chromosome 8. The results obtained from the comparative mapping showed that the isolated *TPP* gene is localized on rice chromosome 8, durum wheat chromosome 20, bread wheat chromosome 3B, oat linkage group E, sorghum chromosome 4 and barley 5H (Figures 14 to 16). This work utilizes a comparative analysis of durum and bread wheat, barley, oat, sorghum and rice based on linkage maps and consensus markers across the genome with the goal of linking the complex wheat genome to simpler diploid species such as barley and rice that serve as references. However, more detailed comparisons are needed to veri-

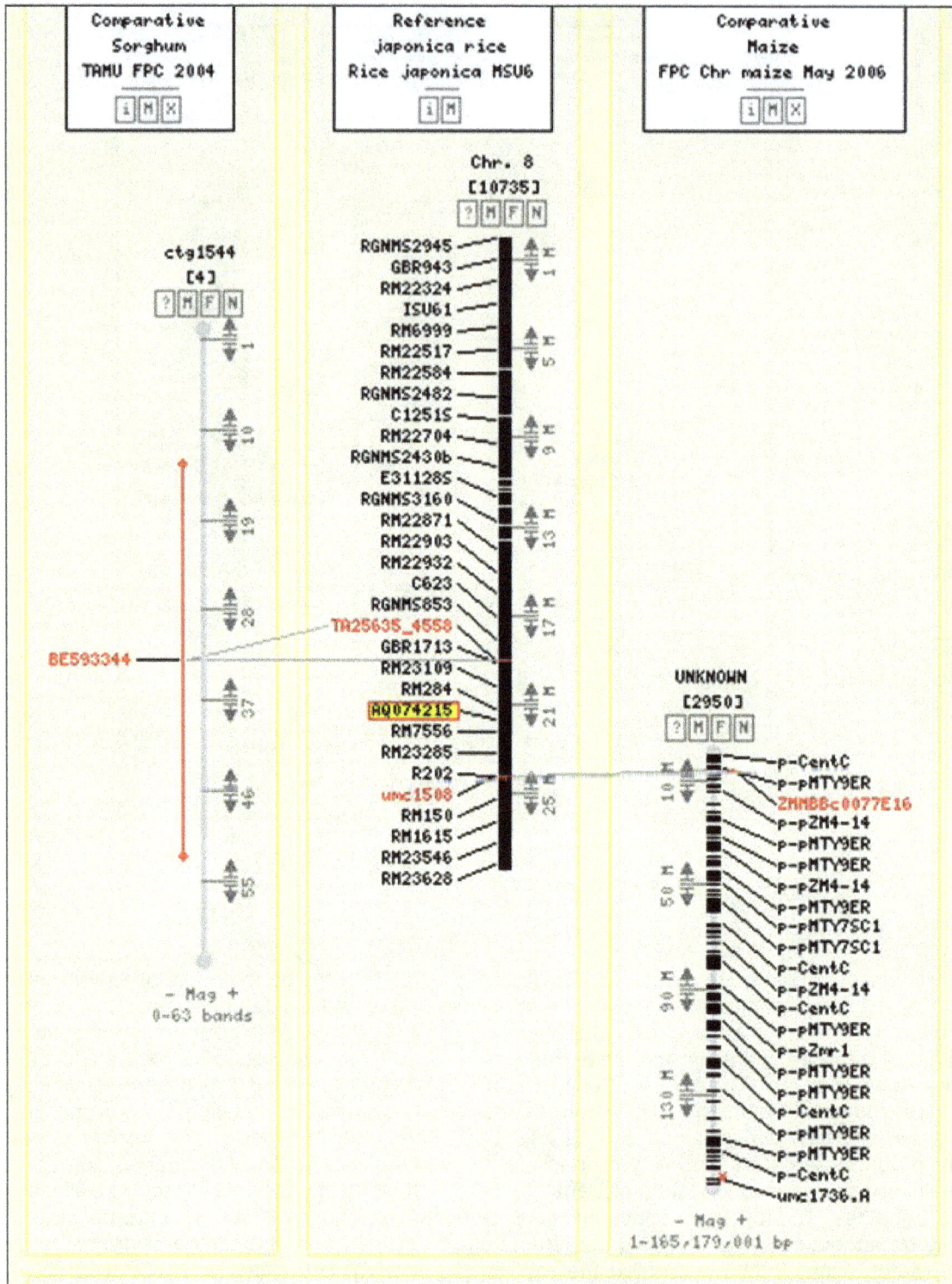

Figure 15. Comparative map showing the locus AQ074215 on rice chromosome 8 and maize and sorghum.

fy conserved regions associated the *TPP* gene.

In this study, the *TPP* gene was isolated, characterized and cloned to be used in the strategic improvement of crops for abiotic stress tolerance through genetic

transformation and the *in silico* comparative mapping of this gene would open the gate for the use of the *TPP* gene for marker assisted selection in breeding programs for abiotic stress tolerant crops.

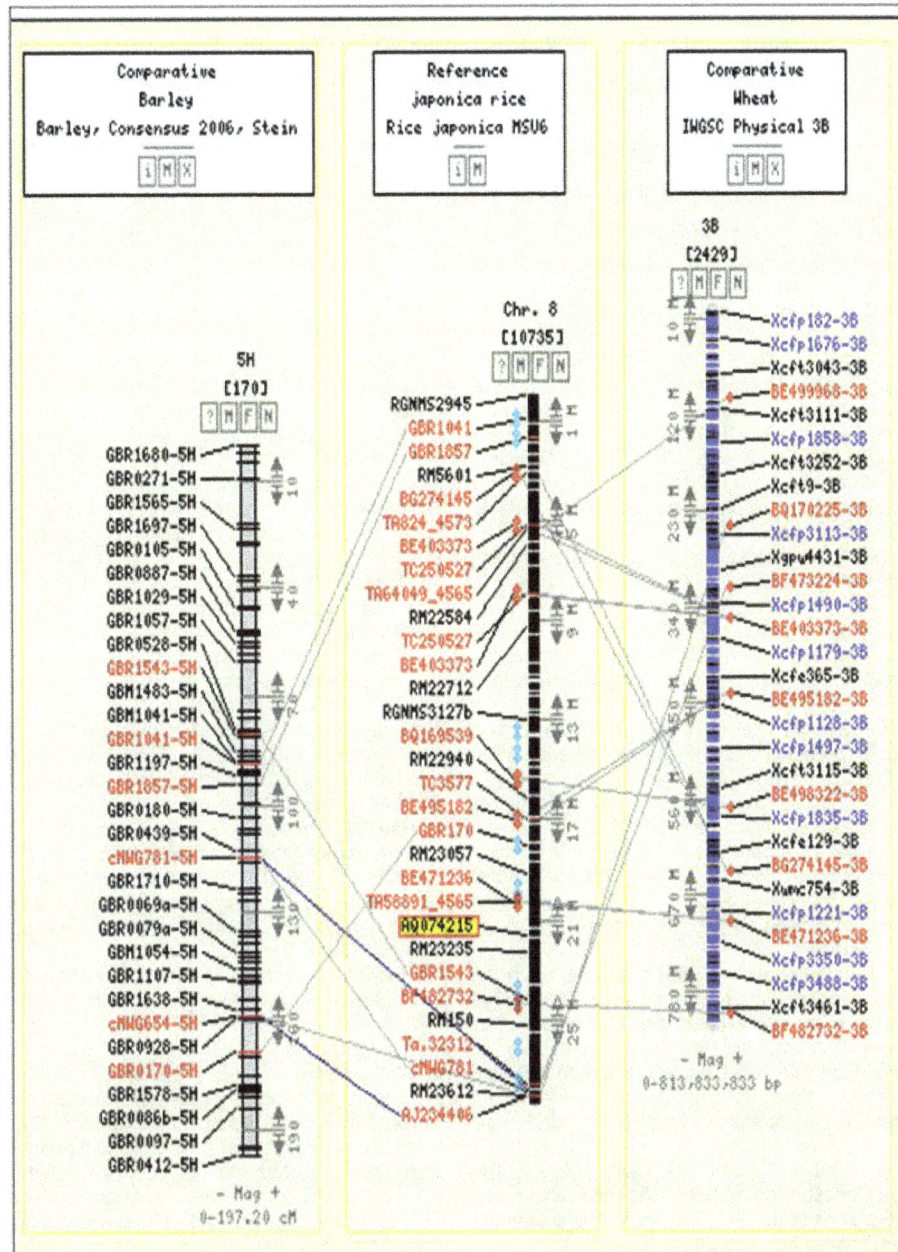

Figure 16. Comparative map showing the locus AQ074215 on Rice chromosome 8. and its linkage to barley and bread wheat.

REFERENCES

Abou Ali R, El-Hefnawy M, Nada A (2009). Isolation and characterization of *Cab8* gene from wild *Vicia cinera* species. Egyptian J. Genet. Cytol. 38:335-346.

Aeschbacher RA, Müller J, Boller T, Wiemken A (1999). Purification of the trehalase GMTRE 1 from soybean nodules and cloning of its cDNA: GMTRE 1 is expressed at a low level in multiple tissues. Plant Physiol. 119:489-496.

Alexandrov NN, Brover VV, Freidin S, Troukhan ME, Tatarinova TV, Zhang H, Swaller TJ, Lu YP, Bouck J, Flavell RB, Feldman KA (2009). Insights into corn genes derived from large-scale cDNA sequencing. Plant Mol. Biol. 69:179-194.

Altschul SF, Madden TL, Schäffer AA, Zhang J, Zhang Z, Miller W, Lipman DL, (1997). Gapped BLAST and PSI-BLAST: A new generation of protein database search programs. Nucleic Acids Res. 25:3389-3402.

Baena-González E, Rolland F, Thevelein JM, Sheen J (2007). A central integrator of transcription networks in plant stress and energy signalling. Nature 448:938-942.

Blazquez MA, Santos E, Flores CL, Martinez-Zapater JM, Salinas J, Gancedo C (1998). Isolation and molecular characterization of the *Arabidopsis TPS1* gene, encoding trehalose-6-phosphate synthase. Plant J. 13:685-689.

Bradford MM (1976). A rapid and sensitive method for the quantification of microgram quantities of protein utilizing the principle of protein-dye binding, Anal. Biochem. 72:248-254.

Brodmann A, Schuller A, Müller J, Aeschbacher RA, Wiemken A, Boller

T, Wingler A (2002). Introduction of trehalose in Arabidopsis plants infected with the trehalose-producing pathogen *Plasmodiphora brassicca*, Mol. Plant Microb. Interact. 15:693-700.

Carpenter JF (1993). Stabilization of proteins during freezing and dehydration: application of lessons from nature. Cryobiology 30:220-221.

Chomczynski P (1993). A reagent for the single-step simultaneous isolation of RNA, DNA and proteins from cell and tissue samples. BioTechniques 15:532-537.

Crowe JH, Crowe LM, Leslie SB, Fisk E (1993). Mechanisms of stabilization of dry biomolecules in anhydrobiotic organisms. In: Close TJ, Bray EA (eds.), Plant Responses to Cellular Dehydration During Environmental Stress. Am. Soc. Plant Physiol. Riverside, CA. pp. 11-20.

Cumino A, Curatti L, Giarrocco L, Salerno GL (2002). Sucrose metabolism: Anabaena sucrose-phosphate synthase and sucrosephosphate phosphatase define minimal functional domains shuffled during evolution. FEBS Lett. 517:19-23.

Diab AA, Ageez A, Abdelgawad BA, Salem TZ (2007). Cloning, sequence analysis and in silico mapping of a drought-inducible gene coding for s-adenosylmethionine decarboxylase from durum wheat. World App. Sci. J. 2(4):333-340.

Eastmond PJ, Graham IA (2003). Trehalose metabolism: a regulatory role for trehalose-6-phosphate? Curr. Opin. Plant Biol. 6:231-235.

Eastmond PJ, Van Dijken AJ, Spielman M, Kerr A, Tissier AF, Dickinson HG, Jones JD, Smeekens SC, Graham IA (2002). Trehalose- 6-phosphate synthase 1, which catalyses the first step in trehalose synthesis, is essential for *Arabidopsis* embryo maturation. Plant J. 29:225-235.

El-Bashiti T, Hamamcı H, Oktem HA, Yucel M (2005). Biochemical analysis of trehalose and its metabolizing enzymes in wheat under abiotic stress conditions. Plant Sci. 169:47-54.

Elbein ADYT, PanYT, Pastuszak I, Carroll D (2003). New insights on trehalose: a multifunctional molecule. Glycobiol. 13:17R-27R.

FAO, WFP (2010).. The State of Food Insecurity in the World, Addressing food insecurity in protracted crises. Food and Agriculture Organization of the United Nations Rome.

Frohman MA (1994). On beyond classic RACE (rapid amplification of cDNA ends). Meth. Appl. 4:S40-58.

Gaber BP, Chandrasekhar I, Pattabiraman N (1986). The interaction of trehalose with the phospholipid bilayer:a molecular modeling study. In: Leopold AC (ed.), Membranes, metabolism and dry organisms. pp. 231-241.

Garg AK, Kim JK, Owens TG, Ranwala AP, Choi YD, Kochian LV, Wu RJ (2002). Trehalose accumulation in rice plants confers high tolerance levels to different abiotic stresses. Proc. Natl. Acad. Sci. USA 99:15898-15903.

Ge LF, Chao DY, Shi M, Zhu MZ, Gao JP, Lin HX (2008). Overexpression of the trehalose-6-phosphate phosphatase gene Os*TPP*1 confers stress tolerance in rice and results in the activation of stress responsive genes. Planta 228:191-201.

Goddijn O, Smeekens S (1998). Sensing trehalose biosynthesis in plants. Plant J. 14:143 146.

Gomez LD, Baud S, Gilday A, Li Y, Graham IA (2006). Delayed embryo development in the *Arabidopsis* Trehalose- 6-phosphate synthesis 1 mutant is associated with altered cell wall structure, decreased cell division and starch accumulation. Plant J. 46:69-84.

Heid CA, Stevens J, Livak KJ, Williams PM (1996). Real time quantitative PCR. Genome Res. 6(10):986-994.

Higo A, Katoh H, Ohmori K, Ikeuchi M, Ohmori M (2006). The role of a gene cluster for trehalose metabolism in dehydration tolerance of the filamentous cyanobacterium Anabaena sp. PCC 7120. Microbiol. 152(4):979-987.

Jang IC, Oh SJ, Seo JS, Choi WB, Song SI, Kim CH, Kim YS, Seo HS, Choi YD, Nahm BH, Kim JK (2003). Expression of a bifunctional fusion of the *Escherichia coli* genes for trehalose-6-phosphate synthase and trehalose-6-phosphate phosphatase in transgenic rice plants increases trehalose accumulation and a biotic stress tolerance without stunting growth. Plant Physiol. 131:516-524.

Katoh H, Asthana RK, Ohmori M (2004). Gene expression in the cyanobacterium Anabaena sp. PCC 7120 under desiccation. Microb. Ecol. 47:164-174.

Leyman B, Van Dijck P, Thevelein JM (2001). An unexpected plethora of trehalose biosynthesis genes in *Arabidopsis thaliana*. Trends Plant Sci. 6:510-513.

Liang P, Pardee AB (1995). Recent advances in differential display. Curr. Prot. Mol. Biol. 7:274-280.

Livak KJ, Schmittgen TD (2001). Analysis of relative gene expression data using real-time quantitative PCR and the 2(-Delta Delta C(T)) Method Meth. 25:402-408.

Martı́nez-Barajas E, Delatte T, Schluepmann H, Jong G, Nunes SC, Primavesi LF, Coello P, Mitchell RAC, Paul MJ (2011). Wheat Grain Development Is Characterized by Remarkable Trehalose 6-Phosphate Accumulation Pregrain Filling: Tissue distribution and relationship to SNF1-Related Protein Kinase1 Activity. Plant Physiol. 156(1):373-381.

Mellor RB (1992). Is trehalose a symbiotic determinant in symbioses between higher plants and microorganisms? Symbiosis 12:113-129.

Mitchell RAC, CastellsBrooke N, Taubert J, Verrier PJ, Leader DJ, Rawl ings CJ (2007). Wheat estimated transcript server (WhETS): a tool to provide best estimate of hexaploid wheat transcript sequence. Nucleic Acids Res. (Web Server issue) 35:W148-151.

Müller J, Aeschbacher RA, Wingler A, Boller T, Wiemken A (2001). Trehalose and trehalase in Arabidopsis. Plant Physiol. 125:1086-1093.

Müller J, Staehelin C, Mellor R, Boller T, Wiemken A (1992). Partial purification and characterization of trehalose from soybean nodules, J. Plant Physiol. 140:8-13.

Murphy D, Carter D (1990). Vasopressin gene expression in the rodent hypothalamus: transcriptional and posttranscriptional responses to physiological stimulation. Mol. Endocrinol. 4(7):1051-1059.

Ozturk NZ, Talame V, Michalowski CB, Gozukirmizi N, Tuberosa R and Bohnert HJ, (2002). Monitoring large-scale changed in transcript abundance in drought-and salty-stressed barley. Plant Mol. Biol. 48:551-573.

Paul M, Pellny T, Goddijn O (2001). Enhancing photosynthesis with sugar signals. Trends Plant Sci. 6(5):197-200.

Penna S (2003). Building stress tolerance through over-producing trehalose in transgenic plants. Trends Plant Sci. 8:355-357.

Potts M (1994). Desiccation tolerance of prokaryotes. Microbiol. Rev. 58:755-805.

Pramanik MH, Imai R (2005). Functional identification of a trehalose 6-phosphate phosphatase gene that is involved in transient induction of trehalose biosynthesis during chilling stress in rice. Plant Mol. Biol. 58:751-762.

Saitou N, Nei M (1987). The neighbor-joining method: a new method for reconstructing phylogenetic trees. Mol. Biol. Evol. 4:406-425.

Sambrook J, Fritisch EF, Maniatis T (1989). Molecularc loning: a laboratory manual, 2nd ed. Cold Spring Harbor Laboratory Press.

Sanger F, Nicklen S, Coulson AR (1977). Biochemistry DNA sequencing with chain-terminating inhibitors (DNA polymerase/nucleotide sequences/bacteriophage 4X174). Proc. Nati. Acad. Sci. 74(12):5463-5467.

Satoh-Nagasawa N, Nagasawa N, Malcomber S, Sakai H, Jackson D (2006). A trehalose metabolic enzyme controls inflorescence architecture in maize. Nature 441:227-230.

Schluepmann H, Pellny T, Van Dijken A, Smeekens S, Paul M (2003). Trehalose 6-phosphate is indispensable for carbohydrate utilization and growth in *Arabidopsis thaliana*. Proc. Natl. Acad. Sci. USA 100:6849-6854.

Schluepmann H, Van Dijken A, Aghdasi M, Wobbes B, Paul M, Smeekens S (2004). Trehalose mediated growth inhibition of *Arabidopsis* seedlings is due to trehalose-6-phosphate accumulation. Plant Physiol. 135:879-890.

Shima S, Matsui H, Tahara S, Imai R (2007). Biochemical characterization of rice trehalose-6-phosphate phosphatases supports distinctive functions of these plant enzymes. FEBS J. 274:1192-1201.

Shinozaki K, Yamaguchi-Shinozaki K (2007). Gene networks involved in drought stress response and tolerance. J. Exp. Bot. 58:221-227.

Strom AR, Kaasen I (1993). Trehalose metabolism in *Escherichia coli*: stress protection and stress regulation of gene expression. Mol. Microbiol. 8:205-210.

Talame V, Ozturk NZ, Bohnert HJ, Tuberosa R (2006). Barley transcript profiles under dehydration shock and drought stress treatments: a comparative analysis. J. Exp. Botany Integr. Approaches Sustain Improve Plant Prod. Drought Stress Special Issue. pp. 1-12.

Vandercammen A, Francois J, Hers HG (1989). Characterization of trehalose-6-phosphate synthase and trehalose-6-phosphate phosphatase of Saccharomyces cerevisiae. Eur. J. Biochem. 182:613-620.

Van Dijck P, Mascorro-Gallardo JO, De Bus M, Royackers K, Iturriaga G, Thevelein JM (2002). Truncation of Arabidopsis thaliana and Selaginella lepidophylla trehalose-6-phosphate synthase unlocks high catalytic activity and supports high trehalose levels on expression in yeast. Biochem. J. 366:63-71.

Vogel G, Aeschbacher RA, Müller J, Boller T, Wiemken A (1998). Trehalose-6-phosphate phosphatases from *Arabidopsis thaliana*: identification by functional complementation of the yeast *TPS2* mutant. Plant J. 13:673-683.

Wiemken A (1990). Trehalose in yeast, stress protectant rather than reserve carbohydrate. *Antonie* Van Leeuwenhoek 58:209-217.

Wingler A (2001). Trehalose metabolism in *Arabidopsis*: occurrence of trehalose and molecular cloning and characterization of trehalose-6-phosphate synthase homologues. J. Exp. Bot. 52:1817-1826.

Wolf A, Krämer R, Morbach S (2003). Three pathways for trehalose metabolism in *Corynebacterium glutamicum* ATCC 13032 and their significance in response to osmotic stress. Mol. Microbiol. 49:1119-1134.

Xue GP, McIntyre LC, Glassop D, Shorter R (2008). Use of expression analysis to dissect alterations in carbohydrate metabolism in wheat leaves during drought stres. Plant Mol. Biol. 67:197-214.

17

Lung function decline: Screening of alpha-1 antitrypsin gene in a population exposed to coal dust

M. Bhattacharjee[1], B. G. Unni[1]*, S. Das[1], M. Deka[2] and P. G. Rao[1]

[1]Biotechnology Division, CSIR-North-East Institute of Science and Technology, Jorhat-785006, Assam, India.
[2]Department of Biological Sciences, Institiute of Science and Technology, Gauhati University, Guwahati, 781014 Assam, India.

Alpha-1 antitrypsin (AAT) deficiency is an inherited disorder that causes low levels of, or no alpha-1 antitrypsin in the blood. Most commonly, it is associated with chronic obstructive pulmonary disease (COPD). COPD includes chronic bronchitis and emphysema chronic bronchitis - inflammation of the lining of the bronchial tubes emphysema - permanent destruction of the alveoli. Mutations in the PI gene, located on chromosome 14, are associated with this genetic disorder. The Z protein is due to a single amino acid substitution of 342 glutamine to lysine .Chronic respiratory diseases have a pre-eminent role in the health conditions of people residing near coalmine areas with implications for morbidity and excess mortality from specific causes. We screened 412 individuals (COPDs and Non-COPDs) for carriers of deficient ZZ allele of AAT gene at the coal mine site, Assam. DNA extraction was done by standard phenol chloroform method and amplification for Alpha-1-antitrypsin gene was done by site directed mutagenesis PCR method. Coal dust exposure was a potential factor in development of COPD. AAT deficiency was not found to be present in our study population.

Key words: COPD, A1AT gene, Coal Dust, ZZ type, air pollution.

INTRODUCTION

A widely accepted definition from Global Initiative for Obstructive Lung Disease (GOLD) defines Chronic Obstructive Pulmonary disease (COPD) as "a disease state characterized by airflow limitation that is not fully reversible. The airflow limitation is usually both progressive and associated with an abnormal inflammatory response of the lungs to noxious particles or gases" (Pauwels et al., 2001). One of the most significant breakthroughs in the field of COPD in the past

30 years was the discovery of a close association between an inherited deficiency of a protein in the blood called the α1- antitrypsin (AAT). It is the only known genetic disorder that leads to COPD. AAT deficiency accounts for less than 1% of COPD in USA (ATS, 1995). Alpha -1- Antitrypsin deficiency is one of the mostcommon serious hereditary disorders in the world, because it effects all major racial subgroups, and there is an estimated 120.5 million carriers and deficient subjects in the 58 countries surveyed worldwide (de Seres, 2002). Alpha -1-Antitrypsin (AAT) is a 52 kDa alpha-1-glycoprotein, composed of 394 amino acid residues and three asparagines-linked complex carbohydrate side chains (Brantly et al,, 1988). It is produced mainly by hepatocytes and secreted into the blood, where it acts as a circulating serine protease inhibitor whose principal substrate is neutrophil elastase (NE) (Pierce, 1988). The AAT gene is located on the long arm of chromosome 14, has been mapped to chromosome 14q31 - 32.3 (Byth et al, 1994). The normal gene is designated PiM and about 100 normal and defective genetic variant is recognizable

---

*Corresponding author. E-mail: bgunni@rrljorhat.res.in, bgunni@yahoo.com.

List of Abbreviations: COPD: Chronic Obstructive Pulmonary Disease; **A1AT/AAT:** Alpha 1 Antitrypsin; **GOLD:** Global Initiative for Obstructive Lung Disease; **DALY(s):** Disability Adjusted Life Year(s); **GBDS:** Global Burden of Disease Study; **ATS:** American Thoracic Society; **API:** Antiprotease Inhibitor; **OR**: Odds Ratio; **FEV1/FVC:** Forced Expiratory Volume in one Second/Forced Vital Capacity.

are recognizable by isoelectric focusing (IEF) (WHO, 1996). Intermediate and severe alpha -1-antitrypsin deficiency is almost entirely caused by the Z and S alleles as opposed to the wild-type M allele in the alpha - 1-antitrypsin gene: individuals with six different genotypes, ZZ, SZ, MZ, SS, MS, and MM, have relative plasma alpha -1-antitrypsin concentrations of almost 16, 51, 83, 93, 97, and 100% respectively (Brantly et al., 1991). The families of normal α_1-antitrypsin alleles is referred to as M (M1, M2, and M3) and are found in approximately 90% of the population. The most deficiency allele associated with emphysema is the Z allele. The specific mutations responsible for many forms of α_1-antitrypsin deficiency have been identified. The abnormal Z allele is associated with replacement glutamic acid by lysine at position 342 as a result of a single base mutation from GAG to AAG (Nukiwa et al.,1986).This substitution results in alteration of the three dimensional configuration of the molecule.

Chronic Obstructive Pulmonary Disease (COPD) is one of the leading causes of mortality. By 2020, it is expected to rise to the third position as a cause of death and at fifth position as the cause of disability adjusted life years (DALYs) as per projections made in Global Burden of Disease study (GBDS) (Jindal, 2006). Chronic irritation of the airways by inhaled substances such as cigarette smoke is the major known risk factor for COPD. However it has recently been estimated that 15 to 19% of COPD in smokers may be attributed to occupational exposures (Balmes et al., 2003; Hnizdo, 2002). Very less information on the disease is available from India. Screening of AAT deficiency in every community is essential for enhancing awareness of this disorder among health-care givers and the general public (WHO 1996, Carrell and Lomas, 1982) and also for planning health policy and financial medical resources and to their utilization by the scientific community, governments and the pharmaceutical industry. The diagnosis of this deficiency is relatively simple, but many population studies have indicated that AAT deficiency is under diagnosed and delay in its diagnosis is frequent (Mc Elvaney et al., 1997). The importance of dust exposure specifically, mineral dusts in the underground miners for the development of respiratory symptoms, airflow obstruction, and COPD has been well established (Becklake, 1989; Oxman et al., 1993).The present study was undertaken to screen the A1AT gene encoding the Antiprotease Inhibitor (API) protein, alpha 1 antitrypsin amongst the COPDs or non-COPDs indentified at open-cast coal mine site, Ledo.

MATERIALS AND METHODS

Study design

Prior to survey in the study site, we had conducted air analysis for parameters such as Respirable Suspended Particulate Matter

(RSPM), SO_2 and NO_2 and we established that all these parameters were very high throughout the year 2009 and 2010 (Bhattacharjee, 2011) thus, indicating that the people living near the coal mine had a cumulative exposure to coal dust, SO_2 and NO_2. The individuals categorized as COPDs and non-COPDs had participated in an earlier study (Bhattacharjee, 2011) conducted by us in the vicinity of the open-cast coal mine area at Ledo, near Tirap in Assam (Latitude 27°13' to 27°23'N and Longitude 95°35' to 96'00E); with the help of doctors at the local Primary Health Centre during the period of January 2009 to December 2010. Our statistical population shows that these people have been living there for 11 to 35 years within this 1.5 km of the mine area. We drew 412 people randomly as the sample for our study irrespective of age, sex, and livelihood. The status of lung function was confirmed through pre and post –bronchodilator spirometry as described earlier (Bhattacharjee, 2011). Blood samples (3 ml) were collected with informed consent from each subject by standard venipuncture method and stored at 4ºC until DNA extraction. The study was carried out at CSIR (NEIST), Jorhat after ethical clearance from Institutional Ethics Committee NEIST, Jorhat. The authors declare that they have no conflicts of interest.

DNA extraction and PCR amplification

Genomic DNA was extracted from whole blood using GeNei™ Whole Blood DNA extraction Kit, Bangalore Genei, India. PCR amplification for Alpha-1-antitrypsin gene was done by site directed mutagenesis PCR method as described by (Tazellar et al., 1992) with slight modifications using the primers thus highlighted:

Forward primer: ATAAGGCTGTGCTGACCATCGTC
Reverse primer: CTTTTCACCACTTAGGGTGGGTT

All amplifications were started in a 50 µl reaction volume containing 25 µM deoxynucleotide triphosphate (Bangalore GeNei, India), 2 mM MgCl₂ (Bangalore GeNei, India), 20 pM of each primer synthesized by Sigma Aldrich , USA), 200 ng DNA, 2.5 U Taq Polymerase (Sigma Aldrich , USA). An initial denaturation was carried out at 94ºC for 10 min, amplification was carried out for 30 cycles, each cycle consisted of 2 min denaturation time at 94ºC, 2 min annealing time at 52ºC and 3 min extension time, at 64ºC followed by a final extension of 10 min at 72ºC. The PCR reaction was done in a thermal cycler (Applied Biosystems, USA, model 2720).

Restriction enzyme digestion and electrophoresis

10 µl of PCR products were then digested with digested with Taq I (50U) restriction enzyme (Sigma Aldrich, USA) diluted with 1 × endonuclease buffer (Sigma Aldrich, USA) and made the volume up to 20 µl. This digestion mixture was incubated at 65ºC for 2 h according to manufacturer's instructions. Finally, the digested products were analyzed in a 3% agarose (Amresco Superfine Resolution Grade, USA) gel in 89 mM Tris-borate buffer containing 1 mM ethylenediamine tetraacetic acid, pH 8.3 at constant 120 volts for 1 h.

Sequencing of A1AT gene

A few PCR products of COPD (smoker and non to smoker) group and non-COPD (smoker and non - smoker) group were sequenced at Vimta labs, Hyderabad, India. The DNA sequencing samples were processed using ABI 3130 (4 capillary) or 3730XI (96 capillary) electrophoresis instruments. We have used EMBOSS

Table 1. Demographic and lung function observations in the subjects participating in our study at open cast coalmine site (Ledo, Assam).

Variable	Study Population (coal mine area)			
	COPD smokers (N = 92)	COPD non-smokers (N = 194)	Non-COPD non-smokers (N = 84)	Non-COPD smokers (N = 42)
[a]Male	85	140	73	38
[b]Female	7	54	11	4
Sex ratio (F/M)	0.08	0.38	0.15	0.10
[c]Smoking (pack years ± S.D)	15.91± 7.62			16.21 ± 7.31
[d]Coal dust exposure (years ± S.D)	30.40 ± 13.76	30.03 ± 13.16	15.17 ± 7.2	15.16 ± 3.83
[e]Mean age (± S.D)	40.5 ± 10.58	35.53 ± 13.14	34.17 ± 12.84	41.0 ± 11.99
[f]FEV1/FVC(% predicted) (± S.D)	53.15 ± 8.93	56.68 ± 9.56	112.75 ± 16.97	107.78 ± 15.11
[g]post bronchodilator FEV1 (liters)	0.87 ± 0.09	0.85 ± 0.13		

Demographic observations and status of lung function of study subjects at open-cast coal mine , Ledo, Assam. X^2 test was done for categorical variables and unpaired t- test for continuous variables (Bhattacharjee, 2011). [a] COPD Smokers vs. Non COPD Smokers: X^2 = 0.14, p = 0.70, OR = 0.29 (1.28 to 5.27); [a] COPD Non Smokers vs. Non COPD Non Smokers*: X^2 = 7.08, p = 0.007, OR = 0.39 (0.18 to 0.83); [b] COPD Smokers vs. Non COPD Smokers: X^2 = 0.14, p = 0.70, OR = 0.29 (1.28 to 5.27); [b] COPD Non Smokers vs. Non COPD Non Smokers* : X^2 = 7.08, p =0.007, OR = 0.39 (0.18 to 0.83); [c] COPD Smokers vs. Non COPD Smokers: t = 0.21, df = 132, p = 0.8; [d] COPD Smokers vs. Non COPD Smokers*: t = 7.03, df = 132, p = 0.009; [d] COPD Non Smokers vs. Non COPD Non Smokers*: t = 9.71, df =276, p = 0.02; [e] COPD Smokers vs. Non COPD Smokers: t = 0.28, df = 132, p = 0.77; [e] COPD Non Smokers vs. Non COPD Non Smokers: t = 0.796, df = 276, p = 0.42; [f] COPD Smokers vs. Non COPD Smokers*: t = 27.17, df = 132, p = 0.006; [f] COPD Non Smokers vs. Non COPD Non Smokers*: t = 34.53, df = 276, p = 0.003; *Statistically significant.

Needle (global) pairwise sequence alignment algorithm to find the mutation if any at the position Glu342 GAG ➔Lys AAG of the sequence.

Statistical analysis

Data were tabulated and classified as per the study variables mentioned in 'Material and Methods'. Chi-square test with Yates correction was applied to test significant difference in the number between smoker and non-smokers, males and females amongst symptomatic and asymptomatic subjects. Unpaired' test was applied to test for significant difference in continuous variables and X^2 test was used to test the significance in categorical variables.

RESULTS

Lung function, smoking status and sorting out of the subjects

The symptomatic smokers (n = 92) and non-smokers in the (n = 194) coal mine showed obstructive pattern of lung function in both males and females. In the coal mine site, amongst COPD non-smokers (n = 194) and Non-COPD non-smokers (n = 84), significantly, more number of symptomatic males [χ^2 = 7.08, p = 0.007, OR = 0.39 (0.18 to 0.83)] and symptomatic females [χ^2 = 7.08, p = 0.007, OR = 0.39 (0.18 to 0.83)] was recorded. Detailed observations on demographic variables are shown in Table 1 (Bhattacharjee, 2011).

The coal dust exposure years also significantly differed amongst COPD smokers versus non-COPD smokers (t = 7.03, df = 132, p = 0.009); and COPD non-smokers versus Non-COPD non smokers (t = 9.71, df = 276, p = 0.02).

Genotyping and sequencing

PCR amplification showed characteristic 179 bp band (Figure 1) indicating the presence of homozygous 'MM' type in all the samples. On restriction digestion, a band was observed at 157 bp in all the samples. As such, there was no ZZ mutation in these subjects in their A1AT gene. Since all the samples were homozygous 'MM' type, our data did not fit the 'Hardy-Weinberg equation'. Sequencing of the A1AT gene also agreed to the findings of PCR and Restriction enzyme analysis. There was no change in position of Glu342 GAG➔Lys AAG (highlighted in yellow) of the sequences indicating normal MM type in all categories of the individuals studied (Samples 1 to 4).

DISCUSSION

Amongst many risk factors of COPD, the genetic deficiency of A1AT attributed to ZZ type is the best documented reasons (Carp, 1978). Phenotype M is the normal variant phenotypes, S and Z are the two most frequent abnormal variants (Hutchinson, 1998). Calculated values of PiZZ prevalence are approximately: 1:1000 to 1:45,000 in Western and Northern Europe, 1:45,000 to 1:10,000 in Central Europe; and 1: 10,000 to 1:90,000 in Eastern Europe and in Southernmost and Northern areas of the continent. In the white population of USA, Canada, New Zealand, PIZZ phenotype prevalence ranges from 1: 2000 to1:7000 individuals (Andolfatto et al., 2003). In our population subset, we found that all the subjects were having the normal MM type (Figure 1) which was confirmed through site directed mutagenesis PCR and restriction digestion. Our investigation suggests

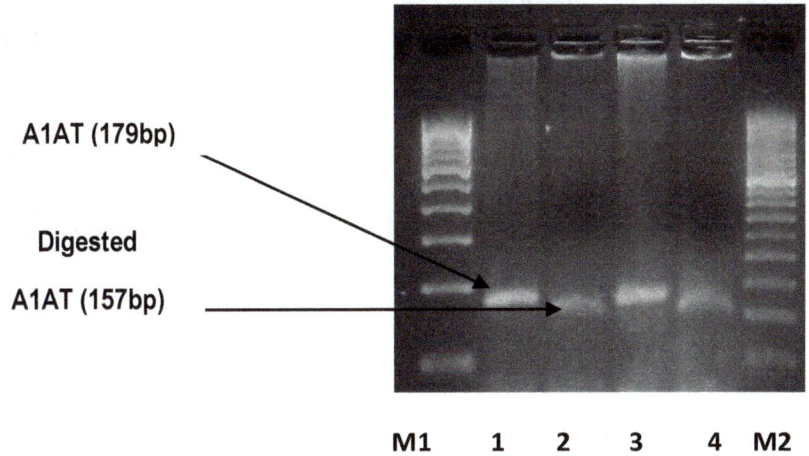

Figure 1. Detection of A1AT gene by site directed mutagenesis PCR method. Lanes 1, 2 A1AT gene179 bp normal (MM type). The primers used to amplify the sequence that included the Z mutation site yielded a product of the correct size (179 bp) in all cases. Lane 1, 2 (Non-COPD); lanes 3, 4 (COPD); M1 = 100bp ladder, M2 = 50bp ladder. Subsequently, PCR products were digested with *Taq* I enzyme. The normal fragment was 157 bp long. Lane 1, 2 (Non-COPD); lanes 3, 4 (COPD); M1 = 100bp ladder, M2 = 50 bp ladder.

```
TCCGCCATCAATATCGAGTTCTGGTCATCATTAAGAAGACAAAGGGTTTGTTGAACTTGACCTCGGG
GGGGATAGACATGGGTATGGCCTCTAAAAACATGGCCCCAGCAGCTTCAGTCCCTTTCTCGACGAGG
GTCAGCACAGCCTAAAAAA
```

```
EMBOSS_001      1 -------TGCATAAGGCTGTGCTGACCATC-----GACGAGAAAGGG---
35
                   |..|||..| .||.|||..|||    ||.||.||||||
EMBOSS_001      1 TCCGCCATCAATATCG-AGTTCTGGTCATCATTAAGAAGACAAAGGGTTT
49

EMBOSS_001     36 ACTGAAGCTG--CTGGGGCCATGTTTTTAGAGGCCATACCCATGTCTATC
83
                   ..|||||..|| ||.||| ||..|||..||||..|||.
EMBOSS_001     50 GTTGAACTTGACCTCGGG-------------GGGGATAGACATGGGTATG
86

EMBOSS_001     84 CCC-------------CCCGAGGTCAAGTTCAACAAACCCTTTGTCTTCT
120
                   .|| |||.|| ||..|| ||..||||||.||
EMBOSS_001     87 GCCTCTAAAAACATGGCCCCAG--CAGCTT---CAGTCCCTTTCTC----
127

EMBOSS_001    121 TAATGATTGAACAAAATACCAAGTCTCCCCTCTTCATGGGAAAAGTGGTG
170
                                                        ||..|| |||
EMBOSS_001    128 ---------------------------------------GACGAG-GGT-
136

EMBOSS_001    171 AATCCCACCCAAAAATAACTGCCTCTCGCTCCTCAACCCCTCCCCTCCAT
220
                   .|.|.||.||.|||||.|
EMBOSS_001    137 CAGCACAGCCTAAAAAA---------------------------------
153

EMBOSS_001    221 CCCTGGCCCCCTCCCTGGATGACATTAAAGAAGGGTTGAGCTGG      264

EMBOSS_001    154 -------------------------------------------      153


#-------------------------------------
#-------------------------------------
```

Sample 1. (COPD smoker).

GAGAAGGGGGGGACGCACCGCCCCTGTTTTTAGAGGCCATACCCATGTCTATCCCCCCCGAGGTCAAGTTCAACA
AACCCTTTGTCTTCTTAATGATTGACCAAAATACCAAGTCTCCCCTCTTCATGGGAAAAGTGGTGAATCCCACCCAA
ACATC

```
EMBOSS_001        27  GAGAA---AGGGACTGAAGCTGCTGGGGCCATGTTTTTAGAGGCCATACC
73

                      |||||   .|||||          ||...|.||.|||||||||||||||||

EMBOSS_001         1  GAGAAGGGGGGGAC-------GCACCGCCCCTGTTTTTAGAGGCCATACC
43

EMBOSS_001        74  CATGTCTATCCCCCCCGAGGTCAAGTTCAACAAACCCTTTGTCTTCTTAA
123

                      ||||||||||||||||||||||||||||||||||||||||||||||||||

EMBOSS_001        44  CATGTCTATCCCCCCCGAGGTCAAGTTCAACAAACCCTTTGTCTTCTTAA
93

EMBOSS_001       124  TGATTGAACAAAATACCAAGTCTCCCCTCTTCATGGGAAAAGTGGTGAAT
173

                      ||||||||.|||||||||||||||||||||||||||||||||||||||||

EMBOSS_001        94  TGATTGACCAAAATACCAAGTCTCCCCTCTTCATGGGAAAAGTGGTGAAT
143

EMBOSS_001       174  CCCACCCAAAAAT      186
                      |||||||||||.||
EMBOSS_001       144  CCCACCCAAACAT      156


#------------------------------------
#------------------------------------
```

Sample 2. (COPD non-smoker).

TGAGGTGAGTGAACCCCCCCAATGTTTTTAGAGGCCATACCCATGTCTATCCCCCCCGAGGTCAAGTTCAACAAAC
CCTTTGTCTTCTTAATGATTGAACAAAATACCAAGTCTCCCCTCTTCATGGGAAAAGTGGTGAATCCCACCCAAAAC
CCCGT

```
EMBOSS_001        27  GAGAAAGGGACTGAAGCTGCTGGGGCC-ATGTTTTTAGAGGCCATACCCA
75

                      |||    |.||.||||.|..|        ||  |||||||||||||||||||

EMBOSS_001         2  GAG---GTGAGTGAACCCCC-----CCAATGTTTTTAGAGGCCATACCCA
43

EMBOSS_001        76  TGTCTATCCCCCCCGAGGTCAAGTTCAACAAACCCTTTGTCTTCTTAATG
125

                      ||||||||||||||||||||||||||||||||||||||||||||||||||

EMBOSS_001        44  TGTCTATCCCCCCCGAGGTCAAGTTCAACAAACCCTTTGTCTTCTTAATG
93

EMBOSS_001       126  ATTGAACAAAATACCAAGTCTCCCCTCTTCATGGGAAAAGTGGTGAATCC
175

                      ||||||||||||||||||||||||||||||||||||||||||||||||||

EMBOSS_001        94  ATTGAACAAAATACCAAGTCTCCCCTCTTCATGGGAAAAGTGGTGAATCC
143

EMBOSS_001       176  CACCCAAAA      184
                      |||||||||
EMBOSS_001       144  CACCCAAAA      152


#------------------------------------
#------------------------------------
```

Sample 3. (Non-COPD smoker).

```
TGTTATCCCTCTTCGCCTCCTTATCCATCATTAAGAAGACAAAGGGTTTGTTGAACTTGACCTCGGGGGGGATAGA
CATGGGTATGGCCTCTAAAAACATGGCCCCAGCAGCTTCAGTCCCTTTCTCGACGATGGTCAGCACAGCCTTATAA
A


EMBOSS_001      13 TGCTGA-CCATC-----GACGAGAAAGGG---ACTGAAGCTG--CTGGGG
51

                   |.||.| |||||    ||.||.||||||  ..|||||..|| ||.|||
EMBOSS_001      18 TCCTTATCCATCATTAAGAAGACAAAGGGTTTGTTGAACTTGACCTCGGG
67


EMBOSS_001      52 CCATGTTTTTAGAGGCCATACCCATGTCTATCCCC------------CC
88

                        ||..|||..||||..|||..||           ||
EMBOSS_001      68 ------------GGGGATAGACATGGGTATGGCCTCTAAAAACATGGCC
104


EMBOSS_001      89 CGAGGTCAAGTTCAACAAACCCTTTGTC---TTCTTAATGAT--TGAACA
133

                   |.|| ||..|||| |||  ||||..|.||| |.|.||
EMBOSS_001     105 CCAG--CAGCTTCA----------GTCCCTTTCTCGACGATGGTCAGCA
141


EMBOSS_001     134 AA-----ATA     138
                    .|       |||
EMBOSS_001     142 CAGCCTTATA     151


#-------------------------------------
#-------------------------------------
```

Sample 4. (Non-COPD non-smoker).

that A1AT deficiency is not prevalent in our population subset. As already documented, the ZZ allele is rarely present in most of the populations; the reasons for occurrence of COPD in our study subset could be attributed to other factors. We already established earlier, that the study area was considerably polluted due to high RSPM, SO_2 and NO_2 and also that there was increase incidence of COPD cases when the coal dust exposure years was more (Bhattacharjee, 2011). It was important to also screen the AAT gene for ZZ carriers and to observe whether the population was genetically susceptible to COPD, as in most parts of the world, the disease is under diagnosed (McEvlaney, 1997). A recent report by the World Health Organization (WHO) and the recent guidelines of the American Thoracic Society (ATS)/European Respiratory Society (ERS) for the management of COPD, has recommended the detection programmes of AAT deficiency (Anonymous, 1997; ATS/ERS statement, 2003). Several longitudinal, epidemiological and associative studies have established that acute episodes of atmospheric pollution causes increased risk of adverse pulmonary events (Dockery et al., 1993; Laden et al., 2006). Significantly, more number of symptomatic that is, people with COPD were recorded than asymptomatic that is, people without COPD in our study site ($p < 0.01$).

There are about 1.00 billion tonnes coal reserves estimated in the coal bearing zone of North-East (NE) India, which is 0.5% of the country's total reserve of about 200 billion tonnes (Chaoji, 2002). Pressure to increase coal mining is likely to intensify because of concerns about energy and power generation and exploration of natural resources. Increased coal demand may exacerbate negative health effects of coal mining activities including occupational health hazards of open-cast coal mining and community exposure (Scott et al., 2004; Coggon and Taylor, 1998).

Hitherto, we can interpret that the lung function decline in the population is mostly due to the effect of respirable mixed coal dust as per our earlier findings (Bhattacharjee, 2011) and there is no deficiency of alpha 1 antitrypsin in the population.

Moreover, COPD is a complex polygenic disease (Mehrotra et al, 2010). It is also possible that certain other genes might play a role and these genes need to be studied to find out whether individual susceptibility due to genetic factors or external risk factors such as smoking, occupational exposure etc are responsible for the disease.

ACKNOWLEDGEMENT

The authors are thankful to CSIR, Government of India, New Delhi for financing the network project.

REFERENCES

American Thoracic Society (1995). Medical Section of The American Lung association. 1995. Standards for the Diagnosis and Care of Patients with Chronic Obstructive Pulmonary Disease. Am J. Crit. Care Med. 152: S77-S120.

American Thoracic Society (ATS)/European Respiratory Society (ERS) Statement (2003): Standards for the diagnosis and management of individuals with Alpha 1 Antitrypsin Deficiency.Am J. Respir. Crit. Care Med. 168: 818-900.

Anonymous (1997). Alpha 1 Antitrypsin Deficiency: memorandum from a WHO meeting. Bulletin of the World Health Organisation .75: 397:41.

Balmes J, Becklake M, Blanc P, Henneberger P, Kreiss K, Mapp C (2003). American Thoracic Society Statement: Occupational Contribution to the burden of airway disease. Am. J. Respir. Crit. Care Med. 167: 787-97.

Becklake MR (1989). Occupational Exposures: evidence for a casual association with chronic obstructive pulmonary disease Am. Rev. Respir. Dis., 140: S85-91.

Bhattacharjee M, Unni BG, Das S, Deka M, Wann SB (2011). IJES (In Press).

Brantly ML, Wittes JT, Vogelmier CF, Hubbard RC, Fells GA, Crystal RG (1991). Use of a highly purified Alpha -1-Antitrypsin standard to establish ranges for the common normal and deficient Alpha -1-Antitrypsin phenotypes. Chest. 100: 703-708.

Brantly MT, Nukiwa T, Crystal RG (1988). Molecular Basis of Alpha-1-Antitrypsin Deficiency. Am. J. Med. 84: 13-31.

Byth BC, Billingsley GD, Cox DW (1994).Physical and genetic mapping of the serpin gene cluster at 14q32.1: allelic association and unique haplotype associated with alpha-1-antitrypsin deficiency. Am. J. Hum. Genet., 55: 126-133.

Carp HJA (1978).Possible mechanisms of emphysema in smokers. In Vitro suppression of elastase-inhibitory capacity by fresh cigarette smoke & its prevention by antioxidants. Am. Rev. Respir. Dis. 118: 617-621.

Carrell RW, Lomas DA (1982). Alpha -1-Antitrypsin deficiency – a model for conformational diseases. N. Engl. J. Med. 346: 45-53.

Chaoji SV (2002).Environmental challenges and the future of Indian Coal. J. Mines, Metals Fuels. p.257.

Coggon D, Taylor AN (1998). Coal mining and Chronic Obstructive Pulmonary Disease: A review of the evidence. Thorax. 53: 398-407.

de Seres FJ (2002)Worldwide racial and ethnic distribution of alpha-1-antitrypsin deficiency: details of an analysis of published genetic epidiomol surveys. Chest .122: 1818-1829.

Dockery DW, Pope CA, Xu X, Spengler JD, Ware JH (1993).An Association between air Pollution and Mortality in six US Cities. N. Engl. J. Med. 329: 1753-1759.

Hnizdo E, Sullivan PA, Bang KM, Wagner G (2002). Association between Chronic Obstructive Pulmonary disease and employment by industry and occupation in the US population: a study of data from the Third National Health and Nutrition Examination survey. Am. J. Epidemiol. 156: 738-746.

Hutchinson DC (1998).Alpha 1 Antitrypsin Deficiency in Europe: Geographical distribution of Pi types S and Z. Respir. Med. 92: 367-377.

Jindal KS (2006). Emergence of Chronic Obstructive Pulmonary Disease as an epidemic in India. Ind. J. Med. Res. 124: 619-630.

Laden F, Schwartz J, Speizer FE, Dockery DW (2006). Reduction in Fine Particulate Air Pollution and Mortality: extended follow-up of the Harvard Six Cities Study. Am. J. Respir. Crit. Care Med. 173: 667-672.

Mc Elvaney NG, Stoller JK, Buist AS, Prakash UBS, Brantly M, Schluchter M, Crystal RG (1997). Baseline characteristics of enrollees in the National Heart, Lung and Blood Institute Registry of Alpha-1-Antitrypsin deficiency. Chest. 111: 394-403.

Mehrotra S, Sharma A, Kumar S, Kar P, Sardana S, Sharma J.K (2010). Polymorphism of glutathione S-transferase M1 and T1 gene loci in COPD. Int. J. Immunogenet. 37: 263-267.

Nukiwa T, Brantly ML, Ogushi F, Fells AG, Crystal RG (1988). Characterization of the Gene and Protein of the Common Al-Antitrypsin Normal M2 Allele. Am. J. Hum. Genet, 43: 322-330.

Oxman AD, Muir DC, Shannon HS, Stock SR (1993). Occupational Dust Exposure and Chronic Obstructive Pulmonary Disease.A systematic overview of the evidence. Am. Rev. Respir. Dis.148: 38-48.

Pauwels RA, Buist AS, Calverly PM (2001). Global Strategy for the diagnosis, management, and prevention of Chronic Obstructive Pulmonary Disease, NHLBI/WHO Global Initiative for Chronic Obstructive Lung Disease(GOLD)Workshop summary. Am. J. Respir. Crit. Care Med. 163: 1256-76.

Pierce JA (1988). Antitrypsin and Emphysema.Perspective and prospects. JAMA. 259:2890-2895.

Andolfatto S, Namour F, Garnier AL, Chabot F, Gueant JL, Aimone-Gastin I (2003). Genomic DNA extraction from small amounts of serum to be used for A1- Antitrypsin genotype analysis. Eur. Respir. J. 21: 215-219.

Scott DF, Grayson RL, Metz EA (2004). Disease and illness in U.S. mining, J. Occup. Environ. Med., 46: 1272-1277

Tazellar JP, Friedman KJ, Kline RS, Guthrie ML, Farber RA (1992). Detection of Alpha 1- Antitrypsin Z and S mutations by polymerase chain reaction Mediated Site- directed Mutagenesis. Clin. Chem., 38/8: 1486-1488.

WHO (1996). AAT Deficiency. Geneva: World Health Organization.

18

Genetic variability among *Coleus* sp. studied by RAPD banding pattern analysis

Govarthanan M.[1]*, Guruchandar A.[2], Arunapriya S.[3], Selvankumar T.[1] and Selvam K.[1]

[1]Department of Biotechnology, Mahendra Arts and Science College, kalippatti, Namakkal, Tamilnadu-637501, India.
[2]Department of Molecular Biology, School of Life sciences, University of Skovde, Sweden.
[3]Department of Biotechnology, Vysya Arts and Science College, Salem, Tamilnadu-636 103, India.

Genetic improvement of medicinal plants depend on the existence, nature and extent of the genetic variabiles available for manipulation. Genetic analysis with random amplification of polymorphic DNA markers has been extensively used to determine genetic diversity among *Coleus* sp. and to identify the best quality for human consumption and its medicinal purpose. The objectives of the present study were to assess molecular variation among *Coleus amboinicus*, *Coleus aromaticus* and *Coleus forskohlii* and to determine the level of genetic similarity among them. Analyses carried out include random amplification of polymorphic DNA (RAPD) on three strains of *Coleus* sp. and random primers were used for PCR. Electrophoresis on denaturing acrylamide gels improved RAPD reproducibility and increased the band number. In this study, the primer OPW 6 and OPW 7 gave reproducible results and the band profiles.

Key words: RAPD, genetic diversity, conservation, primers, genetic polymorphism.

INTRODUCTION

Medicinal plants play a vital role to preserve our health. The genus, *Coleus* consists of herbs, that are widespread in all over India and represents highly valuable plant species having therapeutic and nutraceutical importance. Genetic variation is essential for long term survival of species and it is a critical feature in conservation. For efficient conservation and management, the genetic composition of the species in different geographic locations needs to be assessed. In recent years, fingerprinting systems based on RAPD analysis have been increasingly utilized for detecting genetic polymorphism in several plant genera.

Coleus sp. is one of the important medicinal plants extensively used by traditional practiceoners in India for its medicinal value. *Coleus forskohlii* has been traditionally used to treat high blood pressure. Other benefits include help in losing weight, improving digestion and nutrient absorption, fighting cancer, and immune system support (Gilbert et al., 1999). RAPD analysis (Williams et al., 1990; Welsh and McClelland, 1990) is capable of detecting differences among strains of a single species. The simplicity and fast sample processing of RAPD technique makes it useful for assessing population genetic parameters such as within-population and between-population genetic diversity. An additional advantage is that knowledge of the DNA sequences is not necessary to apply this technique (for a review on this technique see Weising et al., 1995). PCR based RAPD markers have been widely used in assessing genetic variation with in a species by measuring genetic diversity in many species, including medicinal plants (Rosa et al., 2005).

Since variability is a prerequisite for selection programme, it is necessary to detect and document the amount of variation existing within and between populations. DNA marker based fingerprinting can distinguish species rapidly using small amounts of DNA and therefore can assist to deduce reliable information on their phylogenetic relationships. DNA markers are not typically influenced by environmental conditions and

---

*Corresponding author. E-mail: gova.muthu@gmail.com.

Table 1. Primers used in RAPD-PCR reactions.

S. No	Primer	Sequence (5'-3')
1	OPW 6	ACGCCCGATG
2	OPW 7	CTGGACGTCA
3	OPW 8	GACTGCCTCT
4	OPW 9	GTGACCGAGT
5	OPW 10	TCGCATCCCT
6	OPU 15	ACGGGCCAGT
7	OPU 16	CTGCGCTGGA
8	OPU 17	ACCTGGGGAG
9	OPU 18	GAGGTCCACA
10	OPU 19	GTCAGTGCGG

therefore can be used to describe patterns of genetic variation among plant populations and to identify duplicated accessions within germplasm collections (Mohamad et al., 2009).

Molecular markers have been shown to be useful for genetic variation of plant species. These markers based on the polymerase chain reaction (PCR) technique which is the most commonly used for these purposes. Several different PCR- based techniques have been developed during the last decade, each with specific advantages and disadvantages. The randomly amplified polymorphic DNA (RAPD) markers techniques is quick, easy and requires no prior sequence information; it detects nucleotide sequence polymorphisms using single primer of arbitrary nucleotide sequence (Williams et al., 1990). RAPD marker have been extensively used for DNA fingerprinting (Moreno et al., 1998) (Gilbert, 2001), (Gilbert et al., 1999) and (Gilbert et al., 2006), genetic diversity studies (Hoz et al., 1996).Population genetic studies (Wolfe et al., 1998; Wu, 2005).

Among the different types of molecular markers available, random amplified polymorphic DNA (RAPD) are useful for the assessment of genetic diversity because of their simplicity, speed and relatively low cost compared to other molecular markers (William et al., 1990; Rafalski and Tingey, 1993).in this we identify the genetic variability of *Coleus* sp. medicinal plants.

MATERIALS AND METHODS

Plant sample

Medicinal plants such as *Coleus amboinicus*, *Coleus aromticus* and *C. forskohlii* were collected from botanical garden of Kollihills, Namakkal, Tamilnadu (India).

DNA extraction

Two leaves of 2-week-old seedlings were ground in a mortar with liquid nitrogen. Then 1.5 ml of extraction buffer (Tris-HCl 100 mM, pH 7; EDTA 100 mM, pH 7; NaCl 3 M) and SDS at a final concentration of 1% were added. Each sample was shaken and 1.5 ml of equilibrated phenol was added. The phases were separated by centrifugation at 13,000 rpm for 30 min; 500 ml of supernatant was transferred to 1.5 ml eppendorfs and 2 volume of 100% ethanol and 50 ml of 3 M sodium acetate were added to the supernatant for DNA precipitation. The pellets were washed with 70% ethanol and re-suspended in 200 ml of Tris EDTA. Samples were treated with RNAse, and the DNA purity was estimated by measuring the OD (optical density) at 260 to 280 nm (Martins et al., 2006).

RAPD analysis

The random primers (OPW 6-10 and OPU 15-19) were purchased from chromous Dsiotech, Bangalore, India (Table 1). RAPD reactions were carried out in a 20 µl reaction volumes containing 1 µl of template DNA solution (about 50 mg), 0.25 µM primers, 100 µM dATP, dGTP, dCTP, dTTP, respectively, 2.5 mM Mg^{2+}, 1X polymerase buffer and 1.0 U Taq DNA polymerase (Chromous Biotech Bangalore). PCR amplification reaction was conducted with Masteral Thermal Cycler (Corbett Research Australia). The cycling parameters were 94°C, 5 min and 45 cycles to denature at 94°C, 30 s; anneal at 36°C, 1 min; extend at 72°C, 2 min. the amplification products were separated on the 1.2% agarose gel for 1 to 2 h at 80 V, and recorded with the Alpha digital imager gel documentation system (Alpha innotech, Australia) after staining with ethidium bromide (0.5 µg/ml). All tests were repeated twice.

Electrophoresis and silver stain

The RAPD-PCR reactions were electrophoresed in a 12% horizontal acrylamide gel (0.5 mm) refrigerated at 15°C (200 V/ 23 mA/5 W for 10 min followed by 600 V/30 mA/18 W for 2 h). Lambda Marker was used to obtain the band sizes. The bands were visualized by silver staining with a Silver Xpress kit (Bio-Rad), based on fixing with acetic acid, silvering with $AgNO_3$, and developing with Na_2CO_3-sodium thiosulphate. The gel was finally impregnated with acetic acid-glycerol and air-dried.

Quantitative analysis of DNA

The purity and quantity of the isolated DNA was determined by Nanodrop spectrophotometry. The extinction ratio (260/280 nm) was found between 1.4 to 1.8 that indicated the DNA was pure enough for RAPD analysis.

Table 2. Total number of bands.

S No.	Samples	Operon code					Total
		OPW 6	OPW 7	OPW 8	OPW 9	OPW 10	
1.	Coleus amboinicus	8	7	0	0	0	15
2.	Coleus aromaticus	0	0	0	0	0	0
3.	Coleus forskohlii	2	0	4	4	3	13
							28

S No.	Samples	Operon code				
		OPU 15	OPU 16	OPU 17	OPU 18	OPU 19
1.	Coleus amboinicus					
2.	Coleus aromaticus			NO BANDS		
3.	Coleus forskohlii					

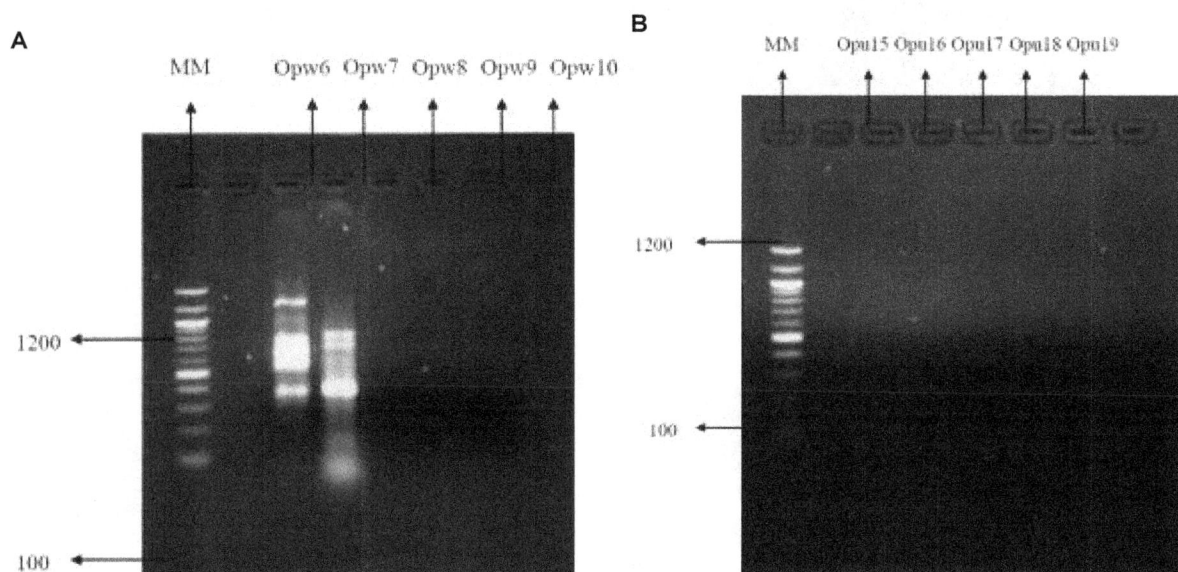

Figure 1. RAPD-PCR products of *Coleus amboinicus*.

RESULTS AND DISCUSSION

The RAPD technique was used to find out the extent of genetic diversity in *Coleus* sp. and primers OPW (6-10) and OPU were used for amplification. Maximum number of bands were observed in *C. amboinicus* with primer OPW 6 followed by OPW 7 (Table 2). A sum total of 28 bands were amplified with respect to all the 5 primers (OPW 6-10). By using OPW (6-10) total number of bands amplified were 28, where as there was no amplification in OPU (15-19).So that variation among *Coleus* sp. was identified from OPW but not in OPU primers. In *C. amboinicus* totally 15 bands were amplified, in *C. aromaticus* there was no amplification, *C. forskohlii* 13 bands were amplified.

The maximum numbers of bands were identified from *C. amboinicus* collected from botanical garden of Gandhi

Krushi Vignana Kendra (GKVK) and Biocentre in Bangalore, followed by the *C. forskohlii*. The bands obtained ranged in size from 100 - 1200 bp. In the present study genetic variability among the three *Coleus* sp. were determined by RAPD technique by using the random primers OPW (6-10) and OPU (15-19) series. The PCR products obtained were analyzed (Figures 1a, b, 2a, b, 3a, and b).

The purity and quantity of the isolated DNA was determined by Nanodrop spectrophotometry. The extinction ratio (260/280 nm) was found between 1.4 to 1.8 that indicated the DNA was pure enough for RAPD analysis (Figure 4a, b, c). Phylogenetic variation were determined in *Coleus* sp. by converting RAPD data into a frequency similarity and analysed by Unweighted Pair Group Method with Arithmetic mean (UPGMA) (Figure 5 and 6).

A

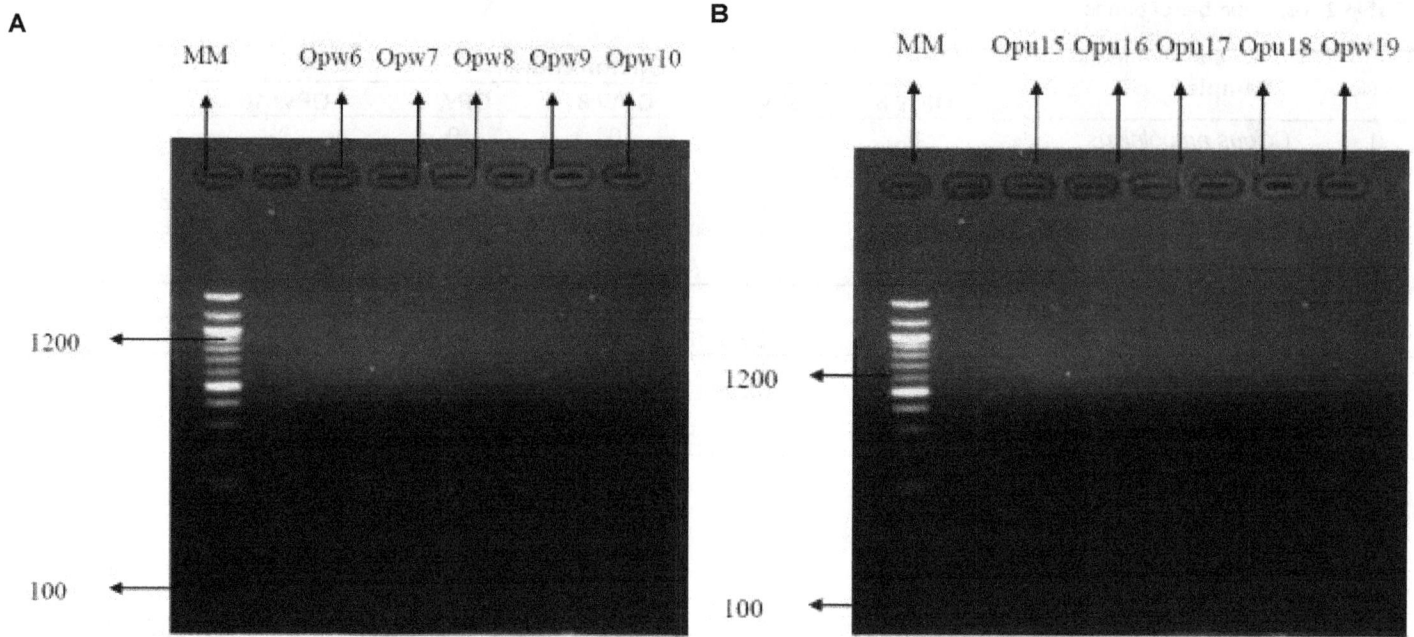

B

Figure 2. RAPD-PCR products of *Coleus aromaticus*.

A

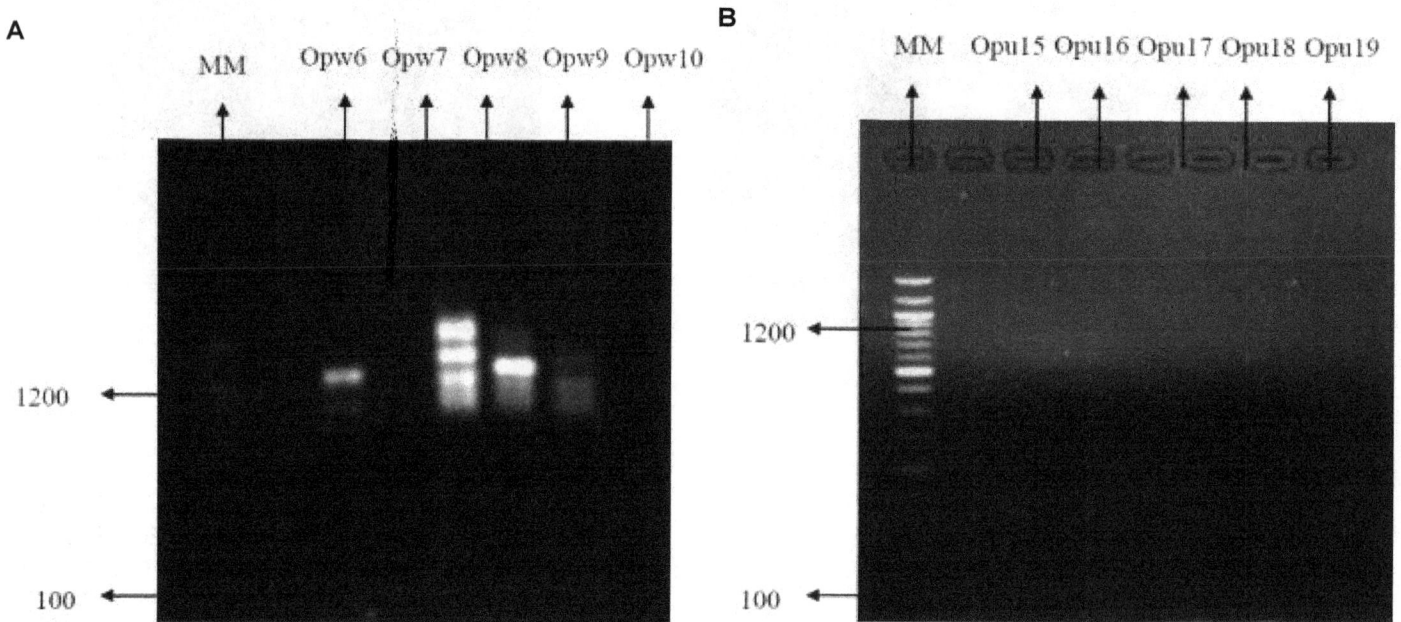

B

Figure 3. RAPD-PCR products of *Coleus forskohlii*.

RAPD can be considered to be essential tool in cultivar identification (DNA typing), assessment of genetic variability and relationships management of genetic resources and biodiversity, studies of phylogenetic relationships and in genome mapping (Hasibe et al., 2009). An attempt was made to analyse genetic variability among coleus species by Randomly Amplified Polymorphic DNA (RAPD) technique. DNAs of three

coleus species were isolated in Phenol chloroform method and tested by Qualitative and Quantitative methods in order to obtain good quality and DNA for analysis.

The result provided further evidence that RAPD technique offers reliable method for characterizing variations among species, within a species and among population and is in agreement with the findings of

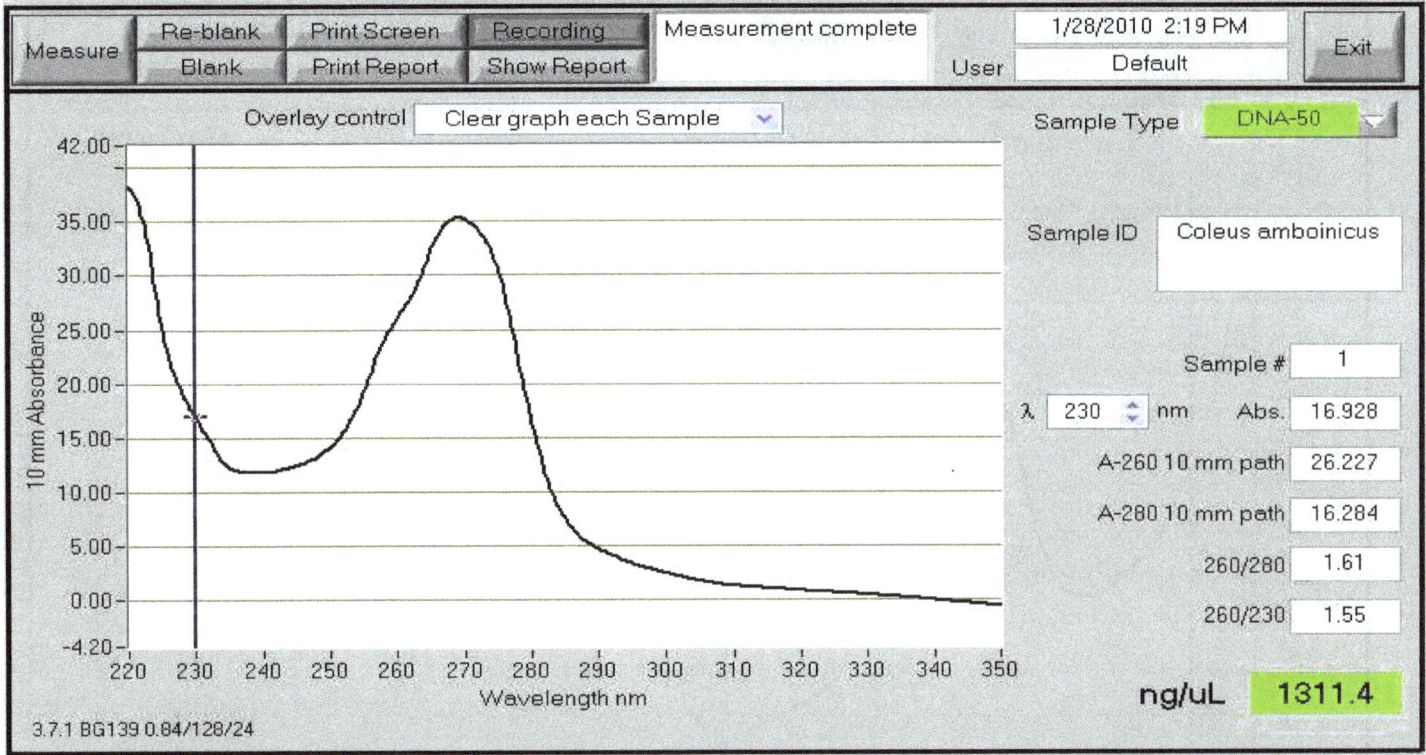

Figure 4a. Quantitative analysis of *Coleus amboinicus* DNA by Nanodrop spectrophotometry.

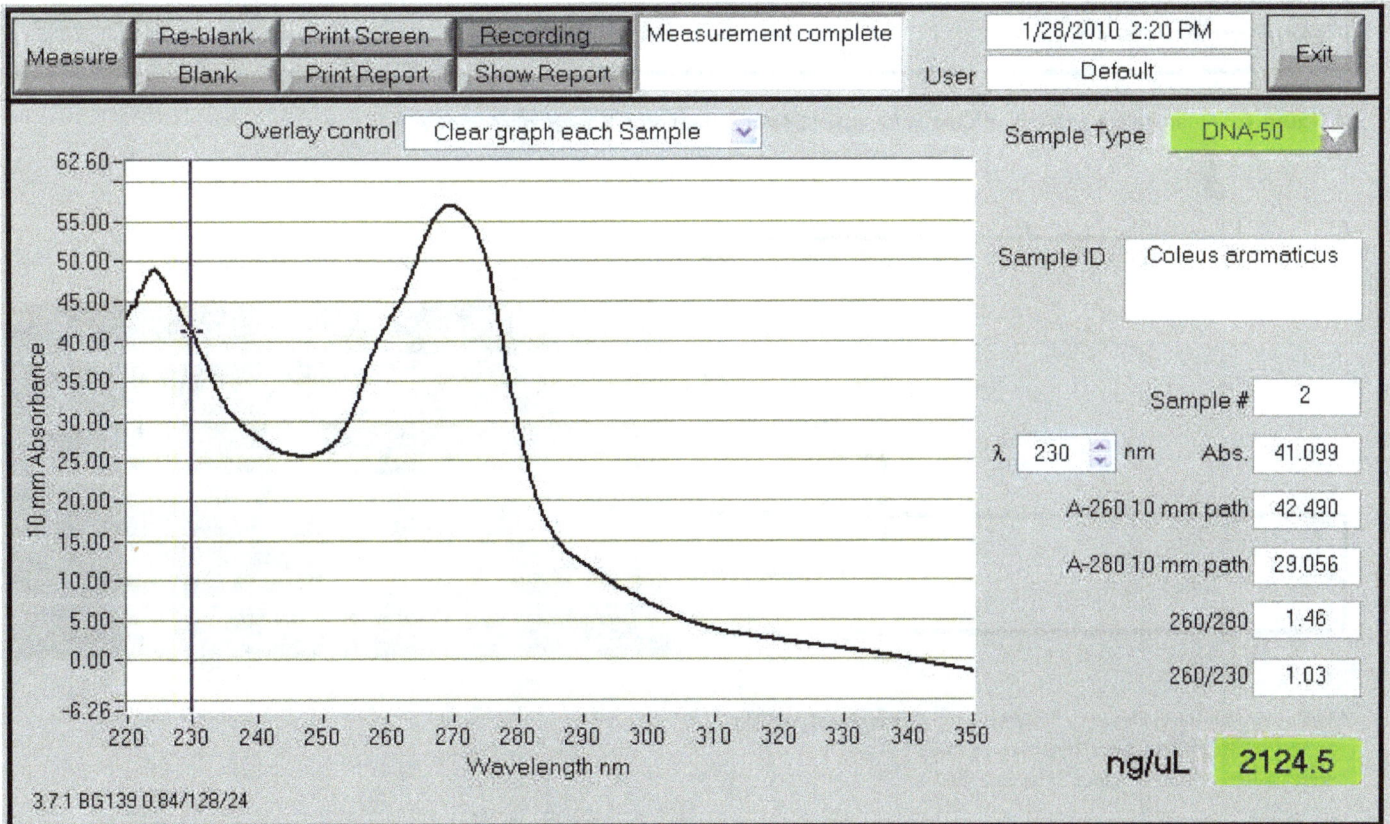

Figure 4b. Quantitative analysis of *Coleus aromaticus* DNA by Nanodrop spectrophotometry.

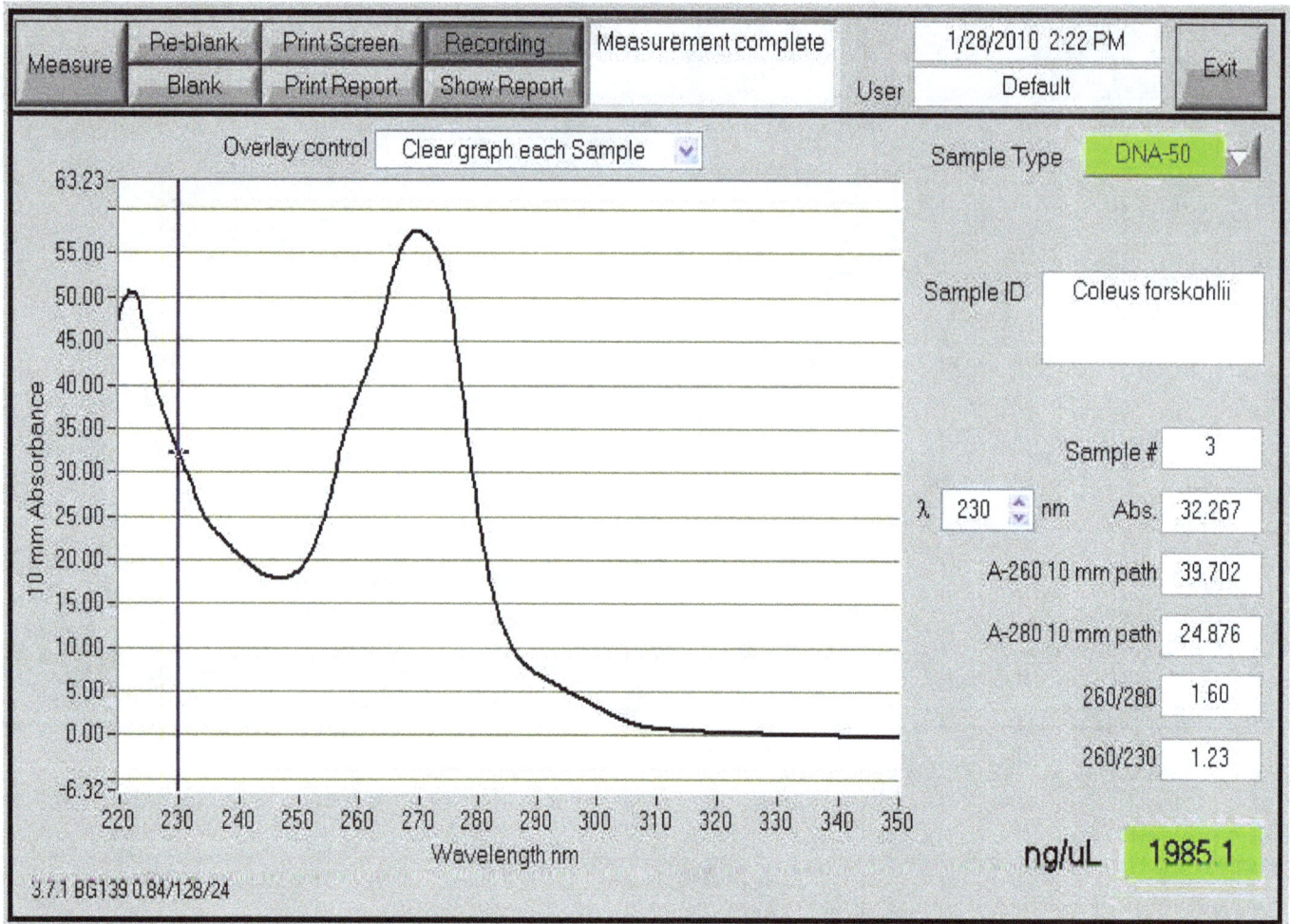

Figure 4c. Quantitative analysis of *Coleus forskohlii* DNA by Nanodrop Spectrophotometry.

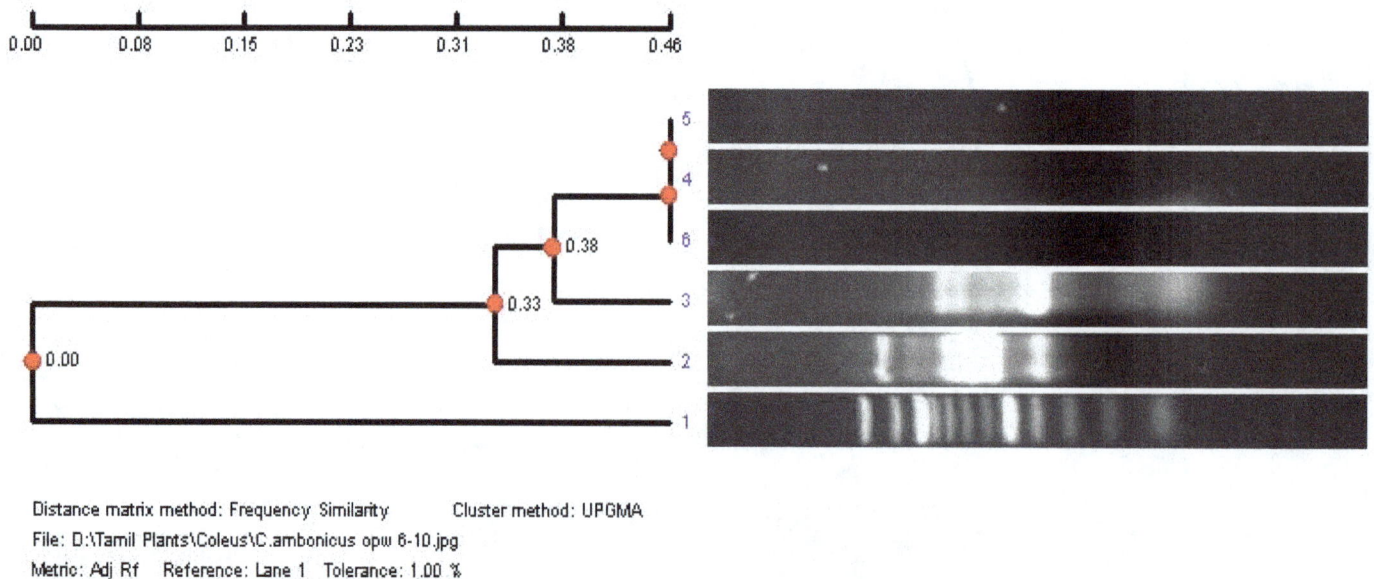

Distance matrix method: Frequency Similarity   Cluster method: UPGMA

File: D:\Tamil Plants\Coleus\C.ambonicus opw 6-10.jpg

Metric: Adj Rf   Reference: Lane 1   Tolerance: 1.00 %

Figure 5. Unweighted pair group method with arithmetic mean (UPGMA) analysis of *Coleus* sp.

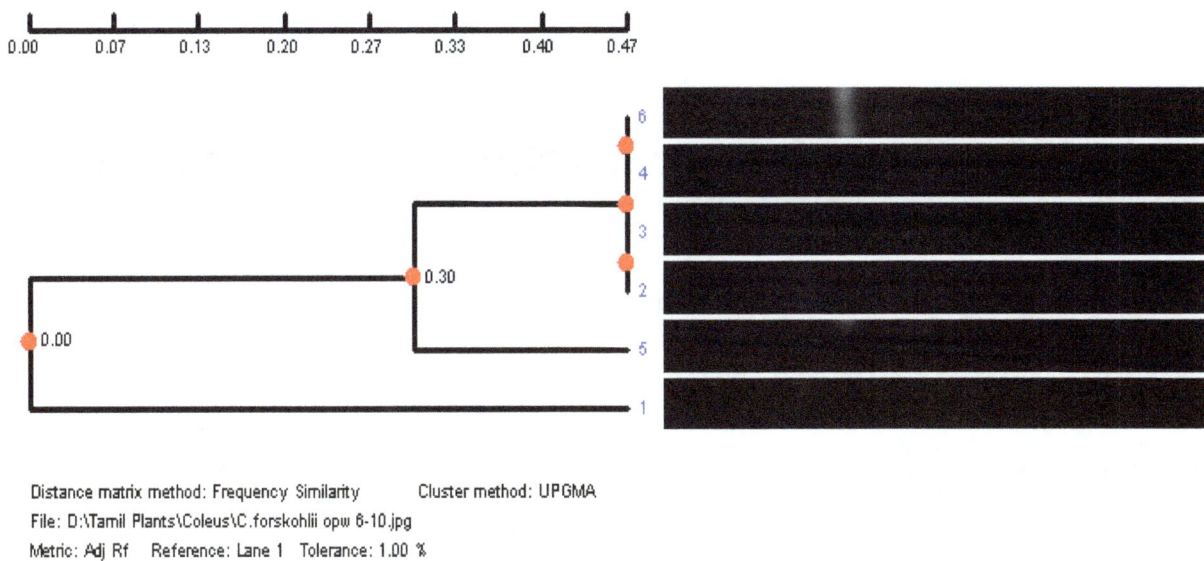

Distance matrix method: Frequency Similarity   Cluster method: UPGMA
File: D:\Tamil Plants\Coleus\C.forskohlii opw 6-10.jpg
Metric: Adj Rf   Reference: Lane 1   Tolerance: 1.00 %

Figure 6. Unweighted pair group method with arithmetic mean (UPGMA) analysis of *Coleus* sp.

Neelambra et al. (2009).

Ten random primers were used in this study these primers generated polymorphic amplification fragments that were clear and highly reproducible but showed different intensity after illustrated under UV.

The results of RAPD showed that we need to take individuals from more different populations so as to preserve their diversity for the future. The study also confirms the suitability of RAPD as a reliable, simple, easy to handle and elegant tool in molecular diagnosis of different accessions of an important medicinal plant. Concurrently, it is also proved that the entries which were found to be similar in taxonomical classification based on morphological characters.

REFERENCES

Gilbert J (2001). Comparison of Canadian *Fusarium graminearum* isolates for aggressiveness, vegetative compatibility, and production of ergosterol and mycotoxins. Mycopathology, 153: 209-215.

Gilbert JE, Lewis RV, Wilkinson MJ, Caligari PD (1999). Developing an appropriate strategy to assess genetic variability in plant collections. Theor. Appl. Genet., 98: 1125-1131.

Gilbert JE, Lewis RV, Wilkinson MJ, Caligari PD (2006). Heterogeneityof three molecular data partitionphylogenies of Mints related to M. Piperita.

Hasibe CV (2009).Genomic DNA isolation from aromatic and medicinal plants growing in Turkey. Sci. Res. Essay, 4(2): 59-64   .

Hoz SD, Davila JA, Loarce Y, Ferrer E (1996). Sample sequence repeat primers used in polymerase chain reaction amplification to study genetics diversity in barley. Genome, 39: 112-117.

Martins SR, Vences FJ, Saenz de Miera LE, Barroso MR, Carnide V (2006). A RAPD analysis of genetic diversity among and within Portuguese landraces of common white bean (Phaseolus vulgaris L.) J. Sci. Horticult., 108: 133-142.

Mohamad abdulla jubera BS, Janagoudar DP, Biradar RL, Ravikumar RVK, Patil SJ (2009). Genetic diversity analysis of elite Jatropha curcas (l.) genotypes using randomly amplified polymorphic DNA markers.Karnataka J. Agric. Sci., 22(2): 293-295.

Moreno S, Martin JP, Ortiz JM (1998). Inter- simple sequence repeat PCR for characterization of closely related grapevine germplasm. Euphylica, 101: 117-125.

Neelambra V, Koche V, Tiwari KL, Mishra SK (2009). RAPD analysis reveals genetic variation in different populations of *Trichodesma indicum*-A perennial medicinal herb Afri. J. Biotechnol., 8(18): 4333-4338.

Rafalski JA, Tingey SV (1993). Genetic diagnosis in plant breeding: RAPDs microsatellites and machines. Trends Genet. 9: 275–280.

Rosa Martinez, Carolina A, Sonsoles F (2005) Genetic variability among Alexandrium tamarense and Alexandrium minutum strains studied by RAPD banding pattern analysis. J. Harmful Algae, (5): 599-607.

Weising K, Nybom H, Wolff K, Meyer W (1995). DNA Fingerprinting in Plants and Fungi. CRC Press, Boca Rato.

Welsh J, McClelland M (1990). Fingerprinting genomes using PCR with arbitrary primers. Nucleic Acids Res., 18: 7213–7218.

Williams JGK, Kubelik AR, Livak KJ, Rafalski JA, Tingey SV (1990). DNA polymorphisms amplified by arbitrary primers are useful as genetic markers. Nucleic Acids Res., 18: 6531–6535.

Wolfe AD, Liston A (1998). Contributions of PCR-based methods to plant systematics and evolutionary biology. In D. E. Soltis, P. S. Soltis, and J.J. Doyle (Eds.), Molecular Systematics of Plants II, DNA Sequencing. Kluwer Aca.

Wu W, Zheng YL, Chen L, Wei YM, Yang RW, Yan ZH (2005)s. Evolution of genetic relationships in the genus Houttuynia thumb in china based on RAPD and ISSR markers. Biochem. Syst. Ecol., 33: 1141-1157.

19

Microbial bioinoculants and their role in plant growth and development

Sujata Dash and Nibha Gupta*

Microbiology laboratory, Regional Plant Resource Centre, Bhubaneswar -751015 Orissa, India.

The increase in the world's population, coupled with the limitations in the world's supply of natural resources and widespread degeneration of the environment presents a major challenge to agriculturalists. Chemical fertilizer is used to give the plant nutrient requirements within a short period to get faster results. Newly improved varieties of crops need high proportions of fertilizer. But chemical fertilizer has certain limitations and entails a lot of disadvantages. No doubt, the application of chemical fertilizer provides nutrition in high concentration in the soil and plants. When chemical fertilizer is applied, the entire contents would not be absorbed by the plants and the remaining parts would react in the soil. Part of it would be washed away and would contaminate water and some part of it would evaporate to the atmosphere; thereby the environment is polluted. Intense activity is involved in efforts to create plants that by themselves are able to fix nitrogen from the atmosphere – that is, convert nitrogen gas into nitrates that can be used by plant's metabolic machinery. At present, only certain plants called legumes are naturally able to do so and even legumes require the aid of symbiotic bacteria. Such a development would drastically curtail the amount of fertilizer required by agricultural crops (Tilak, 2001). The main factor is the price, biotechnology could provide an alternative to technologies that have harmful effects on the environment and it would have the potential of enhancing production on a sustainable basis (Al- Garni, 2006). Microorganisms are useful for biomineralization of bound soil and make nutrients available to their host and/or its surroundings. Nitrogen and phosphorus are major plant nutrients which occupy a key place in balanced use of fertilizer. Phosphorus is an important requirement of legumes for their nitrogen fixation process (Huda et al., 2007). All tropical legumes fix the atmospheric nitrogen by Rhizobium which requires optimum level of phosphorus in plant tissue. Their seedlings establish better in presence of mineral solubilizers because more of the tropical soils are phosphate fixing and make it unavailable to the plants (Dabas and Kaushik, 1998; Sahgal et al., 2004; Tilak et al., 2005; Hameeda et al., 2008; Gupta et al., 2007). It is due to the phosphate solubilizing organisms those solubilize the bound form of phosphorus and AM fungi acts as up-taker of phosphorus and make it available to the host plants. Microorganisms facilitate plant mineral nutrition by changing the amounts, concentrations and properties of minerals available to plants. These changes lead to change in growth, development and chemical composition of plant that are common and substantial enough to encourage the exploitation of plant microbe interaction for improvement of crop productivity. Possible approaches include both the introduction of foreign microorganisms and capitalization on the indigenous microflora. There are various groups of organisms that can be solubilize and/or leaching of phosphate, iron and other mineral metals. Since the chemical fertilizers are becoming important ingredient of the agricultural farming and production, need based technology should be given priority. As the production and manufacture cost of the chemical fertilizer are very high, its availability and uses are also becoming imperative. Biomineralizing phenomenon of the microbes is very important in this regard. Plant microbe interaction is an important phenomenon and also useful in the development of most suitable bioinoculant which may be able to improve the plant productivity under adverse condition. A large number of literatures are available regarding the microbial interaction and beneficial uses in plants of agriculture, horticulture and forestry. Keeping this in view, some important information regarding the biofertilizing potential of some important group of microbes and their application for the development of sustainable technology has been reviewed here.

Key words: Bioinoculants, phosphate solubiliser, nitrogen fixing bacteria, mycorrhiza, PGBR.

INTRODUCTION: TYPES OF BIOINOCULANTS AND THEIR ROLE

Phosphate solubilizing microorganisms

Phosphorus is added to soil in the term of phosphate fertilizer, part of which is utilized by the plant and the remainder converted into fixed and insoluble forms of phosphorus (Kumar et al., 2004; Afzal et al., 2005; Mehrvarz et al., 2008). There is nevertheless enough scope to use finely ground rock phosphate directly on the farm, especially in acidic soils. Since P availability from the phosphoric reserves; that is rock phosphate under neutral and alkaline conditions is scarce/negligible, the phosphate solubilizing microbes dissolving imprisoned/interlocked phosphates appear to have an important implication in Indian agriculture. A wide range of microorganisms from autotrophs to heterotrophs, diazotrophs to phototrophs, fungi and actinomycetes including mycorrhizas are known to have the ability to solubilize inorganic P from insoluble sources (Patil et al., 2002; Mehrvarz and Chaichi, 2008; Reis et al., 2008). From time to time the ability of different microorganisms to solubilize bound phosphates incorporated in agar soil or liquid media has been demonstrated. These reactions take place in rhizosphere and because solubilizing microorganisms dissolve more phosphates than they require for growth and metabolism, the surplus can be absorbed by plants. The mechanisms of conversion of the insoluble phosphorus by these organisms to available form include: altering the solubility of organic compounds to the ultimate soluble form through production of acids and H_2S, under aerobic and anaerobic conditions respectively, mineralizing organic compounds with release of inorganic phosphate (Gaind and Gaur, 1991; Bijaya et al., 2003; Dubey, 2000). Phosphate solubilizing fungi improve the growth performance of the plants (Kehri et al., 2002; Manoharachary et al., 2008). A study showed among five fungi such as Aspergillus niger strain 1, A. niger strain 2, A. niger strain 3, Aspergillus flavus and Aspergillus fumigates when tested for their efficiency in solubilizing phosphate (that is tricalcium phosphate). The fungal isolate A. niger strain 1 found to be efficient in phosphate solubilization (Mahalingam and Thilagavathy, 2008). Vazquez et al. (2000) concluded that most of the bacterial species produced more than one organic acid whereas A. niger produced only succinic acid. The production of organic acids by rhizosphere microorganisms is possibly involved in the solubilization of insoluble calcium phosphate. Mechanical plant compost from Calcutta and an alluvial soil (Inceptisol) from West Bengal, India were used for isolation of P-solubilizers. Fourteen each of P-solubilizing bacteria and fungi were isolated from both sources. All the bacterial isolates were spore-forming Bacilli. Penicillium and Aspergillus were the predominant fungal genera. P-solubilizers isolated from carbon rich compost exhibited higher solubilizing ability than soil isolates (Kole and Hajra, 1998).

Kumar et al. (2004) studied the influence of phosphate solubilizing microorganisms on Mussoorie rock phosphate (MRP) and 'aluminum phosphate' under two reactions (pH 5.1 and 6.2) categories of acid soils from upland and lowland areas soils were inoculated using suspension of fresh cultures of test isolates of bacteria and fungi, respectively. Fungi were superior to bacteria in releasing P from acidic soils. Ramesh et al. (1998) studied the influence of sources of phosphate, FYM and phosphate solubilizing microorganisms on yield and quality parameters of irrigated groundnuts was investigated at Hebbal, Bangalore. Pod, seed, oil and protein yields were highest when P was applied as 75% Mussoorie rock phosphate and 25% superphosphate, plus application of 9 tons farmyard manure/ha and seed inoculation with Aspergillus awamori. Narloch et al. (2002) studied the effect of phosphate-solubilizing fungi MSF-044, MSF-062 (Penicillium sp.) and MSF-087 (Aspergillus sp.) and soluble phosphate $[Ca(H_2PO_4)_2.2H_2O]$ at 0.0, 4.5, 9.0, 17.5 35.0, 70.0 and 140.0 mg; kg (-1) P on dry matter production and P uptake of radish in non-sterilized soil under green house conditions. Isolates differed in their capacity to stimulate dry matter production of plants depending upon the P level. Isolates MSF-044 and MSF-062 were more efficient under low P (4.5 to 17.5 $mg/kg^{-1}P$), while the isolate MSF-087 only stimulated dry matter production in the highest P level (140.0 $mg/kg^{-1}P$). Plants inoculated with the isolate MSF-062 at 17.5 $mg/kg^{-1}P$ presented no differences in dry matter compared to non-inoculated plants with 70.0 $mg/kg^{-1}P$. Phosphorus uptake by inoculated plants did not differ between treatments. Illmer and Schinner (1995) studied that two species Penicillium aurantiogriseum and Pseudomonas sp. (PI 18/89) with proven ability to solubilize inorganic phosphates were tested for their efficiency to dissolve calcium phosphates under non-sterile conditions. In laboratory experiments plant available phosphorus was only increased when phosphate solubilizing microorganisms disposed off sufficient nutrient supply. Field studies on forest soils in Australia indicated that phosphate-solubilizing microorganisms were too vulnerable and unreliable to be used in agriculture. Two solubilizing fungi identified were Aspergillus terreus and A. niger (Casanova et al., 2002). Reyes et al. (2006) studied the biodiversity of phosphate-solubilizing microorganisms (PSM) of rock phosphate mine in Tachira, Venezuela, a larger number of PSM were found in the rhizospheric than in the bulk soil.

*Corresponding author. E-mail: nguc2003@yahoo.co.in.

Six fungal strains belonging to the genus *Penicillium* and with high hydroxyapatite dissolution capacities were isolated from bulk soil of colonizer plants. Five of these strains had similar phenotypes to *Penicillium rugulosum* IR-94MF1 but they solubilized hydroxyapatite at different degrees with both nitrogen sources. From 15 strains of 'gram-negative bacteria' isolated from the rhizosphere of colonizer plants, 5 were identified as diazotrophic free-living encapsulated *Azotobacter* species able to use ammonium and/or nitrate to dissolve hydroxyapatite with glucose, sucrose and/or mannitol. Different nitrogen and carbohydrate sources are parameters to be considered to further characterize the diversity of PSM. Xiao et al. (2008) did the optimization for rock phosphate solubilization by phosphate-solubilizing fungi isolated from phosphate mines. The optimization for rock phosphate (RP) solubilization by phosphate-solubilizing fungi (*Candida krissii*, *Penicillium expansum* and *Mucor ramosissimus*) isolated from phosphate mines (Hubei, PR China) was investigated. The content of soluble phosphorus (P) released by these isolates was tested on the National Botanical Research Institute's phosphate growth medium (NBRIP) containing RP as sole P source. Results showed that the optimum conditions were at temperature of 32°C, shaking speed of 160 rpm, RP concentration of 2.5 g/l, EDTA concentration and 0.5 mg/ml. The content of soluble P gradually decreased with a larger particle size of RP with all the isolates. The content of soluble P was the highest when the initial pH for RP solubilization was 5.5 in the medium inoculated with *C. Krissii* which was different from that of 7.0 in the medium inoculated with P. The bacterial isolate *Micrococcus* spp. were found to be efficient in phosphate solubilization (Mahalingam and Thilagavathy, 2008). Phosphorus solubilization and immobilization data indicated that bacterial isolates solubilized more P than fungi, which in turn immobilized more phosphate (Alam et al., 2002). Wu et al. (2007) studied the biodegradability and mechanical properties of polycaprolactone composites encapsulating phosphate-solubilizing bacterium *Bacillus* sp PG01. The outcome of this work seems to suggest that by proper manipulation of composite compositions, the controlled release of the bacterial fertilizer (that is, *Bacillus* sp. PG01 cells) might be achievable. Mechanical plant compost from Calcutta and an alluvial soil (inceptisol) from West Bengal, India were used for isolation of P-solubilizers. Fourteen P-solubilizing bacteria were isolated from both sources. All the bacterial isolates were spore-forming *Bacilli* (Kole and Hajra, 1997). Line et al. (2006) used *Burkholderia cepacia* CC-AI74 with a high ability for solubilizing tricalcium phosphate (TCP) to study the P-solubilization mechanism. They collected filtrates able to solubilize TCP from the cultures of strain CC-AI74 and demonstrated that the P-solubilization increased 200 times during exponential growth when the pH decreased from 8 to 3. Wang et al. (2007) stated that 'inoculation'

with phosphate-solubilizing fungi diversifies the bacterial community in rhizospheres of maize and soybean. Application of phosphate-solubilizing microorganisms (PSMs) has been reported to increase P uptake and plant growth. However, no information is available regarding the ecological consequences of the inoculation with PSMs. Sharma and Prasad (2003) conducted a field experiment at Indian Agricultural Research Institute, New Delhi during 1996 to 1997 to 1998 to 1999 to study the effect of phosphate solubilizing bacteria (PSB) and incorporation of wheat and rice residue on the relative efficiency of diammonium phosphate (DAP) and Mussoorie rock phosphate (MRP) in three cycles of rice-wheat cropping system. The results of the present study; therefore indicate that low grade rock phosphate such as MRP can be advantageously utilized in rice-wheat cropping system when applied with PSB inoculation and incorporation of rice and wheat residues. Chen et al. (2006) studied the ability of a few soil microorganisms to convert insoluble forms of phosphorus to an accessible form is an important trait in plant growth-promoting bacteria for increasing plant yields. The use of phosphate solubilizing bacteria as inoculants increases the P uptake by plants. Mirik et al. (2008) used phosphate solubilizing bacteria three *Bacillus* species to control bacterial spot disease. Hameeda et al. (2008) studied the growth promotion of maize by phosphate-solubilizing bacteria isolated from composts and macrofauna. Five bacterial strains with phosphate-solubilizing ability and other plant growth promoting traits increased the plant biomass (20 to 40%) by paper towel method. Rivas et al. (2006) studied the biodiversity of populations of phosphate solubilizing *Rhizobia* that nodulates chickpea in different Spanish soils.

Within *Rhizobia*, two species nodulating chickpea, *Mesorhizobium ciceri* and *Mesorhizobium mediterraneum* are known as good phosphate solubilizers.

Nitrogen fixing bacteria

Nitrogen is a necessary component which is used for the growth of the plant. Plants need a limited amount of nitrogen for their growth. The type of crops also determines the level of nitrogen. Some crops need more nitrogen for their growth while some crops need fewer amounts. The type of soil also determines which type of biofertilizers is needed for this crop. Nonsymbiotic nitrogen fixation is known to be of great agronomic significance. The main limitation to nonsymbiotic nitrogen fixation is the availability of carbon and energy source for the energy intensive nitrogen process. This limitation can be compensated by moving closer to or inside the plants, namely: in diazotrophs present in rhizosphere, rhizoplane or those growing endophytically. *Azotobacter* not only help in nitrogen fixation but also capable of producing antibacterial and antifungal compounds, hormones and sidorophores (Tilak, 1993). Biological nitrogen fixation is estimated to contribute 180×10^6 metric tons/year globally of which 80% comes from symbiotic associations and the rest from free-living or associative systems. The ability to reduce and siphon out such appreciable amounts of nitrogen from the atmospheric reservoir and enrich the soil is confined to bacteria and archaea. These include symbiotic nitrogen fixing forms namely: *Rhizobium*, the obligate symbionts in leguminous plants and *Frankia* in non-leguminous trees and non-symbiotic (free-living, associative or endophytic) N_2- fixing forms such as *cyanobacteria, Azospirillum, Azotobacter, Acetobacter diazotrophicus, Azoarcus*, etc. (Tilak et al., 2005). The family Azotobacteriaceae comprises of two genera namely: *Azomonas* (non-cyst forming) with three species (*A. agilis, A. insignis* and *A. macrocytogenes*) and *Azotobacter* (cyst forming) comprising of 6 species, namely: *A. chroococcum, A. vinelandii, A. beijerinckii, A. nigricans, A. armeniacus* and *A. paspali*. *Azotobacter* is generally regarded as free-living aerobic nitrogen fixer. Yield improvement is attributed more to the ability of *Azotobacter* to produce plant growth promoting substances such as phytohormones IAA and siderophore azotoactin, rather than to diazotrophic activity (Tchan, 1984; Tchan and New, 1984). Members of the genus *Azospirillum* fix nitrogen under microaerophilic conditions, and are frequently associated with root and rhizosphere of large number of agriculturally important crop and cereals. Due to their frequent occurrence in the rhizosphere, these are known as associative diazotrophs (Okon, 1985; Tilak and Subba, 1987).

After establishing in the rhizosphere, *Azospirilla* usually, but not always, promote the growth of plants. Despite of their N_2 fixing capacity, the increase in yield is mainly attributed to improved root development due to the production of growth promoting substances and consequently increased rates of water and mineral uptake. *Azospirilla* can contribute towards long term goal of improving plant microbes interactions for salinity affected fields and crop productivity. *Azospirillum* plays a major role in osmoadaption through increase in osmotic stress (Basan and Holguin, 1997). Two groups of nitrogen-fixing bacteria, that is *Rhizobia* and *Frankia* have been found. *Frankia* forms root nodules on more than 280 species of woody plants from 8 different families (Schwintzer and Tjepkema, 1990). However, its symbiotic relationship is not understood. Species of Alnus and Casuarina are globally known to form effective symbiosis with *Frankia* (Wheeler and Miller, 1991; Huss-Danell, 1990; Werner, 1992; Dommergues and Marco-Bosco, 1998). 'Rhizobia' are defined as bacteria capable of forming root nodules on legumes, mediated by *nod* genes. The nitrogen-fixing symbiotic relationship has been exploited in agriculture to enhance crop and pasture growth without the addition of nitrogen fertilizers. For this reason, the majority of research in this field has focused on herbaceous crop and forage legumes of agricultural

significance (Jinturkar and kale, 2005). In contrast, few studies have been made of rhizobial associations among non-crop legumes, despite the fact that they may be ecologically important in the natural landscape (Boring et al., 1988). Worldwide, there are an estimated 17000 to 19000 legume species, although nodulating bacterial species have only been identified for a small proportion of these. *Rhizobia* are soil inhibiting bacteria with the potential for forming specific root structures called nodules. In effective nodules the bacteria fix nitrogen gas (N_2) from the atmosphere into ammonia (O'Gara and Shanmugam, 1976) which is assimilated by the plant and supports growth particularly on nutrient deficient soils. In nutrients [predominantly decarboxylic acids (Lodwig and Poole, 2003) and are protected inside the nodule structure (van Rhin and Vanderleyden, 1995)]. In infective nodules, no nitrogen is fixed, yet *Rhizobia* are still supplied with nutrient and in this situation the *Rhizobia* could be considered parasitic (Denison and Kiers, 2004).

The nitrogen fixing symbiotic relationship has been exploited in agriculture to enhance crop and pasture growth without the addition of nitrogen fertilizers. For this reason, the majority of research in this field and focused on herbaceous crop and forage legumes of agricultural significance. In contrast, few studies have been made of rhizobial associations among non-crop legumes, despite the fact that they may be ecologically important in the natural landscape (Boring et al., 1988). The first and foremost example of bioinoculant is the *Rhizobium* or legume inoculants which was first marketed in U.S.A during the early part of the century. In the present scenario legume – *Rhizobium* association has gained importance in the study of microbial inoculant. Biological nitrogen fixation (BNF) is an important attribute to symbiotic association of legume host with *Rhizobia*. To achieve maximum BNF out of any legume- *Rhizobium* association it is necessary to properly characterize and identify *Rhizobia* before they are made commercially available for field experiment (Sahgal and Johri, 2003). Symbiosis between leguminous plants and soil bacteria commonly referred to as *Rhizobia* is of considerable environmental and agricultural importance, since they are responsible for an estimated 180×10^6 tons per year of biological nitrogen fixation worldwide. Rapid industrialization is associated with land degradation. The available statistics reveals that the situation is alarming. Legume-*Rhizobium* associations have potential application in ecological restoration of such degraded land considering the potential if legume-*Rhizobium* symbiosis. Rhizobial inoculants have been used to improve plant and soil health for more than a century now. Inherent with the use of bioinoculants is the problem of variability in field performance and successful establishment of introduced strain on account of competition with the indigenous rhizobacterial population. Bioinoculant formulations are usually based on laboratory screening followed by appropriate trials in the field (Sahgal and Johri, 2003). *Rhizobia* live in a mutualistic symbiotic relationship with legumes- a relationship that has existed and coevolved for 10 millions of years (Sprent, 1994). The nodulation process includes a complex array of signaling molecules, molecular recognition, and regulation, legumes secrete secondary metabolites known as flavonoid into the soil, *Rhizobia* which are motile are attached to these flavonoids and attached to the rhizoplane. The flavonoids also induce the bacteria to secrete specific signal molecules known as Nod factors (Werner, 2004). The nod factors bind to a receptor in the root -hair cell, and cause root hair curling and eventual penetration of the acterium into the root-hair cell. Nod factors are critical molecules for nodule formation. After entering the root hair, bacteria travel down an infection thread a plant structure made specifically for this purpose (Gage and Margolin, 2000). The growing infection thread branches as it reaches the developing nodule primodium formed by dividing cortical cells. This growth is also initiated by Nod factor which reactivate the cell cycle (Patriarea et al., 2004). In most cases *Rhizobia* then differentiate morphologically to form bacteroids which are usually larger than the free living bacteria and have altered cell walls, bacteroids are released from the infection thread and form symbiosomes in nodule cells (Oke and long, 1999).

Bacteroids are the nitrogen fixing cells and are incapable of cell division and further reproduction (Perret et al., 2000). A compactable Nod factor is not the only requirement for effective nodulation. Bacterial cell surface components such as lipopolysaccharides (LPS) cyclic ß-glucans, exopolysaccharides (EPS), capsular proteins and K-antigens are also recognized by the plant and help determine host specificity (Spaink, 2002; Fraysse et al., 2003; Mathis et al., 2005). If these components are not recognized by the host, then the process is disturbed to various degrees for example, of infection threads fail to form non fixing empty nodules (Nod[+] Fix [+]) may result (Perret et al., 2000). The *Rhizobium* legume interactions have been reported to be very specific in nature. One of the major factors contributing for this specificity is the activity of plant root lectins. It has been speculated that the agrichemicals, accumulated due to extensive application to soil may protect the *Rhizobium* recognition sited on the root surface of legumes. As a result, the biological nitrogen fixation and consequently the yield of leguminous crops will be decreased due to reduced nodulation. Worldwide, there are an estimated 17000 to 19000 legume species (Martinez- Romero et al., 1991), although nodulating bacterial species have only been identified for a small production of these. Competition among the microorganisms determines the outcome of many biological events in nature and yet competitiveness is poorly understood. The *Rhizobium* legume symbiosis provides a good model system with which to study the molecular basis of bacterial competitiveness because

nodulation competitiveness is a ready quantifiable trait. About 170 million tons nitrogen is contributed annually through biological nitrogen fixation by many bacteria and some actinomycetes in association with some legumes and non-legumes. The latter techniques involves use of species *Rhizobia* that shows loose association with roots of non-legumes or enzymatic removal of cell wall of root cells that facilitates *Rhizobial* infection into the root cortex of treatment of seedling-root with phytohormones as 2, 4-D, NAA, IBA or cytokinins along with inoculation by *Azorhizobium cauinodans* or treatment with the signal compound (flavonoids) that are involved in nodule development and colonization by *Rhizobia*.

Rhizobia may invade the roots of non-legumes by specialized crack entry at the ruptured corners of the root that occur due to emerging lateral roots. The nitrogen fixation is attributed to presence of nod, nif and fixes genes that code for nitrogenase complex and other accessory proteins needed for proper functioning of the nitrogenase complex. A feeble nitrogen fixation was observed in paranodules of non-legumes since they lack the important O_2 scavenger and leghaemoglobin. Therefore, other possible sites of colonization as xylem vessels should be studied and possible techniques of transfer and expression of this gene in non- legumes should be perused. Though induced nodulation in non-legumes holds possibility of extending *Rhizobial* symbiosis and biological nitrogen fixation to cereals but this arena remains to be sheltered by numerous similar cascades of reactions present in legumes to be deciphered in non-legumes (Kalia and Gupta, 2002). Flavonoids released from plants play an important role as signal molecules in early stages of legume- *Rhizobium* symbiosis and in regulating the activities of soil microorganism. Flavonoids not only regulate transcription of *Rhizobial* nod genes, but also promote chemotaxis, growth, metabolism and symbiotic efficiency of *Rhizobia*; important effects on other soil microorganisms are evident form studies form how plants regulate *Rhizobia* and other soil microorganisms with natural plant products offer a basis of defining new concepts of rhizosphere ecology (Jain and Nainawatee, 2002). Rhizobial systematics has been determined from the beginning by association to the host plant. By 1980 the species names reflected those of their hosts (Skerman et al., 1980). Nevertheless, there were many strains that were unclassified, or indeed unclassifiable under this scheme most of these anomalies were included in the 'cowpea' *Rhizobia* group. This group eventually contained isolates from the majority of all nodulated legumes. This situation was widely considered unsatisfactory (Howieson and Brockwell, 2005).

The realization that transmissible genetic element-plasmid and symbiosis islands- could carry genes that conferred nodulation ability (Prakash et al., 1981; Fenton and Jarvis, 1994; Rao et al., 1994; Sullivar and Ronson, 1998) resolved one of the long standing problem of

Rhizobial systematics via strains with identical nodulation profiles could appeared to be different in biochemical and genetic tests and vise versa. *Rhizobium* classification is tentative and controversial since it is based on properties of isolates from less than 10% of known legumes. The genus is divided into 2 major groups (fast and slow growers) on morphological, physiological, symbiotic (cross- inoculation), and serological properties. All currently known *Rhizobia* are in the phylum proeobacteria, most in the class Alphaproteobacteria, which contains six Rhizobial families in a single order Rhizobiales. There are a number of species present in these rhizobial genera that have not been observed to form nodules, and therefore do not fit the functional definition of *Rhizobia*. These include many of the species that were formerly known as *Agrobacterium* for example *R. larrymoorei*, *R. rubi* and *R. vitis* (Young et al., 2001, 2004). However, there is recent evidence that other species formerly classified as *Agrobacterium* are capable of nodulation. For example *R. radiobacter* nodulates *Phaseolus vulgaris*, *Campylotropis* spp., *Cassia* spp. (Han et al., 2005) and *Wisteria sinensis* (Liu et al., 2005); both nodules and tumours were formed on *P. vulgaris* by *R. rhizogenes* strains containing a Sym plasmid (Velázquez et al., 2005). There are also other species, although classified within genera commonly considered to be represented entirely by nodulating strains, in fact include strains apparently devoid of nodulation ability. For example *Bradyrhizobium betae* forms tumours on *Beta vulgaris* (beetroot) but is not known to fix N_2 (Rivas et al., 2004). *Mesorhizobium thiogangeticum* is a sulfur-oxidising bacterium, and does not nodulate the tested legumes of *Clitoria ternatea*, *Pisum sativum* and *Cicer arietinum* (Ghosh and Roy, 2006). There are also non-symbiotic strains of *Mesorhizobium* (and other genera) that can become nodulating species by acquiring symbiosis genes.

The genus *Sinorhizobium* was recently reclassified to *Ensifer* on the basis of similarity of DNA sequences (Willems et al., 2003; Young et al., 2004). *Ensifer adhaerens* is a soil bacterium that attaches to other bacteria and may cause cell lysis (Casida, 1982). Although wild type *E. adhaerens* did not nodulate *P. vulgaris* or *Leucaena leucocephala*, it did so when transformed with a symbiotic plasmid from *Rhizobium tropici* (Rogel et al., 2001), demonstrating its capacity to become a rhizobial species. Other *E. adhaerens* strains were subsequently isolated those nodulated legumes naturally. These form a single clade with *Sinorhizobium* in 16S rRNA and *recA* phylogenies leading Willems et al. (2003) to suggest that these strains be reclassified as *Sinorhizobium adhaerens*. However, *Ensifer* (Casida, 1982) is the senior heterotypic synonym and thus takes priority (Young et al., 2001). This means that all *Sinorhizobium* spp. must be renamed as *Ensifer* spp. according to the Bacteriological code (Lapage et al., 1990). Symbiotic nitrogen fixing *Rhizobium* species are

able to specifically interact with leguminous plants. Both bacterial and plant genes are involved in the formation of root nodules on the host plants where atmospheric nitrogen is converted to ammonia (Dusha et al., 1987). In *Rhizobium melliloti*, genes coding for nitrogen fixation enzymes (nif genes) and for other functions of symbiotic nitrogen fixation (fix genes) are localized on a mega plasmid. Bacterial genes determining nodules has been found to be highly host specific for different *Rhizobium* species. Joshi et al. (2008) studied the effect of expression of *Bradyrhizobium japonicum* 61A152 fegA gene in *Mesorhizobium* sp., on its competitive survival and nodule occupancy on *Arachis hypogea*. Nodulation is affected by temperature. It was adversely affected at both 28 and 37°C and the effect was more pronounced during the first third of the nodule formation (Dudeja and Khurana, 1989). Kinselecton could be the main selective force maintaining nodulation and nitrogen fixation in *Rhizobium* bacteria (Olivieri and Frank, 1994). The effect of number of *Rhizobium* supplied on the subsequent numbers of nodules formed was tested. The final nodule numbers are related to the growth of the root system since the final number of nodules per gram of root is independent of the initial dose of inoculum, the effect of which was mainly to cause and increase in root growth (Bhaduri, 1951). The development of nodules involves a highly specific signal exchange between the two symbionts (Verma, 1992) cyclic adenosine 3'5'-monophosphate (cAMP) may have an important role in the sequence of biological events regulating nodule formation and functioning (Terakado et al., 1997).

Pauly et al. (2006) studied several reactive oxygen and nitrogen species are continuously produced in plants as by products of aerobic metabolism or in responds to stresses. This compactable interaction initiated by a molecular dialogue between the plant and bacterial partners, leads to the formation of a novel root organ capable of fixing atmospheric nitrogen under nitrogen limiting conditions. Saha and Haque (2005) studied that the nitrogen fixing bacteria (*Rhizobium*) isolates were subjected to cultural, morphological, biochemical and nodulation test for characterization. Graham (1969) developed a selective medium for selective isolation and growth of *Rhizobium*. Qadri and Mahmood (2003) studied the characterization of *Rhizobia* isolated form some tree legumes growing at Karachi university campus. *Rhizobia* isolates from nodules of *Dalbergia sissoo* were acid – producing and fast growers, utilizes 11 carbon sources and resistant against the antibiotics Amoxicillin, Ampicillin, Cloxacillin, Erythromycin, Neomycin, Sulphamethozazole, Trimethoprim and susceptible to Gentamycin and Tetracycline. The ability of a bacterial strain to form nodules on a particular species of host plant is thus not necessarily related to any particular set of serological characters amenable to the agglutination test (Kleczkowski and Thorton, 1944). Nehra et al. (2007) studied the characterization of heat resistant mutant strains of *Rhizobium* sp (Cajanus) for growth, survival and symbiotic properties. Ahalwat and Dadarwal (1996) studied the Bacteriocin production by *Rhizobium* sp. and its role in nodule competence. *Rhizobium* associated with the tree legumes: Sahgal et al. (2004) studied the selection of growth promotory *Rhizobia* for *D. sissoo* from diverse soil ecosystem. 35 strains were isolated among them, 3 nodulated in *L. leucocephala*. Ultra structure study of root nodules was studied by Qadri et al. (2007). The internal structure of a mature nodule showed an epidermis, cortex, vascular region and a bacteriod region. Vascular bundles were amphicribral. A distinct periderm consisted of scleroid tissue could be observed in the cortex outside the vascular tissue. The bacteriod region contained infected and uninfected (interstitial) cells intermingled with each other. Infected cells of developing nodules as well as of mature nodules were vacuolated. Infection threads were also observed in the bacteriod zone.

Maheshwari and Kumar (1997) studied that although, *Rhizobia* are abundant in soil, all of them are not able to nodulate various types of legumes because of their specificity for each legumes, effective inoculant is required to develop individual crop. *Rhizobial* inoculants (as living fertilizer) composed of microbial inoculants or groups of microorganisms, which are able to fix atmospheric nitrogen required for the development of sustainable system (agriculture/forestry) for uncultivable land areas. The response of particular legume to stress is not a property of the host plant alone, it demands consideration of the inoculant *Rhizobia* and the process of symbiosis. Attempts are made to isolate, characterize, identify the suitable indigenous *Rhizobia* for their use as bioinoculant preparations for certain fast growing fuel wood tree species. Saha and Haque (2005) isolated *Rhizobium* from the healthy nodules. Pure culture of isolated bacteria was subjected to cultural, morphological, biochemical and nodulation tests for characterization. In all cases of inoculation nodule number was higher on plants cultivated on sterile soil mixed with *Rhizobium* than on natural soil. The present finding indicates that the plant growth was related to the number of nodules on roots of plant. Rathi et al. (2009) studied the influence of *Rhizobium leguminosarum* inoculation along with sulphur and micronutrients. Because many soil lack *Rhizobia* or the density of the *Rhizobia* is too low for profitable legume cultivation, legume seeds should be inoculated with large amounts of a host plant specific microsymbiont on their surface just before being sown to ensure efficient legume growth.

Diouf et al. (2003) have done the optimization of inoculation of *L. leucocephala* and *Acacia mangium* with *Rhizobium* under greenhouse conditions. The physiological stage of the bacterial culture had no effect on nodulation and growth of the seedlings of *A. mangium* inoculated and cultivated *in vitro* for four months. For *L. leucocephala*, the number of nodules was significantly higher when the seedlings were inoculated with a bacterial culture in stationary phase. Grange and Hungria (2004) studied the genetic diversity of *Rhizobium* in indigenous common bean (*Phaseolus vulgaris*) in two Brazilian ecosystems and tested its nodulation in *Acacia leucocephala*. Twenty-five different profile combinations were obtained. *Rhizobium etli* was the predominant species, with 73 strains showing similar RFLP profiles; while 12 other strains differed only by the profile with one restriction enzyme. Fifty strains were submitted to sequencing of a 16S rDNA fragment and 34 clustered with *R. etli* including strains with RFLP-PCR profiles similar to those species or differing by one restriction enzyme. However, other strains differing by one or two enzymes were genetically distant from *R. etli* and two strains with identical profiles showed higher similarity to *Sinorhizobium fredii*. Other strains showed higher similarity of bases with *R. Tropici, R. Leguminosarum* and *Mesorhizobium plurifarium*, but some strains were quite dissimilar and may represent new species. Great variability was also verified among the sequenced strains in relation to the ability to grow in YMA at 40°C, in LB, to synthesize melanin *in vitro*, as well as in symbiotic performance including differences in relation to the described species, for example many *R. etli* strains were able to grow in LB and in YMA at 40°C, and not all *R. Tropici* were able to nodulate *Leucaena*. Rahangdale and Gupta (1997) studied the dual inoculation with nitrogen fixing *Rhizobium* and P- uptaker vesicular arbuscular mycorrhizal fungi association. They applied this inoculation to forest tree growing in Chhattisgarh was studied for their symbiotic association with forest trees at laboratory and nursery conditions. A very wide range of forest trees comprising leguminous and non leguminous were found to be associated with *Rhizobium* as well as VAM fungi.

The host specific strains of both organisms (12 strains of *Rhizobium* and 14 isolates of VAM were isolated, purified and maintained in laboratory and glasshouse conditions. These inoculants were evaluated for their efficiency in increasing growth, nodulation, % colonization, phosphate-content, protein and sugar level of 4 different host plants (*Albizzia lebbeck, D. sissoo, Albizzia procera* and *Acacia nilotica*) at individual and dual inoculation level.

Arbuscular mycorrhizal fungi

The arbuscular mycorrhiza (AM) fungi are important rhizospheric microorganisms. They can increase the plant uptake of nutrients especially relatively immobile elements such as P, Zn and Cu (Ryan and Angus, 2003), and consequently, they increase root and shoot biomass and improve plant growth. It has been indicated that AM fungi can colonize plant roots in metal contaminated soil, while their effects on metal uptake by plants are conflicting. They are ubiquitous in geographic distribution occurring with plant growing in all environmental conditions. VAM fungi occur over a broad ecological range from aquatic to desert environments. These fungi belongs to the genera *Glomus, Gigaspora, Aculospora, Scutellospora, entrophosphora, Modicella* and Sclerocystis are obligate symbionts and have not been cultured on nutrient media using standard microbiological techniques. They are multiplied in the roots of host plants and the inoculums is prepared using infected roots of host plants and the inoculums is prepared using infected roots and soil. Crop responses to VAM inoculation are governed by soil type, host variety, VAM strains, temperature, moisture, cropping practices and soil management practices. In general, field experiments with VAM have been the inability to produce clean pure inoculum on large scale. Field trails indicated that VAM inoculation increased yields at certain locations and the response varied with soil type, soil fertility particularly with available P status of soil and VAM culture (Tilak, 1993). In Fungal Biology in 21st century- 'roots' of nearly 90% plants form a symbiotic association with fungi called Mycorrhiza, (fungus root) contrary to popular belief, the luxuriance of rain forests is not because the rain forests soil is more fertile (as torrential rains over millennia leach out soluble minerals), but because the roots associate with fungi whose spreading hyphae increase the area of absorption of scarce nutrients and transport these to the plant in return for photosynthetically fixed carbon.

In the symbiotic interaction, the fungus enters the root cells to form specialized haustoria called arbuscules because they are highly branched tree like structure. Arbuscular mycorrhizal fungi also develop an extensive hyphal network external to the plant root which provides the physical link between the soil and root, drawing phosphorus and other minerals from the soil and translocating them to the root. The mechanisms that are responsible for the increased uptake from soil and transfer to host trough the interface need to be identified, a proteome analysis based on separation of protein by two dimensional electrophoresis and their identification by mass spectrometry has been initiated to identify proteins involved in mycorrhizal development and functioning (Mamatha and Bagyaraj, 2002; Mamatha et al., 2002; Manoharachary, 2002). Mahmood et al. (2004) found some observations on mycorrhizae of *L. leucocephala* (Lam.) De Wit. In young roots of *L. leucocephala* infection hyphae formed appresorium, sending branches in different directions to colonize the host cell. The mycorrhizal fungus was non-septate and branched. Fungal cells (pelotons) were observed in the root cortex. Characteristic arbuscules commonly formed by VAM fungus were visible. The main hyphae branched and the branched hyphae terminated in an arbuscule. The cortical cells of the root showed large number of vesicles which were either globular or elliptical in shape. AM fungi also act as biocontrol agent for plant pathogen

(Raman et al., 2002). An investigation was undertaken to test the hypothesis that amending peat to increase its P buffer capacity and optimizing the P concentration of the amended medium for mycorrhizal activity will enhance its usefulness for raising mycorrhizal seedlings of tree species. The approaches entailed mixing a small quantity of soil of high P adsorbing capacity with peat and constructing a P sorption isotherm for the medium in order to establish solution P concentration near-optimal to optimal for mycorrhization of seedlings. A P sorption isotherm based on incubating the medium with graded amounts of P at 50% of available water-holding capacity was developed. Target solution P concentrations established using the approach enabled us to identify the optimal solution P concentration for mycorrhizal development on roots of our indicator plant *L. leucocephala* grown in the medium. Arbuscular mycorrhizal colonization, host growth, and P status of *L. leucocephala* pinnules measured at target solution P concentrations ranging from 0.12 to 1.0 mg/l revealed that AM fungal activity and symbiotic effectiveness was maximum at solution P concentration of 0.2 mg/l. Medium solution P concentrations in excess of 0.2 mg/l tended to depress AM fungal colonization, but colonization level did not decline below 43% (Peters and Habte, 2001).

Osorio and Habte (2001) studied the synergistic influence of an arbuscular mycorrhizal fungus and a P solubilizing fungus on growth and P uptake of *L. leucocephala* in an oxisol. An investigation was carried out to assess the role that P solubilizing microorganisms play in the P nutrition of mycorrhizal and mycorrhiza-free *L. leucocephala* (Lain.). Soil microorganisms able to solubilize rock phosphate were isolated from the rhizosphere of *L. leucocephala* naturally, growing on three different soils of Hawaii. The isolates were screened for their ability to solubilize rock phosphate in culture medium. The highest activity was observed with one of the fungal isolates, which was identified as *Mortierella* sp. It was multiplied and further evaluated with or without the mycorrhizal fungus *Glomus aggregatum* in a highly, weathered soil for its effectiveness to enhance P uptake and growth of *L. leucocephala*. Phosphorus status of *L. leucocephala* pinnules monitored as a function of time revealed that plants colonized by both microorganisms had the highest P content followed by plants inoculated with the mycorrhizal fungus alone. Inoculation of soil with *Mortierella* sp. alone did not influence P content of plants measured at the time of harvest. However, *Mortierella* sp. increased the P content of mycorrhizal plants by 13% in the unfertilized soil and by 73% in the soil fertilized with rock phosphate. Shoot dry weight measurements showed that *Mortierella* sp. stimulated growth of nonmycorrhizal by 22%, while it stimulated the growth of mycorrhizal plants by 29% regardless of P fertilization.

The results suggest the existence of synergistic interaction between P solubilizing microorganisms and mycorrhizal fungi, although the degree of synergism was more pronounced in terms of P uptake than in terms of growth.

Plant growth promoting microbes

Numerous species of soil bacteria which flourish in the rhizosphere of plants, but which may grow in, on, or around plant tissues, stimulate plant growth by a plethora of mechanisms. These bacteria are collectively known as PGPR (plant growth promoting rhizobacteria). The search for PGPR and investigation of their modes of action are increasing at a rapid pace as efforts are made to exploit them commercially as biofertilizers. PGPR belongs to several genera for example *Actinoplanes, Agrobacterium, Alcaligens, Amorphosporangium, Arthrobacter, Azotobacter, Bacillus, Bradyrhizobium, Cellulomonas, Enterobacter, Erwinia, Flavobacterium, Pseudomonas, Rhizobium, Streptomyces* and *Xanthomonas*. PGPR are believed to improve plant growth by colonizing the root system and pre-emptying the establishment of or suppressing 'deleterious rhizosphere micro organisms' (DRMO) on the root (Schroth and Hancock, 1981). Inoculating planting material with PGPR presumably prevents or reduces the establishment of pathogens (Suslow, 1982). Production of sidorophores is yet another mechanism through which microbes influence plant growth. Siderophores are low molecular weight high affinity iron chealators that transport iron into the bacterial cells and are responsible increased plant growth by PGPR (Kloepper et al., 1988). Under iron deficient conditions, fluorescent siderophore iron complex (Hohnadel and Meyer, 1986) which creates an iron deficient environment deleterious to fungal growth. Plant growth promoting rhizobacteria (PGPR) are considered to promote plant growth directly or indirectly. PGPR can exhibit a variety of characteristics responsible for influencing plant growth. The common traits include production of plant growth regulators (auxin, gibberellins and ethylene), siderophores, HCN and antibiotic production. Other microorganisms that are known to be beneficial to plants are the plant growth promoting rhizobacteria (PGPR). In addition to supplying combined nitrogen by biological nitrogen fixation, certain bacteria affect the development and function of roots by improving mineral and water uptake.

Considerable research is underway globally to exploit the potential of one such group of bacteria that belong to fluorescent *pseudomonad* (FLPs). This bacteria help in maintaining soil health, protect crop from pathogens and are metabolically and functionally (Lata et al., 2002; Lugtenberg and Dekkers, 1999). Most P corrugate, a form that grows at 4°C under laboratory conditions (Pandey and Palni, 1998), produces antifungals such as diacetylphloroglucinil and phenazine compounds that aid in phosphate solubilization. According to Gaur et al.

(2004), 50 to 60% of flurescent *pseudomonads* recovered from the rhizosphere and endorhizosphere of wheat grown in Indo-Gangetic plains were antagonistic towards *Helminthosporium sativum*. Field trials of pseudomonad strain (GRP3) lead to yield increase from 5.6 to 18%. Rangarajan et al. (2001) analysed populations of *Pseudomonas* species have the potential to suppress the bacterial leaf and sheath blight causing pathogens. Recently, concern was showed on the use of FLPs in crop plants as the antifungal substances released by the bacterium, particularly 2, 4-diacetylphloroglucinol (DAPG) could affect the arbuscular mycorrhizal fungi (Andrade et al., 1995). Gaur et al. (2004) confirmed that DAPG producing *Pseudomonads* recovered from wheat rhizosphere did not adversely affect AM colonization. However, given the toxicity of DAPG, such an inhibition may probably be dependent on the amounts released by the bacterium. Filamentous fungi are widely used as producer of organic acids, particularly black *Aspergillus* and some species of *Penicillium*; these species were tested for solubilization of rock phosphate and have been reported for various properties of biotechnological importance such as biocontrol, biodegradation, phosphate solubilization and phosphate fertilization (Richa et al., 2007; Pandey et al., 2008; Manoharachary et al., 2005; Yadav et al., 2011). Environmental and health hazards associated with chemical fertilizer further make their possible use difficult. In the other hand success in agroforestry requires various inputs leading to proper establishment and growth of tree species. Selection of low demanding and fast growing tree species, fertilization, disease and pest management and proper protection of plants are some vital factors. Agroforestry ecosystem after a few years becomes self sustainable because it follows a 'feedback regulating system'.

The nutrient budget in such ecosystem is polycyclic in nature. Weathering of litter and biological cycle are some important processes of nutrient management. The contribution of nutrient from plant to soil varies from species to species and this is an important factor for selecting species for agroforestry. Opencast mining is generally practiced for commercial exploitation of coal. In this process, overburden materials that is, the overlying soil layer with existing vegetation are removed and deposited in another fresh area. Thus, the deposition of million tons of overburdens in the forms of rocks, shale, course tailing results in barren, biologically inert overburden dumps, called mine spoils. Revegetation of these mine spoils is essential for conservation of environment, biodiversity and to make the land productive. But, the revegetation of these mine spoils is difficult because, they are deficient in nutrients such as nitrogen (N), phosphorus (P) and devoid of organic matter with adverse physio-chemical properties. Root symbionts such as *Rhizobium* and arbuscular mycorrhizal fungi (AMF) are also reduced or absent in mine sites

(Noyd et al., 1995). High acidity in mine spoils due to oxidation of residual elemental or iron sulphur is also a constraint to revegetation which hampers root-growth of plant and reduces the population of beneficial microorganisms such as free-living N-fixers (Alexander, 1964; Arminger et al., 1996; Barnishel, 1977; Choudhury, 1996). In recent times, increased ecological awareness among researchers have resulted in search for innovative approaches for revegetation of mine area in India and abroad (Dugaya et al., 1996; Gupta et al., 1994; Kumar and Jena, 1996; Pandya et al., 1997; Prasad and Mahammad, 1990; Sonkar et al., 1998). Mining causes the destruction of natural ecosystems through removal of soil and vegetation and burial beneath waste disposal sites. The restoration of mined land in practice can largely be considered as ecosystem reconstruction- the re-establishment of the capability of the land to capture and retain fundamental resources (Jha et al., 2000). Mining is an extreme form of land degradation having devastating effects on flora and fauna and causing drastic changes in landform and hydrology. Because of large scale destruction of natural areas due to mining operations, a restoration strategy is needed as a part of the overall mining management plan. In restoration, emphasis is given first to build soil organic matter, nutrients and vegetation cover to accelerate natural recovery process. Tree plantations can be used as a tool for mine spoil restoration as they have ability to restore soil fertility and ameliorate microclimatic conditions.

Revegetation of mining overburden dump posses an ecological challenge. Overburden dumps are unstable and become polluted under unmanaged condition. It affects the original and habitat, increase heavy metal concentration and pollute the environment (Sengupta, 1993; Maiti, 2007). Phytostabilization is one way to restore the overburden dumps. As *Acacia auriculiformis* is able to adapt to stress environment, and presumed to be suitable for vegetating heavy metal contaminated soils, a large number open cast coal and iron mine operate in Orissa state, India and phytoremediation measure could be an ideal way to reforest and improve the soil condition of overburden dumps. This species can be used for the revegetation of overburden dumps of open cast mines of Orissa. Soil microorganisms play an important role in ameliorating soil condition that facilitates plant productivity. The effect of soil AM fungi on establishment and growth of *A. auriculiformis* in waste land soil is also reported (Giri et al., 2004). Legume is a natural mini-nitrogen manufacturing factory in the field and can play a vital role in increasing indigenous nitrogen production. Legume help in solubilizing insoluble P in soil, improving the soil physical environment, increasing soil microbial activity, and restoring organic matter and also has smothering effect on weed. The carryover of N derived from legume grown, either in crop sequence or in intercropping system for succeeding crops is also important. In a country like India, where the average consumption of plant nutrients from chemical fertilizers on national basis is very low, the scope for exploiting direct and residual fertility due to legumes has obviously a great potential (Ghosh et al., 2007). It is known that a large number of seedlings of forest species useful for afforestaion in dry deciduous conditions are required for raising plantations.

Microbial application at the nursery stage was also found to be useful in enhancing productivity in some forest trees like *Albizzia, Acacia* and *Dalbergia* etc. (Rahangdale and Gupta, 1998; Sahgal et al., 2004; Thatoi et al., 1993; Verma, 1995). *A. auriculoformis, A. nilotica* and *A. leucocephala* are three legumes suitable for agroforestry because of their litter contains more than 2% nitrogen (Puri, 1960).

These trees can easily be grown at an elevation of 0 to 500 m receiving rainfall 200 to 1400 MM annually (Mishra et al., 2004). These three species can easily withstand 4 to 6 months of dry season. It is however equally important to screen some suitable bioinoculants for these species.

An assessment of the current state of bacterial inoculants for contemporary agriculture in developed and developing countries is critically evaluated from the point of view of their actual status and future use. A biotechnological goal is to use a combined inoculation of selected rhizosphere microorganisms to minimize toxic effects of pollutants and maximize plant growth and nutrition. Selected combinations of microbial inocula enhanced the positive effect achieved by each microbial group, improving plant development and tolerance in polluted soils.

The application of bioinoculants like AM fungi and one or two of the plant growth promoting rhizobacteria such as *Azospirillum, Agrobacteria, Rhizobium, Pseudomonas,* several gram positive *Bacillus* species is an environment-friendly, energy efficient and economically viable approach for reclaiming wastelands and increasing biomass production. *Rhizobium* and phosphate solubilizing microbes may affect AM fungi and their plant host by various mechanisms such as effects on the receptivity of the root, effect on the root fungus recognition, effects on the fungal growth, modification of the chemistry of the rhizospheric soil and effects on the germination of the fungal propagules (Al-Garni, 2006). On the other hand AM fungi are known to enhance nodulation and N fixation by legumes (Amora-Lazecario et al., 1998; Johansson et al., 2004). Moreover AM fungi, *Rhizobium* and phosphate solubilisers often act synergistically on infection rate, mineral nutrition and plant growth.

Development of model systems and sensors for effective N and P utilization and newer probes to assess microbial diversity and microbial enzyme in model systems would generate information on the role of organics and biofertilizers in integrated nutrient management in sustainable agriculture (Tilak, 2001).

MICROBIAL CONSORTIUM AND APPLICATION

Raja et al. (2006) concluded the combination of bioinoculants is a major cause for success of both the plant establishment and the sustainability of bioinoculants and confirms the beneficial effects of microbial consortium over conventional single inoculant application method. Fertilizer application enhanced the efficiencies of N, P and K uptake, whereas reduced their usage efficiencies. Though soil type did not affect microbial inoculation response, fertilizer application significantly affected plant response to microbial inoculation (Muthukumar and Udaiyan, 2006). The microbial inoculants were used in single form or in combinations (Prabakaran and Ravi, 1996; Gupta et al., 1999; Amutha and Kannaiyan, 2004; Aseri and Rao, 2005; Zaidi and Khan, 2006; Anil et al., 2007; Gaikwad et al., 2008). The effects of the inoculation of *Rhizobium* and phosphate solubilizing bacterium (PSB; *Bacillus megaterium* var. *phosphaticum*), singly or in combination gave better result than the uninoculated control (Sengupta et al., 2002; Marimuthu et al., 2002; Jat et al., 2003; Mathew and Hameed, 2003; Kashyap et al., 2004; Purbey and Sen, 2005). Singh and Tilak (2001) said that the synergistic effect of combined phosphorus fertilizers and inert sources of natural P (varisite, strengite, fluorapatite, hydroxyapatite, and tricalcium phosphate) along with bioinoculants (namely: phosphate solubilizing rhizobacteria and arbuscular mycorrhizas) is discussed. Sumana and Bagyaraj (2002) studied the interaction between VAM fungus and nitrogen fixing bacteria and their influence on growth and nutrition on different tree species. Effect of three *Penicillium* spp namely: *P. islandicum*, *P. olvicolor*, and *P. restrictum*, one AM *Glomus* sp. and rock phosphate on the growth and yield of wheat was evaluated on pot culture containing sterilsed soil (Gupta and Baig 2001). Soil compaction induced a limitation in root and shoot growth that was reflected by a decrease in the microbial population and activity. Our results show that bacterial population was stimulated by the decrease in soil bulk density (Canbolat et al., 2006). Enhancement of growth, nodulation and N_2 –fixation in *D. sissoo* roxb. by *Rhizobium* and *Glomus fasciculatum* in the form of dual inoculation was studied by Rao et al. (2003). The structure of the plant root system contributes to the establishment of the rhizosphere microbial population.

The interactions of plant roots and rhizosphere microorganisms are based largely on interactive modification of the soil environment by processes such as water uptake by the plant, release of organic chemicals to the soil by the roots, microbial production of plant growth factors and microbial mediated availability of mineral nutrients. Within the rhizosphere, plant roots have a direct influence on the composition and density of the soil microbial community known as rhizosphere effect. Microorganisms in the rhizosphere may undergo successional changes as the plant grows from seed germination to maturity. During plant development, a distinct rhizosphere succession results in rapidly growing, growth factor requiring and opportunistic microbial populations. These successional changes correspond to changes in the materials released by the plant roots, the rhizosphere during plant maturation. Initially, carbohydrate exudates and mucilaginous materials support the growth of large populations of microorganisms within the grooves of the epidermal plant cells, on the root surface and within mucilaginous layers of the roots. Microorganisms need more time to fix and establishing themselves in soil. The present finding showed that phosphate-solubilizing microorganisms can interact positively in promoting plant growth as well as P uptake of maize plants, leading to plant tolerance improving under water deficit stress conditions (Ehteshami et al., 2007). Highest N and P uptake by *V. radiata* was recorded after treatment with a combination of *B. circulans, C. herbarum* and *G. fasciculatum* in the presence of MRP. Generally, the PSM population increased after AM fungus inoculation (Singh and Kapoor, 1999). Varma and Mathur (1989) studied the effects of interactions between *Azotobacter chroococcum* and *Pseudomonas* striata on their growth pattern were studied. Plant roots have a direct effect on the surrounding microbial populations; microorganisms in the rhizosphere have marked an influence on the growth of plants. In the absence of appropriate microbial populations in the rhizosphere, plant growth may be impaired.

Microbial populations in the rhizosphere may benefit the plant in a variety of ways, mineral nutrients; synthesis of vitamins, amino acids, auxins, cytokinins, and gibberellins, which simulate plant growth; and antagonism with potential plant pathogens through competition and development of amnesal relationships based on production of antibiotics. Organisms in the rhizosphere produce organic compounds that affect the proliferation of the plant root system. Microorganisms synthesize auxins and gibberllin-like compounds, and these compounds increase the rate of seed germination and the development of root hairs that aid plant growth. Rhizosphere microorganisms increase the availability of phosphate through solubilization of materials that would be otherwise be unavailable to plants. Plants have been shown to exhibit higher rates of phosphate uptake when associated with rhizosphere microorganisms than in sterile soils. The principal mechanism of increasing phosphate availability is the microbial production of acids that dissolve apatite (a commom mineral group including calcium fluophophate), releasing soluble forms of phosphorous. The immobilization of nitrogen by microorganisms in the rhizosphere accounts for an appreciable loss added nitrogen fertilizer intended for plant use. Although diverse and complex, the majority of interactions on the rhizosphere are mutually beneficial to both plants and microorganisms and are synergistic in character. Response of *L. leucocephala* to inoculation with *Glomus fasciculatum* and/or *Rhizobium* was studied in a phosphorus deficient unsterile soil. *G. fasciculatum* only inoculation improved nodulation by native *Rhizobia* and *Rhizobium* only treatment improved colonization of roots by native mycorrhizal fungi.

Dual inoculation with both the organisms improved nodulation, mycorrhizal colonization, dry weight, nitrogen and phosphorus content of the plants compared to single inoculation with either organism.

ROLE OF MICROBIAL APPLICATION ON GROWTH AND DEVELOPMENT OF TREE LEGUMES

Duponnois et al. (2000) studied ectomycorrhization of six *A. auriculoformis* provenances from Australia, Papua, New Guinea and Senegal in glass house conditions and its effect on plant growth and on the multiplication of plant parasitic nematode. The results suggest the provenances of *A. auriculoformis* that are well adapted to the indigenous environmental conditions must be selected for their mycorrhizal dependency and for their resistance to plant parasitic nematodes. More over the ectomycorrhizal symbiosis has to be considered as an important component of the cultural system. Roux et al. (2009) studied the *Bradyrhizobia* nodulating the *A. mangium* × *A. auriculoformis* interspecific hybrid are specific and differ from these associated with both parental species. Inoculation experiments performed under *in vitro* or green house conditions showed that all strains were infective with their original hosts but exhibited very variable degrees of effectively according to the host plant tested. They concluded that there is still a high potential for isolating and testing new strains from hybrid to be used as inoculants in the context of large scale afforestation. Giri et al. (2009) studied the effect of inoculation of two arbuscular mycorrhizal fungi, *Glomus fasciculatum* and *Glomus macrocarpum*, alone and in combination, on establishment and growth of *A. auriculoformis* in a wasteland soil was studied under nursery and field conditions. Arbuscular mycorrhiza colonized plants showed significant increment in height, biomass production and girth as compared to non mycorrhizal plants. All growth parameters were higher on dual inoculation of *G. fasiculatum* and *G. macrocarpum* as compared to uninoculated plants under both nursery and field condition. Sharma et al. (2001) studied the mycorrhizal dependency and growth responses of *A. nilotica* and *A. lebbeck* to inoculation by indigenous AM fungi as influenced by available soil P levels in a semi-arid Alfisol wasteland. A series of available phosphorus (Olsen) levels ranging from 10 to 40 ppm were achieved in a semi-arid soil.

The influence of the levels of phosphorus on the symbiotic interaction between two subtropical tree \

species, *A. nilotica* and *A. lebbeck* and a mixed inoculum of indigenous arbuscular mycorrhizal (AM) fungi was evaluated in a greenhouse study. The extent to which the plant species depended on AM fungi for dry matter production decreased as the levels of soil P increased, but the degree of this decrease differed in the two species tested. *A. nilotica* colonized by AM fungi showed a significant increase in shoot P and dry matter at a soil P level. Vijaya and Srivasuki (1999) studied the effect of inoculum type and inoculation dose on ectomycorrhizal development and growth of *A. nilotica* seedlings inoculated with *P. tinctorius* in a nursery. Fumigated nursery bed on a sandy loam was inoculated with the ectomycorrhizal fungus *P. tinctorius* and seeded with *A. nilotica*. Two types of inoculum were compared: mycelium grown in a vermiculite/peat mixture and mycelium grown in liquid medium and entrapped in a calcium alginate gel with different quantities of mycelium. At the end of the first growing season the alginate inoculum at the dose of 4 g mycelium (dry weight) per m^2 proved to be the most efficient. The top dry weight of the seedlings in this treatment was 2.0 fold that of the non-inoculated fumigated controls. This inoculation treatment also ensured 90% mycorrhizal infection by *P. tinctorius*. Line et al. (2006) studied the growth of mycorrhized seedlings of *L. leucocephala* (Lam.) de Wit. in a copper contaminated soil. Due to the low infectivity potential of arbuscular mycorrhizal fungi (AMF) in a mining area located at the State of Bahia, Northeastern Brazil, the effect of mycorrhization on the seedlings of *L. leucocephala* was investigated, in order to use this species for revegetation of the area. Proportions of copper contaminated soil higher than 50% inhibited plant growth, development of the AMF in the roots and consequently, the benefit of mycorrhization.

The pattern of Cu and P absorption in *Leucaena*, associated or not with *G. etunicatum* is maintained when the soil is up to 50% contaminated. Barreto and Fernandes (2001) used *Leucaena leucocephala* in alley cropping systems to improve Brazilian coastal tableland soils. Decreases of the soil density and increases of the porosity were also observed in response to legume incorporation. These effects were most noticeable in the superficial layers. Forestier et al. (2001) studied the effect of *Rhizobium* inoculation methodologies on nodulation and growth of *L. leucocephala*. Their aim of the study was to evaluate the effect of five methods of *Rhizobium* inoculum application on nodulation and nitrogen fixation in *L. leucocephala* seedlings cultivated for 6 months in the greenhouse. Plants inoculated with alginate beads were significantly more developed and more nodulated than plants inoculated with the other methodologies used. Bala and Giller (2001) suggested that the host range and specificity is reported of a genetically diverse group of *Rhizobia* isolated from nodules of *L. leucocephala* and other tree species. Nodule number and nitrogen content was measured in seedlings of herbaceous and woody legume species after inoculation with rhizobial strains isolated from tropical soils, to establish symbiotic effectiveness groups for Rhizobial strains and their hosts. The complex nature of cross-nodulation relationships between diverse rhizobial strains and legume hosts is highlighted. Host plants inoculated with effective rhizobial strains showed better nitrogen use efficiency than plants supplied solely with mineral nitrogen. Wang et al. (1999) studied the genetic diversity of *Rhizobia* from *Leucaena leucocephala* nodules in Mexican soils. *Leucaena* species are leguminous plants native to Mexico. Using two *L. leucocephala* cultivars grown in different soils, we obtained 150 isolates from the nodules.

Twelve rDNA types were identified which clustered into groups corresponding to *Mesorhizobium, Rhizobium* and *Sinorhizobium* by restriction fragment length polymorphism (RFLP) of amplified 16S rRNA genes. Types 2, 4, 5, 6, 10, 11 and 12 were distinct from all the defined species. Others had patterns indistinguishable from some recognized species. Most of the isolates corresponded to *Sinorhizobium* 12. The affinities of host cultivars for different Rhizobial groups and the impact of sail cultivation and the soil populations of *Rhizobia* were analysed from the estimation of isolation frequencies and diversity. The results showed differences in Rhizobial populations in cultivated and uncultivated soils and also differences in *Rhizobia* trapped by *L. leucocephala* cv. *cunningham* or *peruvian*. *A. nilotica* nodulate with fast-growing *Rhizobium* (Dreyfus and Dommergues, 1981). The nitrogen-fixing potential of the species is still unknown. *A. auriculiformis* produces profuse bundles of nodules, which suggests a good nitrogen-fixing potential. Sharma and Ramamurthy (2000) conducted studies under controlled pathogenic conditions to assess the compatibility and nitrogen fixing potential of *Rhizobium* and *Bradyrhizobium* strains for *A. auriculiformis*. Of the 17 strains form different sourced tested only 10 were able to produce visible nodules and their nodulation frequency was also different. Strains isolated from other species of *Acacia* were not particularly more compatible with *A. auriculiformis* than strains isolated form other plant species. There was a significant increase in growth and associated parameter in the nodulated seedlings over that in seedlings without root nodules. However, nodules number appeared to be a poor indicator of the nitrogen fixing potential of strains, as it poorly correlated with nitrogenase activity, seedling dry weight and other growth related biometric parameters. *A. leucocephala (L. leucocephala)* has been the focus of a great deal of research in the past few decades for its nitrogen fixing potential. The acetylene reduction method (Hogberg and Kvarnstrom, 1982) and the difference method (Sanginga et al., 1985), which has been used to evaluate nitrogen fixation by *L. leucocephala,* give figures in the range of 100 to 500 kg N_2 ha^{-1} yr^{-1}. The high nitrogen-fixing potential of this tree is related to its abundant nodulation under specific soil conditions, in which the nodule dry weight was reported to reach approximately 51 kg ha^{-1} in a stand of 830 trees ha^{-1} (Hogberg and Kvarnstrom, 1982), and approximately 63 kg ha^{-1} in a stand of 2,500 trees ha^{-1} (Lulandala and Hall, 1986). *L. leucocephala* generally nodulates with *Rhizobium* (Halliday and Somasegaran, 1982) and occasionally nodulates with *Bradyrhizobium* (Dreyfus and Dommergues, 1981).

The *Rhizobium* strain specific to *L. leucocephala* is not generally found in soils. This explains the positive response to inoculation obtained in most soils where the level of nutrients (other than nitrogen) is high enough to satisfy the-tree's requirements. *Leucaena* has often been regarded as a kind of "miracle tree", an appellation that makes sincere scientists wince. However, such names as subabul in India, Its sensitivity to soil acidity and its high nutrient demand are reflected in its poor performance in infertile soils. Purohit et al. (2007) studied the ecology of soil fungi in *A. nilotica* based agroforestry systems of Rajasthan, India. The present study is aimed to quantify the spatial and temporal variations in the biomass and population of soil fungi in the *A. nilotica* sub. *indica* based traditional agro forestry systems. Soil in the vicinity of the tree possessed higher fungal biomass and population. Increase in electrical conductivity and pH towards alkalinity reduced fungal biomass. Thicker hyphae (12 to 16 µ) were predominant. In the areas under canopy and thinner (8 to 12 µ) in open area. Soil moisture controls fungi hyphal growth and density. Soil organic carbon reduces fungal biomass significantly in the vicinity of the tree than in the open area. Soil nitrogen is high in the under canopy area of the tree. Fungal population and biomass were linearly related. Gupta et al. (2011) studied the influence of mineral solubilizing microbes on growth and biomass of *A. auriculiformis* and developed a possible consortium for the species. A good example of the spread and role of *leucaena* in rural development is given by the work of Manibhai Desai, Director of the non-profit Bharatiya Agro Industries Foundation (BAIF) of Pune, India. He has inspired a generation of Indian scientists to dedicate their efforts to improving the livelihood of the rural poor. *Leucaena* is a major tool in this programme of revegetation, water management and animal improvement, a programme that earned Desai the prestigious Magsaysay Award in 1985. Often featured in their excellent publication, The BAIF Journal, leucaena is the subject of extensive development and research activities at BAIF. The giant leucaenas were carefully appraised in many types of management systems prior to large-scale seed increase (to 40 tons by 1986, equivalent to 800 million seeds) and distribution among India's rural community. Desai and his scientific staff have unquestionably inspired tree planting in the past decade on a scale previously believed impossible for the small farmer (Brewbaker, 1987).

Huda et al. (2007) studied effects of phosphorus and potassium addition on growth and nodulation of *D. sissoo* in the nursery. A study was conducted to test the influence of different inorganic fertilizers (phosphorous and potassium) on the nodulation and growth of *Dalbergia sissoo* grown in the nursery. Before seeds sowing, different combinations of P, K fertilizers were incorporated with the nutrient-deficient natural forest soils, and then amended with cow dung (soil: cowdung = 3:1). Nodulation status (nodule number, shape, fresh weight, dry weight and color) in the roots and the plant growth parameters (length of shoot and root, collar diameter, fresh and dry weight of shoot and root) were recorded 60 days after seeds sowing. Nodulation status and growth of the plants varied significantly in the soils amended with fertilizers in comparison to the control. From the study, it is revealed that PK at the rate of 160 kg/hm^2 fertilizer with soil and cowdung mixture is recommended for optimum growth and nodule formation of *D. sissoo* in degraded soils at a nursery level. Bouillet et al. (2008) did the mixed-species plantations of *Acacia mangium* and *Eucalyptus grandis* in Brazil. Nitrogen accumulation in the stands and biological N$_2$ fixation. Dash and Gupta, (2011) analysed the roots of tree legumes grown in different pots inoculated with mineral solubilizing bacteria and fungi to evaluate AM infection and colonization. As compared to fungal inoculation, bacterial inoculation encouraged increase in percent AM colonization in the roots.

ROLE OF MICROBIAL APPLICATION OF ESTABLISHMENT OF TREE LEGUMES IN WASTELANDS AND UNDER STRESS CONDITION

Mining causes the destruction of natural ecosystems through removal of soil and vegetation and burial beneath waste disposal sites. The restoration of mined land in practice can largely be considered as ecosystem reconstruction - the reestablishment of the capability of the land to capture and retain fundamental resources. In restoration planning, It is imperative that goals, objectives, and success criteria are clearly established to allow the restoration to be undertaken in a systematic way, while realizing that these may require some modification later in light of the direction of the restoration succession (Jha et al., 2000; Tilak, 2001). A restoration planning model is presented where the presence or absence of topsoil conserved on the site has been given the status of the primary practical issue for consideration in ecological restoration in mining. Examples and case studies are used to explore the important problems and solutions in the practice of restoration in the mining of metals and minerals. Even though ecological theory lacks general laws with universal applicability at the ecosystem level of organization, ecological knowledge does have high heuristic power and applicability to site-specific ecological restoration goals. The concept of adaptive management and the notion that a restored site be regarded as a long-term experiment is a sensible

perspective. Unfortunately, in practice, the lack of post-restoration monitoring and research has meant few opportunities to improve the theory and practice of ecological restoration in mining. Vegetative material used in reclamation shall consist of grasses, legumes, herbaceous or woody plants, shrubs, trees or a mixture thereof which is consistent with site capabilities such as drainage, pH, soil depth, available nutrients, soil composition and climate. Such vegetation should be designed to provide a cover consistent with the stated land-use objective and which does not constitute a health hazard (Aseri and Rao, 2005). Mining is an extreme form of land degradation having devastating effects on flora and fauna and causing drastic changes in landform and hydrology. Because of large-scale destruction of natural areas due to mining operations, a restoration strategy is needed as a part of the overall mining management plan. In restoration, emphasis is given first to build soil organic matter, nutrients and vegetation cover to accelerate natural recovery process.

Tree plantations can be used as a tool for mine spoil restoration as they have ability to restore soil fertility and ameliorate microclimatic conditions. We discuss here various approaches of ecosystem restoration on mine spoil, criteria for the selection of plantation species and future research needs in this regard (Singh and Bhati, 2005). While *Rhizobia* and mycorrhizal fungi were a ubiquitous component of the soil biota in all undisturbed woodland soils they were absent or poorly represented in the stockpiled topsoils and some of the rudimentary soils formed in waste rock at the mine site. Kahlon et al. (2006) studied the effect of phosphorus, zinc, sulphur and bioinoculants on yield and economics of cowpea. An incubation study was conducted to test the suitability of Bokaro fly ash, alone or in combination with lignite, charcoal and farmyard manure, to act as a carrier for the *Rhizobium* inoculant. *R. leguminosarum* broth containing 108 cells/ml was mixed with the prospective carriers, packed in polythene and maintained at room temperature. Charcoal and lignite produced the highest populations of *Rhizobium*, possibly due to its ability to absorb toxic compounds and aerate the medium. Fly ash in combination with charcoal (1:1 or 1:3) had the highest population. The *Rhizobium* increased up to 45 days and then remained constant. It is concluded that 25% fly ash in combination with lignite or charcoal can be used as a carrier for bioinoculants (Lal and Mishra, 1998). Revegetation research at Ranger mine concludes that *Rhizobium* treatment is of little benefit because acacias are the easiest species to establish on waste rock dumps (Batterham, 1998). Many processes other than initial establishment, however, will govern long-term success of rehabilitated areas (Reddell and Milnes, 1992). Nitrogen deficiency in mine soils can be amended via the establishment of nitrogen fixing legumes and root nodule bacteria that is *Rhizobium*. Legumes are only effective if an association is formed with the appropriate strain of

Rhizobia (Bell, 2002). If local nitrogen fixing species are to be reintroduced to the mined area, their effective inoculation will be catalysed through the replacement of fresh surface soil or through the careful selection, collection and inoculation of the plants with the appropriate strains. The contribution of plant available nitrogen and of organic matter by Acacias to a re-establishing ecosystem through mechanisms such as litter cycling may rely on symbiosis with *Rhizobia* (Coleman et al., 1983).

Revegetation of the mine spoils is essential for conservation of environment, biodiversity and to make the land productive. But, the revegetation of these mine spoils is difficult because, they are deficient in nutrients such as nitrogen (N), phosphorus (P) and devoid of organic matter with adverse physico-chemical properties. Root symbionts such as *Rhizobium* and arbuscular mycorrhizal fungi (AMF) are also reduced or absent in mine sites (Noyd et al., 1995). High acidity in mine spoils due to oxidation of residual elemental or iron sulphur is also a constraint to revegetation, which hampers root-growth of plant and reduces the population of beneficial microorganisms such as free-living N-fixers (Aleaxander, 1964; Arminger et al., 1996; Barnishel, 1977; Choudhury, 1996). In the recent times, increased ecological awareness among researchers, have resulted in search for innovative approaches for revegetation of coal mine area in India and abroad (Dugaya et al., 1996; Gupta et al., 1994; Kumar and Jena, 1996; Pandya et al., 1997; Prasad and Mahammad, 1990; Sonkar et al., 1998). The use of native and indigenous plant species have been emphasized in revegetation programs with a view to maintain essential processes and life support system, preservation of genetic diversity and to ensure sustainable utilization of species and ecosystem (Banerjee et al., 1996). Plant species also emerge naturally on the barren mined land after certain intervals of time from the initiation of dump, but succession of plant species under such situation proceeds at a much slower rate (Bradshaw and Chadwick, 1980; Roberts et al., 1981). Therefore, it is essential to understand the structure and function of an ecosystem with its primary and secondary succession patterns for a successful revegetation programs. A study was conducted to determine the effect of unamended and variously amended tailings on the survival, growth and metal uptake of different plant species for the phytostabilization of Rajpura-Dariba mine tailings in Rajasthan, India. The species studied were: *A. auriculiformis, A. nilotica, Aegle marmelos, A. lebbeck, Bauhinia purpurea, Boswellia serrata, Caesalpinia bonduc, Cassia fistula, Casuarina equisetifolia, Dendrocalamus strictus, Eucalyptus citriodora, Lawsonia inermis, Parkinsonia aculeata, Peltophorum ferrugineum, Stylosanthes hamata, Withania somnifera, Ziziphus mauritiana, Vigna unguiculata, Triticum aestivum* and *Brassica campestris.*

The conditioning was carried out with inorganic fertilizer

(NPK) and farmyard manure and capping with 15% soil. It was found that *A. auriculiformis, L. inermis, E. citriodora, P. aculeata* and *W. somnifera* were the species that survived and grew in the inhospitable tailing surface though after suitable amelioration.

The results indicate that for successful phyto-remediation of metal polluted sites, a strategy should be considered that combines rapid screening of plant species possessing the ability to tolerate and accumulate heavy metals with agronomic practices that enhance shoot biomass production and/or increase/decrease in metal bioavailability in the rhizosphere (Archana and Aery, 2005). Archer and Cladwell (2004) studied the response of six Australian plant species to heavy metal contamination at an abandoned mine site. This investigation was carried out to assess the potential suitability of certain Australian plants for use in the phytoremediation of derelict mine sites. A revegetation trial was conducted to evaluate the feasibility of growing a legume species, *Prosopis juliflora L.*, on fly ash ameliorated with combination of various organic amendments, blue-green algal biofertilizer and *Rhizobium* inoculation. Plants accumulated higher amounts of Fe, Mn, Cu, Zn and Cr in various fly ash amendments than in garden soil. Further, inoculation of the plant with a fly ash tolerant *Rhizobium* strain conferred tolerance for the plant to grow under fly ash stress conditions with more translocation of metals to the above ground parts. The results showed the potential of *P. juliflora* to grow in plantations on fly ash landfills and to reduce the metal contents of fly ash by bioaccumulation in its tissues (Rai et al., 2004).

COMMERCIALLY AVAILABLE BIOFERTILIZER AND APPLICATION

Russian microbiologists who developed a product called phosphobakterin and tested it on various crops between 1940 to 1960 observed noticeable increases in yields and better phosphorus uptake in inoculated plants. Subsequently, several investigators in other countries have repeatedly attempted to test the value of seed inoculation with phosphate dissolving microorganisms on growth and yield of plants. Recently at I.A.R.I, a carrier based 'microphos' bio fertilizer containing *Pseudomonas striata* has been prepared and tested on crops (Poi, 2008; Mehrvarz and Chaichi, 2008). Application of liquid inoculants to seedlings was better than seed inoculation. Odee et al. (2002) recommend that seedlings raised in the nursery should be inoculated with a liquid inoculant immediately or soon after germination. Turan et al. (2006) evaluated the capacity of phosphate solubilizing bacteria and fungi on different forms of phosphorus in liquid culture. In this study, the capacity of phosphate solubilizing bacterial strain, *Bacillus* (FS3) and fungal isolates, *Aspergillus* FS9 and FS11 have been tested in

National Botanical Research Institutes. The result suggested that phosphate solubilizing bacteria FS3 and fungal strain FS9 have great potential for use bio-tertilizer development in agriculture. Several phosphate solubilizers, including bacteria, actinomyctes and fungi have been isolated. For the convenience of culturing, maintenance, preservation and manipulation bacteria are selected for preparation and use as biofertilizers (Fallah et al., 2007; Xiao et al., 2008). The preparation of the biofertiliser containing *Bacillus megaterium* var. *phosphatium* is known as phosphobacterin cultures obtained from USSR, FOSFO 24 a Czechoslovakian culture and an indigenous culture isolated from *Cassia accidentals*.

Laboratory experiments, trials and tests had to be carried out to get suitable end product that would ideally suit the different conditions of the soil.

REFERENCES

Afzal A, Ashraf M, Asad SA, Farooq M (2005). Effect of phosphate solubilizing microorganisms on phosphorus uptake, yield and yield traits of wheat (*Triticum aestivum L.*) in rain fed area. Int. J. Agri. Biol., 7(2): 207-209.

Ahalwat OP, Dadarwal KR (1996). Bacteriocin production by *Rhizobium* sp. *cicer* and its role in nodule competence. Ind. J. Microbiol. 36(1): 17-23.

Alam S, Khalil S, Ayub N, Rashid M (2002). In vitro solubilization of inorganic phosphate by phosphate solubilizing microorganisms (PSM) from maize rhizosphere. Int. J. Agri. Biol. 4(4): 454-458.

Aleaxander M (1964). Introduction to Soil Microbiology. John Wiley and Sons Inc, New York.

Al-Garni SMS (2006). Increased heavy metal tolerance of cowpea plants by dual inoculation of an arbuscular mycorrhizal fungi and nitrogen-fixer *Rhizobium* bacterium. Afr. J. Biotechnol., 5(2): 133-142.

Amora-Lazecario E, Vazquez MM, Azcon R (1998). Response of nitrogen transforming microorganisms to arbuscular mycorrhizal fungi. Biol. Fertil. Soil, 27: 65-70.

Amutha K, Kannaiyan S (2004). Effect of seed inoculation of *Azorhizobium caulinodans*, *Azospirillum brasilense* and *Pseudomonas fluorescens* on their survival in the spermosphere and rhizosphere of cereal crops. Biofertil. Tech. Scientific Publishers (India): 289-292.

Andrade G, Azcon R, Bethlenfalvey GJ (1995). A rhizobacterium modifies plant and soil responded to the mycorrhozal fungus, *Glomus mosseae*. Appl. Soil. Ecol., 2: 195-202.

Anilkumar AS, Nair KH, Sherief AK (2007). Utilization of enriched coir pith-vermicompost in organic mediculture. Plant. Arch. 2: 617-620.

Archana S, Aery NC (2005). Experimental studies for the revegetation of lead-zinc mine tailings. Proceedings of the National Academy of Sciences India. Section B, Biol. Sci., 75: 197.

Archer MJG, Caldwell RA (2004). Response of six Australian plant species to heavy metal contamination at an abandoned mine site. Water. Air. Soil. Pollut. 157(1-4): 257-267.

Arminger WH, Jhons JN, Bennet OL (1996). Revegetation of land disturbed by strip mining of coal in Appalachia. United States Department of Agriculture Research ARS-NE 71. United States Government. Printing Office, Washington, D.C.

Aseri GK, Rao AV (2005). Effect of bioinoculants on seedlings of Indian Gooseberry (*Emblica officinalis* Gaertn.). Ind. J. Microbiol., 44 (2): 109-112.

Bala A, Giller KE (2001). Symbiotic specificity of tropical tree *Rhizobia* for host legumes. New. Phytol., 149 (3): 495-507.

Banerjee SK, Williums AJ, Biswas SC, Manjhi RB, Mishra TK (1996). Dynamics of natural ecorestoration in coal mine overburden of dry deciduous zone of M.P. India. Ecol. Environ. Conser., 2: 97-104.

Barnishel RI (1977). Reclamation of surface mined coal spoils. United States Department of Agriculture/ EPA Interagency Energy Environment Research and Development Programs Report, CSRS-IEPA 600/7-77/093 USEPA Cincinnati, OH.

Barreto AC, Fernandes MF (2001). Use of *Gliricidia sepium* and *Leucaena leucocephala* in alley cropping systems to improve Brazilian coastal tableland soils. Pesquisa. Agropecuaria. Brasileira. 36(10): 1287-1293.

Basan Y, Holguin G (1997). *Azospirillum*-plant relations. Environmental and Physiological advances. Can. J. Microbiol. 43: 103-121.

Batterham RP (1998). Review of revegetation studies at Ranger Mine 1983–98. Report prepared for ERA Ranger Mine by ERA Environmental Services Pty. Ltd, Darwin.

Bell RW (2002). Restoration of Degraded Landscapes: Principles and Lessons from Case Studies with Salt-affected Land and Mine Revegetation. CMU. J., 1(1): 1-21.

Bijaya T, Andhale MS, Singh NI (2003). Distribution of phosphate solubilizing microorganisms in acidic soils of Manipur. J. Mycopathol. Res., 41(2): 167-170.

Bhaduri SN (1951). Influence of the numbers of *Rhizobium* supplied on the subsequent nodulation of the legume host plant. Ann. Bot. 80: 499-503.

Boring LR, Swank WT, Waide JB, Henderson GS (1988). Sources, fates, and impacts of nitrogen inputs to terrestrial ecosystems: Review and synthesis. Biogeochem., 6(2): 119–159.

Bouillet JP, Laclau JP, Goncalves JLM, Moreira M, Trivelin PCO, Jourdan C, Silva EV, Piccolo MC, Tsai SM, Galiana A (2008). Mixed-species plantations of *Acacia mangium* and *Eucalyptus grandis* in Brazil - 2: Nitrogen accumulation in the stands and biological N-2 fixation. For. Ecol. Manag. 255(12): 3918-3930.

Bradshaw AD, Chadwick MJ (1980). The Restoration of Land. Blackwell Scientific Publications, Oxford, London.

Brewbaker JL (1987). Leucaena: a multipurpose tree genus for tropical agroforestry. Agroforestry a decade of development International Council for Research in Agroforestry.

Canbolat MY, Bilen S, Cakmakci R, Sahin F, Aydin A (2006). Effect of plant growth-promoting bacteria and soil compaction on barley seedling growth, nutrient uptake, soil properties and rhizosphere microflora. Biol. Fertil. Soil. 42(4): 350-357.

Casanova E, Salas AM and Toro M (2002). The use of nuclear and related techniques for evaluating the agronomic effectiveness of phosphate fertilizers, in particular rock phosphate, in Venezuela: II. Monitoring mycorrhizas and phosphate solubilizing microorganisms. International Atomic Energy Agency Technical Documents IAEA-TECDOCs. (1272): 101-106.

Casida LE Jr. (1982). *Ensifer adhaerens* gen. nov., sp. nov.: A bacterial predator of bacteria in soil. Int. J. Syst. Bacteriol. 32(3): 339–345.

Chen YP, Rekha PD, Arun AB, Shen FT, Lai WA, Young CC (2006). Phosphate solubilizing bacteria from subtropical soil and their tricalcium phosphate solubilizing abilities. Appl. Soil. Ecol., 34(1): 33-41.

Choudhury SN (1996). India's North–East: Industrial Resources and Opportunities. Laser King. Tinsukia, Assam, India.

Coleman DC, Reid CPP, Cole CV (1983). Biological strategies of nutrient cycling in soil systems. Adv. Ecol. Res., 13: 1–55.

Dabas P, Kaushik JC (1998). Influence of Glomus mosseae, phosphorus and drought stress on the growth of *Acacia nilotica* and *Dalbargia sissoo* seedlings. Ann. Biol., 14(1): 91-94.

Dash S, Gupta N (2011). Impact of microbial inoculants o AM colonization in tree legumes. Micorr. News. 23(1): 12-13.

Denison RF, Kiers ET (2004). Why are most *Rhizobia* beneficial to their plant hosts, rather than parasitic? Microbiol. Infect., 6(13): 1235–1239.

Diouf D, Forestier S, Neyra M, Lesueur D (2003). Optimisation of inoculation of *Leucaena leucocephala* and *Acacia mangium* with *Rhizobium* under greenhouse conditions. Ann. For. Sci. 60(4): 379-384.

Dommergues YR, Macro-Bosco (1998). The contribution of N_2 fixing trees to soil productivity and rehabilitation in torpical, subtropical and Meditrrranean regions. In Microbial interactions in agriculture and forestry (Eds Subba Rao, N. S. and Dommergues, Y. R.). Oxford & IBH, New Delhi: pp. 65-96.

Dreyfus BL, Dommergues YR (1981). Nodulation of *Acacia* species by fast- and slow-growing tropical strains. Appl. Environ. Microbiol., 41: 97-99.

Dubey SK (2000). Effectiveness of rock phosphate and superphosphate amended with phosphate solubilizing microorganisms in soybean grown on Vertisols. J. Ind. Soc. Soil. Sci., 48(1): 71-75.

Dudeja SS, Khurana SK (1989). The Peagonpea-*Rhizobium* sumbiosis as affected by high root temperature: Effect on nodule formation. J. Expt. Bot., 40(4): 469-472.

Dugaya D, Williums AJ, Chandra KK, Gupta BN, Banerjee SK (1996). Mycorrhizal development and plant growth in amended coal mine overburden. Ind. Forest, 19: 222-226.

Duponnois R, Founoune H, Lesueur D, Thioulouse J, Neyra M (2000). Ectomycorrhization of six *Acacia auriculiformis* provenances from Australia, Papua New Guinea and Senegal in glasshouse conditions: effect on the plant growth and on the multiplication of plant parasitic nematodes. Aus. J. Exp. Agricul., 40(3): 443-450.

Dusha P, Kovalenko S, Banflavi Z, Kondorosi (1987). *Rhizobium* melilitic insertion element ISRm2 and its use for identification of the Fix X gene. J. Bactriol., 1403-1409.

Ehteshami SMR, Aghaalikhani M, Khavazi K, Chaichi MR (2007). Effect of phosphate solubilizing microorganisms on quantitative and qualitative characteristics of maize (*Zea mays* L.) under water deficit stress. Pak. J. Biol. Sci., 10(20): 3585-3591.

Fallah AR, Besharati H, Nourgholipour F, Shahbazi K (2007). Effects of rock phosphate, organic matter, sulfur, *Thiobacillus* and phosphate solubilizing microorganisms on yield and quality of corn, Agri. Sci. Tabriz., 17(3): 25-36.

Fenton M, Jarvis BDW (1994). Expression of the symbiotic plasmid from *Rhizobium leguminosarum* biovar *trifolii* in *Sphingobacterium multivorum*. Can. J. Microbiol., 40(10): 873–879.

Fraysse N, Couderc F, Poinsot V (2003). Surface polysaccharide involvement in establishing the *Rhizobium*-legume symbiosis. Euro. J. Biochem., 270(7): 1365-1380.

Forestier S, Alvarado G, Badjel S, Lesueur D (2001). Effect of *Rhizobium* inoculation methodologies on nodulation and growth of *Leucaena leucocephala*. Wor. J. Microbiol. Biotech., 17(4): 359-362.

Gage DJ, Margolin W (2000). Hanging by a thread: invasion of legume plants by *Rhizobia*, Curr. Opi. Microbiol., 3(6): 613–6177.

Gaikwad AL, Deokar CD, Shete MH, Pawar NB (2008). Studies on effect of phyllosphere diazotrophs on growth and yield of groundnut. J. Plant. Dis. Sci., 3(2): 182-184.

Gaind S, Gaur AC (1991). Thermotolerant phosphate solubilizing microorganisms and their interaction with mung bean. Plant. Soil., 133(1): 141-149.

Gaur R, Shani N, Kawaljeet, Johri BN, Rossi P, Aragno M (2004). Diacetyl phloroglucinol-producing *Pseudomonas* do not influence AM fungi in wheat rhizosphere. Curr. Sci., 86: 453-457.

Ghosh PK, Bandyopadhyay KK, Wanjari RH, Manna MC, Misra AK, Mohanty M, Rao AS (2007). Legume effect for enhancing productivity and nutrient use-efficiency in major cropping systems - An Indian perspective: A review. J. Sust. Agricul., 30(1): 59-86.

Ghosh W, Roy P (2006). *Mesorhizobium thiogangeticum* sp. nov., a novel sulfur-oxidizing chemolithoautotroph from rhizosphere soil of an Indian tropical leguminous plant. Int. J. Syst. Evol. Microbiol., 56 (1): 91–97.

Giri B, Kapoor R, Agarwal L, Mukerji KG (2004). Preinoculation with Arbuscular Mycorrhizae Helps *Acacia auriculiformis* grow in degraded Indian Wasteland soil. Comm. Soil. Sci. Plant. Anal., 35 (1 & 2): 193 - 204.

Graham PH (1969). Selective medium for growth of *Rhizobium*. Appl. Microbiol., 17(5): 769-770.

Grange L, Hungria M (2004). Genetic diversity of indigenous common bean (*Phaseolus vulgaris*) *Rhizobia* in two Brazilian ecosystems. Soil. Biol. Biochem., 36(9): 1389-1398.

Gupta BN, Singh AK, Bhowmik AK, Banerjee SK (1994). Suitability of different tree species for copper mine overburden. Ann. Forest. 2: 85-87.

Gupta N, Baig S (2001). Evaluation of synergistic effect of phosphate solubilizing *Penicillium spp* AM fungi and rock phosphate on growth and yield of wheat. Phil. J. Sci., 130 (2): 139-143.

Gupta N, Dash S, Mohapatra AK (2011). Influence of mineral solubilizers on growth and biomass of *Acacia auriculiformis* Cunn. Ex. Berth. grown in nursery condition (Eds M. Miransari). Soil Microbes and Environmental Health, chapter- 11.

Gupta NS, Sadavarte KT, Mahorkar VK, Jadhao BJ, Dorak SV (1999). Effect of graded levels of nitrogen and bioinoculants on growth and yield of marigold (*Tagetes erecta*). J. Soil. Crop., 9 (1): 80-83.

Gupta RP, Kalia A, Kapoor S (2007). Bioinoculants. A step towards sustainable agriculture. Bioinoculants: a step towards sustainable agriculture. New India Publishing Agency, India. p. 306.

Halliday J, Somasegaran P (1982). Nodulation, nitrogen fixation, and *Rhizobium* and strain affinities in the genus *Leucaena*. In *Leucaena* research in the Asian-Pacific region. Ottawa: IDRC.

Hameeda B, Harini G, Rupela OP, Wani SP, Reddy G (2008). Growth promotion of maize by phosphate-solubilizing bacteria isolated from composts and macrofauna. Microbiol. Res., 163 (2): 234-242.

Han SZ, Wang ET, Chen WX, Han SZ (2005). Diverse bacteria isolated from root nodules of *Phaseolus vulgaris* and species within the genera *Campylotropis* and *Cassia* grown in China. Syst. Appl. Microbiol., 28(3): 265–276.

Hogberg P, Kvarnstrom M (1982). Nitrogen fixation by the woody legume *Leucaena leucocephala*. Plant. Soil., 66: 21-28.

Hohandel OB, Meyer JM (1986). Pyoverdine-facilitated iron uptake among fluorescent pseudomonades. Iron siderophores and plant diseases (Eds: T. R. Swinburne). Plenum press, New York, USA: pp.119-129.

Howieson JG, Brockwell J (2005). Nomenclature of legume root nodule bacteria in 2005 and implications for collection of strainsfrom the field (Eds. J. Brockwell), *14th Australian nitrogen fixation conference*,. The Australian Society for Nitrogen Fixation. pp. 17–23.

Huda SMS, Sujauddin M, Shafinat S, Uddin MS (2007). Effects of phosphorus and potassium addition on growth and nodulation of *Dalbergia sissoo* in the nursery. J. For. Res., 18(4): 279–282.

Huss-Danell K (1990). The physiology of actinorrhizal roots. The biology of *Frankia* and Actinorrhizal Plants (Eds Schwintzerr, C. R. and Tjepkema, J. D.). Acacemy Press, San Diegom USA: pp.128-158.

Illmer P, Schinner F (1995). Phosphate solubilizing microorganisms under non sterile condition. Bodenkultur. 46(3): 197- 204.

Jain V, Nainawatee HS (2002). Plant flavonoids: Signals to legume nodulation and soil microorganisms. J. Plant. Biochem. Biotechnol. 11(1): 1-10.

Jat SL, Sumeriya HK, Mehta YK (2003). Influence of integrated nutrient management on content and uptake of nutrients on sorghum (*Sorghum bicolor* (L.) Moench). Crop. Res. Hisar., 26(3): 390-394.

Jha AK, Singh A, Singh AN, Singh JS (2000). Evaluation of direct seeing of tree species as a hand of revegetation of coal mine spoils. Ind. Forest. 126(11): 1217-1221.

Jinturkar RP, kale SB (2005). Isolation of *Rhizobium* from different varieties of chickpea. Asi. J. Microbiol. Biotechnol. Environ. Sci., 7(4): 727-728.

Johansson JF, Paul LR, Finlay RD (2004). Microbial interactions in the mycorrhizosphere and their significance for sustainable agriculture. FEMS Microbiol. Ecol., 48: 1-13.

Joshi F, Chaudhari A, Joglekar P, Archana G, Desai A (2008). Effect of expression of *Bradyrhizobium japonicum* 61A152 fegA gene in *Mesorhizobium* sp., on its competitive survival and nodule occupancy on *Arachis hypogea*. Appl. Soil. Ecol., 40(2): 338-347.

Kahlon CS, Sharanappa, Kumar R (2006). Effect of phosphorus, zinc, sulphur and bioinoculants on yield and economics of cowpea. Mys. J. Agri. Sci., 40(1): 138-141.

Kalia A, Gupta RP (2002). Nodule induction in non-legumes by *Rhizobia*. Ind. J. Microbiol., 42(3): 183-193.

Kashyap S, Sharma S, Vasudevan P (2004). Role of bioinoculants in development of salt resistant saplings of *Morus alba* (var. sujanpuri) in vivo. Sci. Hort., 100(1/4): 291-307.

Kehri SS, Chandra S, Kehri SM (2002). Improved performance of linseed due to inoculation of phosphate solubilizing fungi. *Front.* Microbiol. Biotech. Plant. Pathol., pp.193-198.

Klecekowski A, Thorton HG (1944). A serological study of root nodule bacteria from pea and clover inoculation groups. An article found from jb.asm.org: pp.661-672.

Kloepper JW, Leong J, Teintze P, Arayangkool T, Sintwongse P, Siripaibool C, Wadisirisuk P, Bookerd N (1988). Nitrogen fixation (15

N dilution) with soybeans under Thai field conditions. II. Effect of herbicides and water application schedule. Plant. Soil., 108: 87-92.

Kole SC, Hajra JN (1997). Isolation and evaluation of tricalcium phosphate and rock phosphate solubilizing microorganisms from acidic terai and lateritic soils of West Bengal. J. Interacademicia., 1(3): 198-205.

Kole SC, Hajra JN (1998). Occurrence and acidity of tricalcium phosphate and rock phosphate solubilizing microorganisms in mechanical plant compost of Calcutta and an alluvial soil of West Bengal. Environ. Ecol., 16(2): 344-349.

Kumar PS, Singh SP, Siddesha KNM, Anuroop CP (2004). Role of the acaropathogenic fungus, Hirsutella thompsonii Fisher, in the natural suppression of the coconut eriophyid mite, Aceria guerreronis Keifer, in Andhra Pradesh, India. Bioinoculants for sustainable agriculture and forestry: Proceedings of National Symposium held on February 16-18,-2001. Scientific Publishers (India). pp. 215-221.

Kumar U, Jena SC (1996). Trial on integrated bio-technical approach in biological reclamation of coal mine spoil dumps in South-eastern Coalfields limited (S.E.C.L.) Bilaspur (Madhya Pradesh). Ind. Forest. 122: 1085-1091.

Lal JK, Mishra B (1998). Fly ash as a carrier for Rhizobium inoculant. J. Res. Birsa. Agricul. Univer., 10(2): 191-192.

Lapage SP, Sneath PA, Lessel EF, Skerman VBD, Seeliger HPR, Clark WA (1990). International Code of Nomenclature of Bacteria. American Society of Microbiology, Washington.

Lata SAK, Tilak KVBR (2002). Biofertilizers to augment soil fertility and crop production. Soil Fertility and crop production. (Ed Krishna, K. R.) Science Publishers, USA: pp. 279-312.

Line TF, Huang HI, Shen FT, Young CC (2006). The protons of gluconic acid are the major factor responsible for the dissolution of tricalcium phosphate by Burkholderia cepacia CC-A174. Bioresour. Tech., 97(7): 957-960.

Liu J, Wang ET, Chen WX (2005). Diverse Rhizobia associated with woody legumes Wisteria sinensis, Cercis racemosa and Amorpha fruticosa grown in the temperate zone of China. Syst. Appl. Microbiol., 28(5): 465–477.

Lodwig EM, Poole PS (2003). Metabolism of Rhizobium bacteroids. Crit. Rev. Plant. Sci., 22(1): 37–38.

Lulandala LLL, Hall JB (1986). Leucaena leucocephala's biological nitrogen fixation: a promising substitute for inorganic nitrogen fertilization in agroforestry systems. In Biotechnology of nitrogen fixation in the tropics (BIOnifT), Proceedings of UNESCO Regional Symposium and Workshop, UPM, Malaysia, 25-29 August 1986. Indian Society of Soil Science (India). Chapter viii: 369.

Lutenberg BJJ, Dekkers LC (1999). what makes Pseudomonas bacteria rhizospheric competent. Environ. Microbiol., 1: 9-13.

Mahalingam PU, Thilagavathy D (2008). A study on phosphate solubilizing microorganisms of an orchard ecosystem in Tamilnadu. J. Pure. Appl. Microbiol., 2(1): 219-222.

Maheshwari DK, Kumar H (1997). Rhizobial inoculant and their applications. 49th Annual meeting of the Indian Phytopathological society and National symposium on the biology of plant microbe interaction February 15-17, 1997, RD university, Jabalpur.pp. 28-29.

Maiti SK (2007). Bioreclamation of coalmine overburden dumps- with special emphasis on macronutrients and heavy metals accumulation in tree species. Environ. Monit. Assess., 125: 111-122.

Mahmood A, Iqbal R, Qadri R, Naz S (2004). Some observations on mycorrhizae of Leucaena leucocephala (Lam.) De Wit. Pak. J. Bot., 36(3): 659-662.

Mamatha G, Bagyaraj DJ (2002). Different level of VAM inoculums application on growth and nutrition of Tomato in the nursery. Bioinoculants for sustainable agriculture and forestry. Proceeding of national Symposium held on February 16-18, 2001(Eds. S.M. Reddy, S. Ram Reddy, M. A. Singarachary and S. Girisham). Scientific Publishers (India). pp.43-48.

Mamatha G, Shivayogi MS, Bagyaraj DJ, Suresh CK (2002). Variations in microbial load and their activity in sandalwood growing in hilly zone of Karnataka. Bioinoculants for sustainable agriculture and forestry. Proceeding of national Symposium held on February 16-18, 2001(Eds. S.M. Reddy, S. Ram Reddy, M. A. Singarachary and S. Girisham). Scientific Publishers (India). pp. 119-123.

Manoharachary C (2002). Arbuscular Mycorrhizal fungi and their SEM

aspects. Bioinoculants for sustainable agriculture and forestry. Proceeding of national Symposium held on February 16-18, 2001(Eds. S.M. Reddy, S. Ram Reddy, M. A. Singarachary and S. Girisham). Scientific Publishers (India).pp.37- 42.

Manoharchary C, Sridhar K, Singh RA, Adholeya A, Rawat S, Johri BN (2005). Fungal biodiversity, distribution, conservation and prospecting of fungi from India. Curr. Sci., 89(1): 59-70.

Manoharchary C, Mohan K C, Kunwar I K and Reddy S V (2008). Phosphate solubilizing fungi associated with Casuarina equisetifolia. J. Mycol. Plant. Pathol., 38(3): 507-513.

Marimuthu S, Subbian P, Ramamoorthy V, Samiyappan R (2002). Synergistic effect of combined application of Azospirillum and Pseudomonas fluorescens with inorganic fertilizers on root rot incidence and yield of cotton. Zeit. Pflanzenkrankheiten. Pflanzenschutz. 109(6): 569-577.

Martínez-Romero E, Segovia L, Mercante FM, Franco A A, Graham P, Pardo MA (1991). Rhizobium tropici, a novel species nodulating Phaseolus vulgaris L. beans and Leucaena sp. trees. Int. J. Syst. Bacteriol., 41(3): 417–426.

Mathew MM, Hameed SMS (2003). Effect of microbial inoculants and phosphorus levels on yield attributes, yield and quality characters of vegetable cowpea (Vigna unguiculata subsp. sesquipedalis (L.) Verdcourt) in an oxisol. Sou. Ind. Horticul. 51(1/6): 186-190.

Mathis R, deRycke R, D'Haeze W, Van Maelsaeke E, Anthonio E, Van Montagu M, Holsters M, Vereecke D, Gijsegem FV (2005). Lippopolysaccharides as a communication signal for progression of legume endosymbiosis. Proceeding of National Academy of sciences of United States of America. 102(7): 2655-2660.

Mehrvarz S, Chaichi MR (2008). Effect of phosphate solubilizing microorganisms and phosphorus chemical fertilizer on forage and grain quality of barley (Hordeum vulgare L.). Am. Eur. J. Agri. Environ. Sci., 3(6): 855-860.

Mehrvarz S, Chaichi MR, Alikhani HA (2008). Effects of phosphate solubilizing microorganisms and phosphorus chemical fertilizer on yield and yield components of barely (Hordeum vulgare L.). Am. Eur. J. Agri. Environ. Sci., 3 (6): 822-828.

Mirik M, Aysan Y, Cinar O (2008). Biological control of bacterial spot disease of pepper with Bacillus strains. Turk. J. Agricul. For., 32(5): 381-390.

Mishra A, Sharma SD, Pandey R (2004). Amelioration of degraded sodic soil by afforestation. Arid. Land. Res. Manag. 18(1): 13-23.

Muthukumar T, Udaiyan K (2006). Growth of nursery-grown bamboo inoculated with arbuscular mycorrhizal fungi and plant growth promoting rhizobacteria in two tropical soil types with and without fertilizer application. New. For., 31(3): 469-485.

Narloch C, deOliveira VL, dosAnjos JT, Silva GN (2002) Responses of radish culture to phosphate-solubilizing fungi. Pesquisa. Agropecuaria. Brasileira. 37(6): 841-845.

Nehra K, Yadav AS, Sehrawat AR, Vashishat RK (2007). Characterization of heat resistant mutant strains of Rhizobium sp [Cajanus] for growth, survival and symbiotic properties. Ind. J. Microbiol., 47(4): 329-335.

Noyd RK, Pfledger FL, Norland MR, Sadowasky MJ (1995). Native Prairie grasses and microbial community responses to reclamation of taconite iron Ore tailing. Can. J. Bot., 73: 1645-1654.

O'Gara F, Shanmugam KT (1976). Regulation of nitrogen fixation by Rhizobia. Export of fixed N_2 as NH_4^+. Biochimica. Biophysica. Acta. 437(2): 313–321.

Odee DW, Indieka SA, Lesueur D (2002). Evaluation of inoculation procedures for Calliandra calothyrsus Meisn. grown in tree nurseries. Biol. Fertil. Soil., 36(2): 124-128.

Oke V, Long SR (1999). Bacteroid formation in the Rhizobium–legume symbiosis. Curr. Opin. Microbiol., 2(6): 641–646.

Okon Y (1985). Azosprillum as a potential inoculants for agriculture. Trend. Biotechnol., 3: 223-228.

Oliviery I, Frank SA (1994). The evolution of nodulation of Rhizobium: Altruism in the rhizosphere. J. Heredit., 85 (1): 46-47.

Osorio NW, Habte M (2001). Synergistic influence of an arbuscular mycorrhizal fungus and a phosphate solubilizing fungus on growth and P- uptake of Leucaena leucocephala in an oxisol. Arid. Land. Res. Manag., 15(3): 159-165.

Pandey A, Das N, Kumar B, Rinu K, Trivedi P (2008). Phosphate solubilization by *Penicillium* spp. Isolated form soil samples of Indian Himalayan Region. Wor. J. Microbiol. Biotechnol., 24: 97-102.

Pandey A, Palni LMS (1998). Isolation of *Pseudomonas corrugates* from Sikkim, Himalaya. Wor. J. Microbiol. Biotechnol., 14: 411-413.

Pandya SR, Patil MR, Khara RB (1997). Revegetation of coal spoils by flyash and pulp and paper mill waste. J. Indust. Pollut. Cont., 13: 151-157.

Patil MG, Sayyed RZ, Chaudhari AB, Chincholkar SB (2002). Phosphate solubilizing microbes: a potential bioinoculant for efficient use of phosphate fertilizers. Bioinoculants for sustainable agriculture and forestry: Proceedings of National Symposium held on February 16-18, 2001. Scientific Publishers (India). pp. 107-118.

Patriarea EJ, Tatè R, Ferraioli S, Iaccarino M (2004). Organogenesis of legume root nodules. Inter. Rev. Cytol., 234: 201–262.

Pauly N, Pucciariello C, Mandon K, Innocenti G, Jamer A, Baudouin E, Frendo P, Puppo A (2006). Reactive oxygen and nitrogen species and glutathrion: Key players in the legume-*Rhizobium* symbiosis. J. Exot. Bot., 57 (8): 1769-1776.

Perret X, Staehelin C, Broughton WJ (2000). Molecular basis of symbiotic promiscuity. Microbiol. Mol. Biol. Rev., 64(1): 180–201.

Peters SM, Habte M (2001). Optimizing solution P concentration in a peat-based medium for producing mycorrhizal seedlings in containers. Arid. Land. Res. Manag., 15(4): 359-370.

Poi SC (2008). Diversity of phosphate solubilizing microorganisms in the soils of district Murshidabad of West Bengal. Environ. Ecol., 26(4): 1502-1504.

Prabakaran J, Ravi KB (1996). Influence of bioinoculants on *Cenchrus ciliaris* nursery. Mad. Agric. J., 83(8): 531-533.

Prakash RK, Schilperoort RA, Nuti MP (1981). Large plasmids of fast-growing *Rhizobia*: homology studies and location of structural nitrogen fixation (*nif*) genes. J. Bacteriol. 145(3): 1129–1136.

Prasad R, Mahammad G (1990). Effectiveness of nitrogen fixing trees (NFTs) in improving microbial status of Bauxite and Coal mined out areas. J. Trop. For., 6: 86-94.

Purbey SK, Sen NL (2005). Effect of bioinoculants and bioregulators on productivity and quality of fenugreek (*Trigonella foenum-graecum*). Ind. J. Agri. Sci., 75(9): 608-611.

Purohit U, Mehar SK, Sundarmoorthy S (2007). Ecology of soil fungi in *Acacia nilotica* based agroforestry systems of Rajasthan, India. J. Ind. Bot. Soc., 86(1&2): 86-94.

Qadri R, Mahmood A (2003). Characterization of *Rhizobia* isolated from some tree legumes growing at the Karachi University campus. Pak. J. Bot., 35(3): 415-421.

Qadri R, Mahmood A, Athar A (2007). Ultra structural studies on root nodules of *Pithecellobium dulce* (Roxb.) benth. (Fabaceae). Agri. Concep. Sci., 72(2): 133-139.

Raman N, Gnanaguru M, Srinivasan V (2002). Biocontrol of plant pathogens by arbuscular mycorrhizal fungi for sustainable agriculture. Bioinoculant for sustainable agriculture and forestry: Proceedings of national symposium held on February 16-18, 2001(Eds. S.M. Reddy, S. Ram Reddy, M. A. Singarachary and S. Girisham), Scientific publishers (India). pp.153-175.

Rahangdale R, Gupta N (1998). Selection of VAM inoculants for some forest tree species. Ind. For., 124: 331-341.

Rai UN, Pandey K, Sinha S, Singh A, Saxena R, Gupta DK (2004). Revegetating fly ash landfills with *Prosopis juliflora* L.: impact of different amendments and *Rhizobium* inoculation. Environ. Inter., 30(3): 293-300.

Raja P, Uma S, Gopal H, Govindarajan K (2006). Impact of bio inoculants consortium on rice root exudates, biological nitrogen fixation and plant growth. J. Biol. Sci., 6 (5): 815-823.

Ramesh R, Shanthamallaiah NR, Jayadeva HM, Hiremath RR, Bhairappanavar ST (1998). Yield and quality parameters of irrigated groundnut as influenced by sources of phosphate, FYM and phosphate solubilizing microorganisms. Crop. Res. Hisar., 15(2/3): 144-147.

Rangarajan S, Loganthan P, Saleena LM, Nair S (2001). Diversity of *Pseudomonads* isolated from three different plant rhizospheres. J. Appl. Microbiol., 91: 742-749.

Rangarajan S, Saleena LM, Vasudevan P, Nair S (2001). Biological suppression of rice diseases by *Pseudomonas* spp. Under saline soil conditions. Plant. Soil., 251: 73-82.

Rao JR, Fenton M, Jarvis BDW (1994). Symbiotic plasmid transfer in *Rhizobium leguminosarum* biovar *trifolii* and competition between the inoculant strain ICMP2163 and transconjugant soil bacteria. Soil. Biol. Biochem., pp.339–351.

Rao VM, Niranjan R, Shukla R, Preek R, Jain HC (2003). Enhancement of growth, nodulation and N_2 –fixation in *Dalbergia sissoo* roxb. by *Rhizobium* and *Glomus fasciculatum*. J. Mycol. Plant. Pathol., 33(1): 109-113.

Rathi BK, Jain AK, Kumar S (2009). Influence of *Rhizobium leguminosarum* inoculation along with sulphur and micronutrients on quality aspect of blackgram (*Vigna mungo* (L.) heeper). J. Ind. Bot. Soc., 88(1&2): 67-69.

Reddell P, Milnes AR (1992). Mycorrhizas and other specialised nutrient-aquition strategies: Their occurrence in woodland plants from Kakadu and their role in rehabilitation of waste rock dumps at a local uranium mine. Aus. J. Bot., 40: 223–242.

Reis MR, Silva AA, Guimaraes, AA, Costa MD, Massenssini AM, Ferreira EA (2008). Action of herbicides on inorganic phosphate-solubilizing microorganisms in sugarcane rhizospheric soil. Planta. Daninha. 26(2): 333-341.

Reyes I, Valery A, Valduz Z (2006). Phosphate-solubilizing microorganisms isolated from rhizospheric and bulk soils of colonizer plants at an abandoned rock phosphate mine. Plant. Soil., 287(1/2): 69-75.

Richa G, Khosla B, Reddy MS (2007). Improvement of Maize plant growth by phosphate solubilizing fungi in rock phosphate amended soil. Wor. J. Agricul. Sci., 3: 481-484.

Rivas R, Peix A, Mateos PF, Trujillo ME, MartinezMolina E, Velazquez E (2006). Biodiversity of populations of phosphate solubilizing *Rhizobia* that nodulates chickpea in different Spanish soils. Plant. Soil., 287(1-2): 23-33.

Rivas R, Willems A, Palomo JL, Garcia-Benavides P, Mateos PF, Martínez-Molina E, Gillis M, Velázquez E (2004). *Bradyrhizobium betae* sp. nov., isolated from roots of *Beta vulgaris* affected by tumour-like deformations. Int. J. Syst. Evol. Microbiol., 54(4): 1271–1275.

Roberts RD, Marrs RH, Skeffington RA, Bradshaw AD (1981). Ecosystem development on naturally colonized china clay wastes. Vegetation changes and overall accumulation on organic matter and nutrient. J. Ecol., 69: 153-161.

Rogel MA, Hernández-Lucas I, Kuykendall LD, Balkwill DL, Martínez-Romero E (2001). Nitrogen-fixing nodules with *Ensifer adhaerens* harboring *Rhizobium tropici* symbiotic plasmids. Appl. Environ. Microbiol., 67(7): 3264–3268.

Roux CL, Tentchev D, Prin Y, Goh D, Japarudin Y, Perrineiu MM, Duponnois R, Domergue O, deLajudie P, Galiana A (2009) Bradyrhizobia nodulating the *Acacia mangium* X *A auriculiformis* interspecific hybrid are specific and differ from those associated with both parental species. Appl. Environ. Microbiol., 75(24): 7752-7759

Ryan MH, Angus JF (2003). Arbuscular mycorrhizae in wheat and field pea crops on a low P soil: increased Zn-uptake but no increase in P-uptake or yield. Plant. Soil., 250: 225-239.

Saha AK, Haque MF (2005). Effect of inocutation with *Rhizobium* on nodulation and growth of Bean, *Dolichos lablab*. J. Life. Earth. Sci., 1(1): 71-74.

Sahgal M, Johri BN (2003). The changing face of Rhizobial systematic. Curr. Sci., 84(1): 43-48.

Sahgal M, Sharma A, Johri BN, Prakash A (2004). Selection of growth promontory *Rhizobia* for *Dalbergia sissoo* from diverse soil ecosystems of India. Symbiosis. 36(1): 83-96.

Sanginga N, Mulongoy K, Ayanaba A (1985). Inoculation of *Leucaena leucocephala* (Lam.) de Wit with *Rhizobium* and its nitrogen contribution to a subsequent maize crop. Biol. Agri. Hort., 3: 347-352.

Schroth MN, Hancock JG (1981). Selected topics in biological control. *Ann.* Rev. Microbiol., 35: 453-476.

Schwintzer R, Tjepkema JD (1990). The biology of *Frankia* and *Actinorrhizal* Plants, Academic press inc. San Diego, USA. p. 99.

Sengupta M (1993). Environmental impacts of mining: Monitoring, restoration and control, 430. Bocaraton, Florida: Lewis Publisher.

Sengupta SK, Dwivedi YC, Kushwah SS (2002). Response of tomato (Lycopersicon esculentum Mill.) to bioinoculants at different levels of nitrogen. Veg. Sci., 29(2): 186-188.

Sharma MP, Bhatia NP, Adholeya A (2001). Mycorrhizal dependency and growth responses of *Acacia nilotica* and *Albizzia lebbeck* to inoculation by indigenous AM fungi as influenced by available soil P levels in a semi-arid Alfisol wasteland. New. For., 21(1): 89-104.

Sharma RS, Ramamurthy V (2000). Nodulation adaptability and growth promoting ability of *Rhizobium* and *Bradyrhizobium* strains for *Acacia auriculiformis*. Ind. J. Microbiol., 40: 131-135.

Sharma SN, Prasad R (2003). Yield and P uptake by rice and wheat grown in a sequence as influenced by phosphate fertilization with diammonium phosphate and Mussoorie rock phosphate with or without crop residues and phosphate solubilizing bacteria. J. Agri. Sci., 141(3-4): 359-369.

Singh G, Tilak KVBR (2001). Phosphorus nutrition through combined inoculation with phosphorus solubilising rhizobacteria and arbuscular mycorrhizae. Fertil. News. 46 (9): 33-36.

Singh S, Kapoor KK (1999). Inoculation with phosphate-solubilizing microorganisms and a vesicular arbuscular mycorrhizal fungus improves dry matter yield and nutrient uptake by wheat grown in a sandy soil. Biol. Fertil. Soil. 28(2): 139-144.

Skerman VBD, McGowan V, Sneath PA (1980). Approved lists of bacterial names. Int. J. Syst. Bacteriol., 30(1): 225-420.

Sonkar DS, Singh AK, Banerjee SK (1998). Relative suitability of different nitrogen fixing and non-nitrogen fixing tree species on coal mine overburden of Jayant, Singaruli. Environ. Ecol., 16: 314-317.

Spaink HP (2002). Root nodulation and infection factors produced by Rhizobial bacteria. Annl. Rev. Microbiol., 54 (1): 257-288.

Sprent JI (1994). Evolution and diversity in the legume–*Rhizobium* symbiosis—Chaos theory. Plant. Soil. 161 (1): 1-10.

Sullivan JT, Ronson CW (1998). Evolution of rhizobia by acquisition, of a 500-kb symbiosis island that integrates into a phe-tRNA gene. Proceedings of the National Academy of Sciences of the United States of America. 95 (9): 5145-5149.

Sumana DA, Bagyaraj DJ (2002). Interaction between VAM fungus and nitrogen fixing bacteria and their influence on growth and nutrition of neem (*Azadirachta indica A. juss*). Ind. J. Microbiol., 42(4): 295-298.

Suslow TV (1982). Role of root colonizing bacteria in plant growth. Phytopathogenic prokaryotes (Eds. M. S. Mount and G. A. Lacy). Academic press, London: pp.187-223.

Tchan YT (1984). Family II. Azotobacteriaceae. In Bergey's Manual of Systematic Bacteriology (Eds Krieg, N. R. and Holt, J. G.) Williams and Wikins, Baltimore, 1: 219.

Tchan YT, New PT (1984). Genus I. *Azotobacter beijerinck*. In Bergey's Manual of Systematic Bacteriology. Williams and Wikins, Baltimore: p.220.

Terakado J, Okamura M, Fujihara S, Ohmori M, Yoneyama T (1997). Cyclic AMP in *Rhizobia* and symbiotic nodules. Ann. Bot. 80: 499-503.

Thatoi HN, Sahu S, Mishra AK, Padhi GS (1993). Comparative effect on VAM inoculation on growth, nodulation and Rhizosphere population of Subabul [*Leucaena leucocephala* (lam.) De wit.] grown in iron mine waste soil. Ind. For., 119(6): 481-489.

Tilak KVBR (1993). Association effects of vesicular arbuscular mycorrhizae with nitrogen fixers. *Proc.* Indian. Natl. Sci. Acad., B59(3&4): 325-332.

Tilak KVBR (2001). Biofertilizers: Their role in integrated nutrient management. Bioinoculants for sustainable agriculture and forestry. Proceeding of national Symposium held on February 16-18, 2001 (Eds. S.M. Reddy, S. Ram Reddy, M. A. Singarachary and S. Girisham). Scientific Publishers (India). pp.1- 7.

Tilak KVBR, Ranganayaki N, Pal KK, De R, Saxena AK, Shekhar NC, Mittal S, Tripathi AK, Johri BN (2005). Diversity of plant growth and soil health supporting bacteria. Curr. Sci., 89(1): 136-150.

Tilak KVBR, Subba RNS (1987). Association of *Azospirillum brasilense* with pearlmillet (*Pennisetum americanum* (L.) Leeke). Biol. Fertil. Soil., 4: 97-102.

Turan M, Ataoglu N, Sahin F (2006). Evaluation of the capacity of phosphate solubilizing bacteria and fungi on different forms of phosphorus in liquid culture. J. Sust. Agric., 28(3): 99-108.

Van Rhijn P, Vanderleyden J (1995). The *Rhizobium*–plant symbiosis. Microbiol. Rev., 59(1): 124-142.

Varma S, Mathur RS (1989). Biocoenotic association between nitrogen-fixing and phosphate-solubilizing microorganisms. Curr. Sci., 58(19): 1099-1100.

Vazquez P, Holguin G, Puente ME, Lopez CA, Bashan Y (2000). Phosphate-solubilizing microorganisms associated with the rhizosphere of mangroves in a semiarid coastal lagoon. Biol. Fertil. Soil. 30(5/6): 460-468.

Velázquez E, Peix A, Zurdo-Piñeiro JL, Palomo JL, Mateos PF, Rivas R, Muñoz AE, Toro N, García-Benavides P, Martínez-Molina E (2005). The coexistence of symbiosis and pathogenicity-determining genes in *Rhizobium rhizogenes* strains enables them to induce nodules and tumors or hairy roots in plants. Mol. Soil. Til. Res., 50(2): 159-167.

Verma DPS (1992). Signals in root nodule organogenesis and endocytosis of *Rhizobium*. Plant. Cell. 4: 373-382.

Verma LN (1995). Conservation and efficient use of organic sources of plant nutrients. Organic agriculture. Peekay Tree Crops Development Foundation. India: pp. 101-143.

Vijaya T, Srivasuki KP (1999). Effect of inoculum type and inoculation dose on ectomycorrhizal development and growth of *Acacia nilotica* seedlings inoculated with *Pisolithus tinctorius* in a nursery. Biologia. 54(4): 439-442.

Wang ET, MartinezRomero J, MartinezRomero E (1999). Genetic diversity of *Rhizobia* from *Leucaena leucocephala* nodules in Mexican soils. Mol. Ecol., 8(5): 711-724.

Wang GH, Jin J, Xu MN, Pan XW, Tang C (2007). Inoculation with phosphate-solubilizing fungi diversifies the bacterial community in rhizospheres of maize and soybean. Pedosphere. 17(2): 191-199.

Werner D (1992). Symbiosis of Plants and Microbes, Chapman and Hall, New York: pp. 387-400.

Werner D (2004). Signalling in the *Rhizobia*-legumes symbiosis. Plant. Surface. Microbiol., pp.99–120.

Wheeler CT, Miller JM (1990). Current and potential uses of Actinorrhizal plants in Europe. In the biology of *Frankia* and *Actinorrhizal* plants (Eds Schwintzer, C. R. and Tjepkem, J. D.) Academic Press, San Diego, USA: pp.128-156.

Willems A, Fernández-López M, Munoz-Adelantado E, Goris J, De Vos P, Martínez-Romero E, Toro N, Gillis M (2003). Description of new *Ensifer* strains from nodules and proposal to transfer *Ensifer adhaerens* Casida 1982 to *Sinorhizobium* as *Sinorhizobium adhaerens* comb. nov. Request for an opinion. Int. J. Syst. Evol. Microbiol., 53 (4): 1207-1217.

Wu KJ, Wu CS, Chang JS (2007). Biodegradability and mechanical properties of polycaprolactone composites encapsulating phosphate-solubilizing bacterium *Bacillus* sp PG01. Process. Biochem., 42(4): 669-675.

Xiao CQ, Chi RA, Huang XH, Zhang WX, Qiu GZ, Wang DZ (2008). Optimization for rock phosphate solubilization by phosphate-solubilizing fungi isolated from phosphate mines. Ecol. Eng., 33(2): 187-193.

Xiao LM, Dou XB, Min WZ, RuiJiang Q, Jian C, Xue YT Song HS (2008). Screening and characterization of phosphate solubilizing microorganisms at high temperature. Res. Environ. Sci., 21(3): 165-169.

Yadav J, Verma JP, Tiwari KN (2011). Plant growth promoting activities of fungi and their effect on Chickpea plant growth. Asi. J. Biol. Sci., 4(3): 291-299.

Young JM, Kuykendall LD, Martínez-Romero E, Kerr A, Sawada H (2001). A revision of *Rhizobium* Frank 1889, with an emended description of the genus, and the inclusion of all species of *Agrobacterium* Conn 1942 and *Allorhizobium undicola* de R. undicola and R. vitis. Int. J. Syst. Evol. Microbiol., 51(1): 89–103.

Young JM, Park DC, Weir BS (2004). Diversity of 16S rDNA sequences of *Rhizobium* spp. implications for species determinations. FEMS. Microbiol. Lett., 238(1): 125–131.

Zaidi A, Khan MS (2006). Co-inoculation effects of phosphate solubilizing microorganisms and *Glomus fasciculatum* on green gram *Bradyrhizobium* symbiosis. Tur. J. Agri. For., 30(3): 223-230.

20

Micropropagation of *Allanblackia stuhlmannii*: Amenability to tissue culture technique

Johnstone Neondo[1]*, Joseph Machua[2], Anne Muigai[1], Aggrey B. Nyende[1], Moses Munjuga[3], Ramni Jamnadass[3] and Alice Muchugi[3]

[1]Jomo Kenyatta University of Agriculture and Technology, P.O. Box 62000-00200, Nairobi, Kenya.
[2]Kenya Forestry Research Institute, P. O. Box 20412-00200, Nairobi, Kenya.
[3]World Agroforestry Centre (ICRAF), P. O. Box 30677-00100, Nairobi, Kenya.

Allanblackia stuhlmannii is an endangered forest tree valued for its edible nut oil. Its limited regenerative potential in the wild hinders the sustainable utilization of its products. To achieve mass production of *A. stuhlmannii*, its amenability to micropropagation technique was examined. Explants were best surface sterilized at 8% sodium hypochlorite for 10 min and rinsed using sterile distilled water. Of eight basal nutrient media tested, Lloyd and McCown Woody plant medium (WPM) was the most suitable (88.89% explants survival). Microshoots were induced from apical meristems cultured on WPM supplemented with different concentrations of 6-benzyladenine (BAP), kinetin (KN), Dichlorophenoacetic acid (2, 4 - D), Naphthalene acetic acid (NAA) and Thidiazuron (TDZ), ($P < 0.05$). All responding explants produced a single microshoot irrespective of the type and concentration of PGRs used. 1.2 mg/IBAP and 1.2 mg/IKIN exhibited the most rapid and consistent shoot length increase ($P < 0.05$). Prolonged culture or sub culturing did not promote further shoot proliferation. Callus was induced from leaf explants cultured on WPM fortified with Gamborg's vitamins, 3% sucrose, 1 mg/IKIN combined with 1.2 mg/l 2, 4 - D. No somatic embryos emerged from the callus. The success in explant sterilization and induction of microshoot and callus in this study is a milestone step in the regeneration of *A. stuhlmannii*.

Key words: Sterilization, media plant growth regulators, shoot proliferation, callus induction.

INTRODUCTION

Allanblackia stuhlmannii Eng. (Clusiaceae) locally known as 'Msambu' is a forest tree with high market potential and grows naturally in East and West Usambara forests, Nguru and Uluguru forest mountains in Tanzania (El Tahir and Mlowe, 2002; Meshack, 2004, Van, 2005). During World War I, German soldiers in Tanzania used fat extracted from *A. stuhlmannii* nuts as an alternative edible fat to butter (Saka, 1995). The communities living around the Eastern ArcMountains, particularly farmers, use the oil extracted from *A. stuhlmannii* nuts for food and soap production (Lovett, 1983; Monela et al., 2001; Osemeobo, 2005, Pye-Smith, 2009).

They also use dry leaves of this tree as medicinal tea to treat chest pain and smear heated seed oil on aching joints, rashes and wounds (Meshack, 2004). Phytochemical analysis of *A. stuhlmannnii* crude extracts showed that Guttiferone F, a prenylatedbenzophenone, a compound related to a group of compounds that have been studied for their anti-HIV property, was present (Fuller et al., 2003). This tree species also has a great commercial potential for margarine production from its edible seed oil whose extraction requires less chemical processing and refraction than palm oil (Atangana et al., 2006). Already, the oil from *A. stuhlmannii* has received the approval of the European Union (EU) Novel Food Regulations that certify safe usage as a foodstuff (Hermann, 2009; Ramni et al., 2010).

Regeneration of *A. stuhlmannii* via seed is however

*Corresponding author. E-mail: jonneondo@yahoo.com.

slow and low (low seed fecundity/viability). Germination typically takes 1 to 7 months to begin and a minimum of 18 months to complete after sowing (Mwaura and Munjuga, 2011). Rooting of cuttings is poor while survival rate of grafted materials is dismal (Mwaura and Munjuga, 2011). Under natural conditions, the trees begin to bloom at the age of 12 years. The fruits take over a year to mature and become ripe between November and March and between August and October (Mathayo et al., 2009). Rodents and monkeys eat the fruits, and hence, providing the only mode of seed dispersal (Glynn and Ritzl, 2000). The limited regenerative potential and dispersal powers of A. stuhlmannii in the wild are likely to be a vulnerable element of the local biodiversity (Amanor et al., 2003; Attipoe et al., 2006). Loss of A. stuhlmannii biodiversity is likely to result to decline in the quality of harvested products (Amanor et al., 2003; Cordeiro et al., 2007; Egyir, 2007). Sustainable development of A. stuhlmannii industry will greatly dependent on mass production of A. stuhlmannii. A micropropagation protocol with high multiplication rates will greatly contribute in the domestication and hence, conserve this economically endangered tree species.

Micropropagation offers a rapid means of producing large quantity of clonal planting stocks and propagation of some commercial crops and also tree species that are difficult to establish conventionally (Bonga, 1987; Merkle and Dean, 2000; Thorpe et al., 1990). Micropropagation of a wide range of tree species have been successfully achieved (Pankaj and Toshiyuki, 2001). However, numerous recalcitrant forest trees of economic value are still difficult to establishment in vitro (Anna et al., 2010). There are no documented studies on the micropropagation of A. stuhlmannii or any member of this genus. The use of explants from mature plants is not frequently accomplished, mainly due to the high level of contamination (Drew, 1988), reduced or absence of morphogenetic ability (Bonga, 2010) and poor rooting of the regenerated shoots. The selection of suitable explants and establishment of shoot cultures are two critical factors for cloning of trees while maintaining clonal fidelity and purity (Maynard, 1988). Induction of cellular differentiation in vitro however, depends on genetic totipotency, culture medium formulation, and incubation conditions (Gasper et al., 1996). During the in vitro culture process, undesirable or inhibitory compounds such as excess phenolic metabolites (Carlberg et al., 1983), ethylene (Mensuali-Sodi et al., 1996) and 5-hydroxy methyl furfural?, an inhibitory by-product of autoclaving sucrose, can be produced during medium preparation. These compounds hinder successful in vitro induction of cellular differentiation. As a result, there is need for preliminary experiments for selecting or modifying the known basal media that will be suitable for micropropagation of the plant species (Preece and Compton, 1991). In this paper, we reported on the adoptable explant sterilization protocol, direct organogenesis and callus induction from leaf explants of A. stuhlmannii.

MATERIALS AND METHODS

Collection and management of plant material

Plant materials used in this study were collected in October, 2009 at Amani Nature Reserve (ANR) located in the Southern part of the East Usambara Mountains (4°48 to 5°13'S, 38°32 to 48°E) in Tanzania. The plant materials consisted of mature seeds, seedlings and cuttings from coppices. They were thoroughly washed at sampling site and then stored in cool boxes and later transported to Kenya Forestry Research Institute (KEFRI) at Muguga in Kenya. On arrival, the seedlings were transplanted in potting bags containing well mixed loam soil. The seeds were germinated on growth trays half of which contained sawdust mixed with sand (1:1) and the other half contained sawdust mixed with decomposing manure. Both transplantation and germination were done in a glasshouse at KEFRI. The cuttings were dipped vertically in sawdust mixed with sand (1:1) in a nursery at KEFRI. They were all watered twice a week using a sprinkler. Seeds that germinated were transplanted into similar potting bags. Well adapted seedlings were used as stock plants in subsequent experiments.

Explants preparation and establishment of sterilization protocol

All the glassware and metallic equipment used for this section were sterilized by autoclaving at 121°C at 1.06 kg cm^{-2} pressure for 15 min before use. Young emerging leaves shoot apices and slender branches were harvested from stock plants and used as explants. The explants were placed in a glass jar containing 500 ml of water into which three drops of Tween® 20 and five drops of Dettol® soap detergent had been added. The jar was swirled gently for 15 min before washing the explants with running tap water for 10 min. Under a clean lamina flow hood, half the number of harvested explants was subjected to sodium hypochlorite and the other half subjected to formaldehyde (sterilants) at varying concentrations and exposure times using the following experimental designs: In experiment 1; explants were separately exposed to the two sterilants at four concentration levels that is, 0 10, 15 and 20% with each concentration level having two exposure times (10 and 20 min). In experiment 2; explants were exposed to 2% Redomil® solution for 15 min prior subjection to sodium hypochlorite at three concentration levels (6, 8 and 10%) each subjected to three exposure times (6, 8 and 10 min). In experiments (1 and 2), explants were rinsed three times using sterile distilled water and placed on sterile blotting paper. The shoots were trimmed to a length of 1 cm and leaves cut into squares of 1 cm dimension before culturing in full strength MS medium. All cultures were daily monitored and assessed for fungal and bacterial infection. The number of explants infected by either fungi or bacteria or both (using morphological descriptors) and those that died due to bleaching were recorded after one week for two consecutive weeks (whereby score 1 represented presence of bacteria and fungi and score 0 represented absence of bacteria and fungi).

Effect of different basal media on explants survival

Eight nutrient media namely; Murashige and Skoog medium, Gamborg (B$_5$) medium, Lloyd and McCown's Woody Plant medium (WPM), White's medium, Preece Hybrid medium, Driver and Kiniyuki walnut (DKW)medium, Anderson medium, and Quorin and

Table 1. Comparison of element levels in the various media used to assess explants survival rate.

Concentration (mg/l) Macro elements	Anderson's medium	DKW medium	Lloyd and McCown's woody plant medium	Quorin and Lepoivre medium	Murashige and Skoog medium	White's medium	Gamborg (B5) medium	Preece hybrid medium
NH_4NO_3	400	1416	400	400	1650	0	0	908
KNO_3	480	0	0	1800	1900	80	2500	0
K_2SO_4	0	1559	990	0	0	0	0	1275
$MgSO_4.7H_2O$	180.7	361.49	180.7	175.8	180.5	720	121.6	271.1
KH_2PO_4	0	265	170	270	170	0	170	217.5
$CaCl_2$	332.2	112.5	72.5	0	332	0	113.2	92.5
$Ca(NO_3)_2.4H_2O$	0	1367	386	833.8	0	300	0	876.5
$Na(H_2PO_4).H_2O$	330	0	0	0	0	16.5	130.4	0
$(NH_4)_2SO_4$	0	0	0	0	0	0	134	0
Micro elements								
$CoCl_2.6H_2O$	0.025	0	0	0.025	0.025	0	0.025	0
$CuSO_4.5H_2O$	0.025	0.25	0.25	0.025	0.025	0	0.025	0.25
$FeNa_2EDTA$	36.7	33.8	36.7	36.7	36.7	3.47	36.7	30.83
H_3BO_3	3	4.8	6.2	6.2	6.2	1.5	3	5.5
KI	0.75	0	0	0.08	0.83	0.75	0.75	0
$MnSO_4.H_2O$	10	33.5	22.3	0.76	16.9	5.31	10	27.9
$Na_2MoO_4.2H_2O$	0.25	0	0.25	0.25	0.25	0	0.25	0.32
$ZnSO_4.7H_2O$	2	17	8.6	8.6	8.6	2.67	2	4.3
$NiSO_4$	0	0.005	0	0	0	0	0	0
$Zn(NO_3)_2$	0	17	0	0	0	0	0	8.5
Vitamins								
Glycine	0	0	2	0	2	0	0	1
Nicotinic acid	0	0	0.5	0	0.5	0	1	0.25
Pyridoxine	0	0	0.5	0	0.5	0	1	0.25
Thiamine	0.4	0	0	0	0.2	0	10	0
Myo-inositol	100	0	0.5	0	100	0	100	0
Adenine	80	0	0	0	0	0	0	0
hemisulphate	0	0	0	0	0	0	0	0

Lepoivre medium were tested for their suitability in micropropagation of A. stuhlmannii. Table 1 presents the ion composition of these media. All the media were supplemented with 30 gl^{-1} sucrose and 0.8 g/l agar. The pH value of each media was adjusted to 5.75 using 0.1M HCl or 0.1M NaOH and dispensed in 150 mm (height) by 25 mm (diameter) culture tubes (10 ml of medium per tube). The media were then sterilized in an autoclave at 121°C at 1.06 kg cm^{-2} pressure for 15 min before use. Cultures were grown at 25 ± 1°C and a 16 h photoperiod provided by white fluorescent Philips light bulbs (40 W) in the growth chamber.

Effect of plant growth regulators and their concentrations on proliferation stage

Modified Lloyd and McCown's Woody plant medium (modification: 76 mg/l $CaCl_2$ and 6.0 mg/l H_3BO_3) was used to test the effect of different PGRs at varying concentrations on shoot proliferation. In the single plant growth regulator applications, 1.2 mg/l, 2.4 mg/l and 3.6 mg/l separately for BAP and KIN and 1.2 mg/land 2.4 mg/l for TDZ and a control (media without PGRs) totaling to nine treatments were set. In the combination sets, three different concentrations of BAP (1.2, 2.4 and 3.6) mg/land two concentrations of NAA (0.2 and 0.4) mg/l were combined in six different treatments. In another combination, three different concentrations of KIN (1.2, 2.4 and 3.6) mg/land two concentrations of NAA that is, 0.2

Table 2. Fungal and bacterial contamination levels and explants mortality when exposed to different concentrations of Sodium hypochlorite and Formaldehyde at varying exposure times (Rep = 9).

Sterilant	Concentration (ml) %	Exposure time (min)	Contamination (%)		Explants mortality (%)
			Fungi	Bacteria	
Sodium	0	10	100.00	96.00	0.00
Hypochlorite	0	20	100.00	92.10	0.00
	10	10	16.70	33.30	33.30
	10	20	33.30	33.30	83.30
	15	10	33.30	16.70	100.00
	15	20	0.00	0.00	100.00
	20	10	16.70	16.70	100.00
	20	20	0.00	0.00	100.00
Formaldehyde	10	10	67.00	67.00	67.00
	10	20	33.00	33.00	67.00
	15	10	33.00	33.00	33.00
	15	20	0.00	33.00	100.00
	20	10	33.00	33.00	100.00
	20	20	0.00	0.00	100.00
L.S.D$_{0.05}$			57.71	57.71	36.50

and 0.4 mg/l were combined in six different treatments. In yet another combination, two concentrations of TDZ (1.2 and 2.4) mg/land, two concentrations of NAA (0.2 and 0.4) mg/l were combined in four different treatments. All the media were supplemented with 30 mg/l sucrose and 8 mg/l agar while medium sterilization and culture conditions were effected as described earlier. Induction of microshoots and their length was evaluated at an interval of 4 weeks after initiation.

Effect of plant growth regulators and their concentrations on callus induction stage

Lloyd and McCown's Woody plant medium with two modifications (76 mg/l $CaCl_2$ and 6.0 mg/l H_3BO_3) supplemented with Gamborg's vitamins (McCown and Sellmer, 1987 and Gamborg et al., 1974) was used to test the effect of different plant growth regulators at varying concentrations on induction of callus from leaf discs. Three different concentrations of KIN (0.5, 1 and 2) mg/l and four concentrations of 2, 4-D (1, 1.25, 1.5 and 2) mg/l were combined in five different treatments including a control (media without PGRs). All the media were supplemented with 30 mg/l sucrose and 8 mg/l agar and medium sterilization and culture conditions as described earlier. The percentage proportion of callus induction around the leaf discs was evaluated at an interval of 4 weeks after initiation.

Effect of plant growth regulators and their concentrations on rooting of shoots

In the first experiment, half strength MS and WPM media in which half of each media contained 30 mg/l and the other half contained 1 mg/l sucrose were prepared. All the media were fortified with IBA, IAA and NAA at various concentrations (0.0, 0.01, 0.05, 0.1 and 2.5) mg/l separately and 8 mg/l agar added into each treatment. In the second experiment, a two-step procedure (Bennett et al., 1994; Gasper and Coumans, 1987) was used. Stable shoot were initially cultured in half-strength MS medium (containing 30 mg/l sucrose and 8 mg/l agar) fortified with IBA, IAA and NAA at 0, 0.01, 0.03, 0.05 and 0.1 mg/l concentration levels separately. The shootlets

were then transferred into the same MS medium but without PGRs after three weeks and regularly checked for any sign of rooting.

Data collection and analysis

Data was collected in MS Excel spreadsheets and analysed using Statistical Analysis System (SAS) and Genstat 12th Edition, statistical softwares. Mean number of explants contaminated by bacteria and fungi and those that died as a result of bleaching / scorching was determined. The best sterilant suitable concentration and the preferred time of exposure were deduced from the analyzed means, hence, adopted as the sterilization protocol. The medium with the highest explants survival rate was also determined. ANOVA tests showing the effects of variations and interactions of the various plant growth regulators used and duration (weeks) on induction of microshoots, callus and roots were compared at $P > 0.05$ (Turkey's test).

RESULTS

Sterilization protocol

Results on the effectiveness of sodium hypochlorite and formaldehyde in the sterilization of A. stulhmannii explants show that, increase in concentration and time of exposure, resulted in high explant mortality rate and decrease in bacterial and fungal (although, inconsistent) contamination levels, $P < 0.05$, (Table 2) with sodium hypochlorite being the preferred sterilant. The trend observed in the outcome of experiment 1 was used to design experiment 2 in which the concentration level and exposure time of sodium hypochlorite were optimized with regard to reduction in bacterial and fungal contamination levels and explant mortality rate.

To adequately control fungal contamination, explants

Table 3. Fungal and bacterial contamination levels and mortality of explants exposed to varying concentrations at different exposure durations of Sodium hypochlorite (%), Rep = 9.

Concentration (ml) %	Time (min)	Contamination (%)		Explants mortality (%)
		Fungi	Bacteria	
6	6	66.70	66.70	0.00
6	8	100.00	66.70	0.00
6	10	100.00	66.70	0.00
8	6	66.70	33.30	0.00
8	8	66.70	33.30	0.00
8	10	33.30	0.00	0.00
10	6	66.70	33.30	33.30
10	8	33.30	0.00	66.70
10	10	0.00	0.00	100.00
$L.S.D_{0.05}$		73.82	87.34	46.69

Table 4. Explants survival rate on various nutrient media.

Type of media	No. of cultured explants	No. of live explants	Explant survival (%)	Grade
Anderson's medium	18	4	22.22	Low
White's medium	18	5	27.78	Low
Murashige and Skoog	18	6	33.33	Low
Preece medium	18	10	55.56	Average
Quoirin and Lepoivre	18	11	61.11	Average
Gamborg	18	11	61.11	Average
Driver and kiniyuki	18	12	66.67	Average
Lloyd and McCown's (WPM)	18	16	88.89	High
$L.S.D_{0.05}$			11.27	

were subjected to 2% Redomil® solution (Perez et al., 2009) prior exposure to sodium hypochlorite in the sterilization experiment 2 (Table 3). It was observed that, *A. stuhlmannii* explants were best surface sterilized when exposed to 2% Redomil® solution for 15 min followed by 10 min exposure to 8% sodium hypochlorite and finally rinsed three times using sterile distilled water (Table 3).

Media selection

When explant survival rates on the selected media were compared using $LSD_{0.05}$ (Turkey's test), significant differences were observed (Table 4). The medium with high explants survival rate (Lloyd and McCown's Woody plant medium with 88.89% explants survival rate) was preferred for subsequent experiments. The modification introduced in the selected medium significantly reduced *in vitro* shoot-tip necrosis and phenol exudation by explants.

Direct shoot induction

When the response (mean shoot length of *in vitro* explants) of PGRs treatments were compared using Turkey's test, significant differences were observed.
Treatment 1.2 mg/l KIN exhibited the most significant

increase in shoot elongation compared to any other treatment while 6 mg/lBAP combined with 2 mg/l of 2, 4 - D was the most effective combined PGRs application treatment (Plate 1).

Callus induction

The mean (%) of callus induced from leaf discs of *A. stuhlmannii* differed significantly between the PGRs treatments used (Tables 5 to 6). Callus was induced from leaf disks (especially from the midrib) cultured on modified WPM supplemented with Kinetin combined with 2, 4 – D (Plate 2).

Root induction

No explants rooted in the two-step experimental setup are described in the methodology.

DISCUSSION

Explant sterilization

Surface sterilization of explants is a prerequisite to successful establishment of clean cultures for manipulation. In this study, although, there was a

Plate 1. Direct shoot formation from nodal explants cultured on WPM supplemented with KIN, BAP and TDZ alone or in combination with NAA or 2, 4-D. Key :i) KIN(A)................1.2KIN ii) BAP+2,4-D(d).......2,2,4-D+2BAP; iii) BAP+NAA(e)........3.6BAP+0.4NAA; iv) TDZ+NAA(b).......1.2TDZ+0.2NAA; v) BAP(a)..............2.4BAP; vi) TDZ(c)...............2.4TDZ+0.4NAA; vii) KIN+NAA(b).......2.4KIN+0.2NAA; viii) KIN+NAA(c)........3.6KIN+0.4NAA..

Table 5. The mean shoot length of microshoots induced from nodal explants cultured on WPM supplemented with KIN,BAP, TDZ alone or in combination with NAA and 2,4-D , Rep = 9.

Treatment (mg/l)	Shootlength (cm)		
	Week 4	Week 8	Week 12
Control	1.04 ± 0.010^a	1.11 ± 0.006^a	1.15 ± 0.030^a
3.6KIN + 0.4NAA	1.13 ± 0.021^{ab}	1.21 ± 0.014^{abcd}	1.21 ± 0.014^{ab}
1.2BAP + 0.4NAA	1.14 ± 0.015^{abc}	1.21 ± 0.023^{abcd}	1.21 ± 0.019^{ab}
2.4TDZ + 0.2NAA	1.15 ± 0.015^{abc}	1.36 ± 0.055^{cde}	1.21 ± 0.013^{ab}
1.2KIN + 0.4NAA	1.15 ± 0.017^{bc}	1.21 ± 0.023^{abcd}	1.22 ± 0.018^{abc}
2,2,4-D+4BAP	1.15 ± 0.016^{bcd}	1.22 ± 0.016^{abcd}	1.25 ± 0.020^{abc}
3.6BAP + 0.4NAA	1.15 ± 0.013^{bcd}	1.21 ± 0.013^{abc}	1.21 ± 0.015^{ab}
1.2BAP +0 .2NAA	1.15 ± 0.015^{bcd}	1.22 ± 0.013^{abcd}	1.25 ± 0.013^{abc}
1.2TDZ + 0.4NAA	1.16 ± 0.012^{bcde}	1.42 ± 0.067^e	1.26 ± 0.018^{abcd}
2,2,4-D + 2BAP	1.16 ± 0.014^{bcde}	1.21 ± 0.014^{abc}	1.24 ± 0.010^{abc}
1.2KIN + 0.2NAA	1.16 ± 0.015^{bcde}	1.23 ± 0.016^{abcde}	1.28 ± 0.009^{abcde}
2.4BAP + 0.4NAA	1.19 ± 0.014^{bcdef}	1.24 ± 0.017^{abcde}	1.23 ± 0.016^{abc}
2.4KIN + 0.4NAA	1.19 ± 0.014^{bcdef}	1.24 ± 0.017^{abcde}	1.23 ± 0.019^{abc}
3.6BAP + 0.2NAA	1.19 ± 0.018^{bcdefg}	1.24 ± 0.025^{abcde}	1.25 ± 0.027^{abc}
1,2,4-D + 2BAP	1.19 ± 0.020^{bcdefgh}	1.25 ± 0.021^{abcde}	1.33 ± 0.025^{bcde}
1,2,4-D + 4BAP	1.19 ± 0.024^{bcdefgh}	1.26 ± 0.030^{abcde}	1.30 ± 0.027^{abcde}
1.2KIN	1.20 ± 0.021^{bcdefgh}	1.39 ± 0.071^{cde}	1.44 ± 0.066^e
1.2BAP	1.20 ± 0.021^{bcdefgh}	1.40 ± 0.070^{de}	1.43 ± 0.077^{de}
3.6KIN + 0.2NAA	1.20 ± 0.021^{bcdefgh}	1.24 ± 0.025^{abcde}	1.28 ± 0.016^{abcde}
2.4TDZ	1.20 ± 0.028^{bcdefgh}	1.17 ± 0.015^{ab}	1.35 ± 0.030^{bcde}
2.4KIN	1.20 ± 0.024^{bcdefgh}	1.35 ± 0.058^{bcde}	1.38 ± 0.055^{cde}
2.4BAP	1.21 ± 0.026^{bcdefgh}	1.37 ± 0.058^{cde}	1.36 ± 0.052^{bcde}
1.2TDZ	1.21 ± 0.022^{bcdefgh}	1.24 ± 0.017^{abcde}	1.20 ± 0.022^{ab}
1,2,4-D + 6BAP	1.25 ± 0.021^{cdefgh}	1.34 ± 0.026^{bcde}	1.37 ± 0.025^{bcde}
3.6BAP	1.26 ± 0.026^{defgh}	1.36 ± 0.031^{cde}	1.38 ± 0.035^{cde}
1.2TDZ + 0.2NAA	1.27 ± 0.027^{efgh}	1.11 ± 0.006^a	1.27 ± 0.018^{abcde}
3.6KIN	1.28 ± 0.027^{fgh}	1.36 ± 0.013^{cde}	1.38 ± 0.036^{cde}
2.4TDZ + 0.4NAA	1.28 ± 0.024^{fgh}	1.36 ± 0.032^{cde}	1.36 ± 0.052^a
2.4KIN + 0.2NAA	1.30 ± 0.021^{fgh}	1.36 ± 0.030^{cde}	1.35 ± 0.030^{bcde}
2.4BAP + 0.2NAA	1.30 ± 0.026^{gh}	1.36 ± 0.034^{cde}	1.34 ± 0.032^{bcde}
2,2,4-D + 6BAP	1.30 ± 0.026^h	1.36 ± 0.028^{cde}	1.35 ± 0.041^{bcde}
L.S.D$_{0.05}$	0.058	0.096	0.089

Mean values within a column followed by the same letter are not significantly different by Turkey's test ($P \geq 0.05$). Letters are assigned in ascending order to treatments with higher mean values within the column.

Table 6. Callus induction (mean %) from leaf disk cultured on WPM supplemented with KIN+2, 4-D at various concentration (Re p= 9).

Callus induction (%) Treatment (mg/l)	Week 4	Week 8	Week 12
Control	0.00^a	0.00^a	0.00^a
2KIN + 2 2,4-D[(a)]	0.000^a	0.11^a	0.11^a
0.5KIN + 1 2,4-D[(b)]	8.33^b	8.33^b	8.33^b
2KIN + 1.5 2,4-D[(c)]	10.56^b	10.56^b	10.56^b
1KIN + 1.25 2,4-D[(d)]	32.22^c	32.22^c	32.22^c
L.S.D$_{0.05}$	13.581	13.582	13.582

Mean values within a column followed by the same letter are not significantly different by Turkey's test ($P < 0.05$).

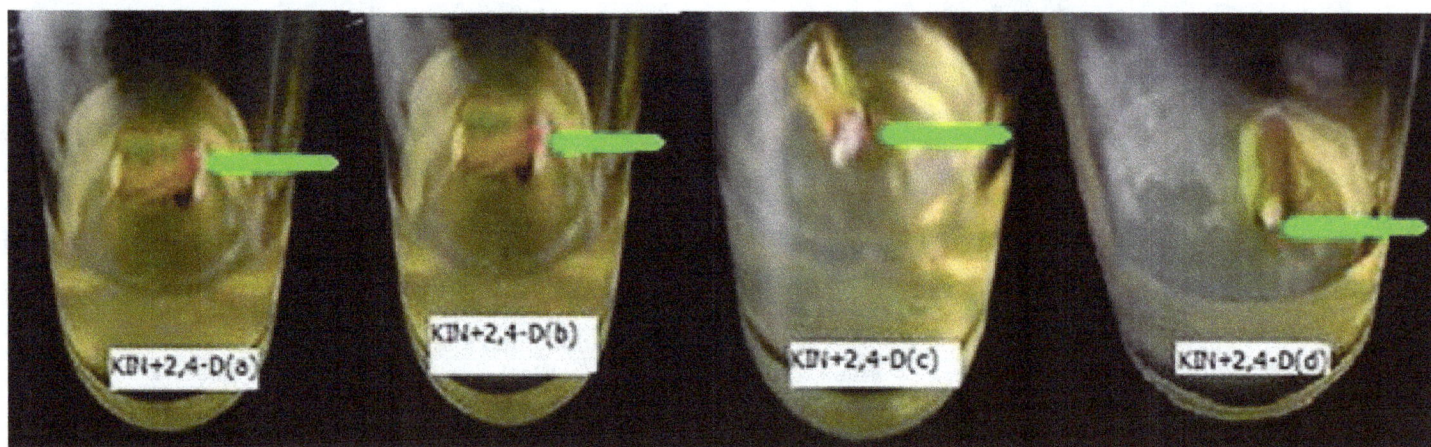

Plate 2. Callus induction from leaf explants (green line indicate the pinkish callus).

significant reduction of fungal and bacterial contamination when the concentrations of the sterilant and exposure time were increased ($P < 0.05$), the mortality rate increased significantly ($P < 0.05$) thereby rendering higher concentration levels and exposure time unsuitable for adoption. The final sterilization experiment showed that, subjection of *A. stuhlmannii* explants to 2% Redomil® solution for 15 min prior to immersion in 8% (ml) sodium hypochlorite at exposure time of 10 min results in clean and live explants than any other concentration level and exposure time used in the experiments (Tables 2 and 3). This is in agreement with Karkonen et al. (1999) findings on *Melaleuca alternifolia*. This was adopted in this research work as the best sterilization protocol for *A. stuhlmannii* seedlings grown in the glasshouse.

Media selection

It is evident that, Lloyd and McCown's Woody plant medium (with 88.89% explants survival) is the best nutrient medium that can used in tissue culture of *A. stuhlmannii* (Table 4).The eventual death of *in vitro* explants on various media such as MS medium,

Anderson's medium, and Quorin and Lepoivre medium was due to production of phenolic compounds and possibly as a result of unsuitability of constituents of individual medium (Preece and Compton, 1991). There is a general tendency of attributing the occurrence of shoot-tip necrosis and subsequent death of *in vitro* explants to the high salt concentration found in some medium for example, MS medium (Bairu et al., 2009). The aforementioned phenomena reduced cell competence and led to eventual loss of their totipotency (Bairu et al., 2009). Alterations of NH_4/NO_3 ratio and sulphur content have been shown to significantly reduce shoot-tip necrosis (Laskshmi and Raghava, 1993). The amounts of calcium and boron in the media also play a critical role in plant tissue culture. Due to its versatility and specificity, calcium plays major structural and functional roles in plants (Hepler, 2005). Plants rely on the unique properties of calcium for their structural, enzymatic and signaling functions and also its role in physiological processes such as cell elongation and cell division (Hirschi et al., 2004). Boron requirements on the other hand differ widely among plant species and are known to have a narrow range between deficiency and toxicity levels when compared to other mineral nutrients (Abdulnour et al., 2000). McCown's woody plant medium was the medium selected for subsequent experiments.

Direct shoot induction

The achievement of uniform and consistent *in vitro* shoot growth is highly problematic for most woody tree species especially those with strong episodic growth characteristic (Gupta et al., 1981; McCown, 2000), such as *A. stuhlmannii*. For many other tree species, *in vitro* clonal propagation is either not possible or the frequency of plant regeneration is too low to be of commercial use (Bonga et al., 2010). Experiments comparing different cytokinin regimes showed that, explants produced between 90 to 100% of responsive (organogenic) explants (Werner et al., 2001). The use of different plant growth regulators either alone or in combinations, significantly affected shoot length of the cultured nodal explants (Zaer, 1982; Nehra et al., 1994; Perez-Tornero and Burgos, 2009). Explants cultured in treatment 1.2 mg/lKIN exhibited significant increase in shoot elongation as compared to any other treatments. The shoot of explants cultured on media with high PGR concentrations, mostly suffered hyperhydricity (vitrification) and hence leaf fall and browning of the shoot apexes (Kataeva et al., 1994). Treatment 6 mg/lBAP combined with 2 mg/ 2, 4- D was the most effective PGRs combination in inducing shoot elongation. Existing reports suggests that, when auxins at lower concentrations are combined with cytokinins, they have critical role in plant regeneration in several systems like *Petasiteshybridus* (Wldi et al., 1998), *Eucalyptus grandis* (Luis et al., 1999), *Hybanthusenneaspermus* (Prakash et al., 1999), *Coleus forskohlii* (Sairam et al., 2001) and *Eleusine indica* (Yemets et al., 2003). Subsequent subculture of explants to either fresh media or media without PGRs resulted in massive loss of explants. All the attempts to induce roots on stabilized *in vitro* shoots failed.

Callus induction

Induction of somatic embryogenesis is a complex phenomenon, which is regulated by numerous factors. In most cases, treatments with exogenous PGRs are required to manipulate cell differentiation (Carman, 1990). Callus was successfully induced from leaf discs (particularly from the midrib) cultured on WPM supplemented with Gamborg's vitamins, 3% (w/v) sucrose and 8 mg/l, 1 mg/l Kinetin combined with 1.2 mg/l 2, 4 - D. Somatic embryogenesis was however not achieved. Induction of callus in *A. stuhlmannii* leaf discs is a possible indication of the potential of this tree species to produce primary somatic embryos that can be made to undergo secondary somatic embryogenesis (SSE). SSE is a phenomenon whereby new somatic embryos are initiated from pre-existing somatic embryos (Raemakers

et al., 1995). Some cultures are able to retain their competence for SSE for many years, and thus provide material for various studies, as described for *Vitisrupestris* (Martinelli et al., 2001). Since new embryos are continually formed from existing embryos, SSE has the potential to produce many plants and once initiated, may continue to produce embryos over a long period of time (Pinto et al., 2002). Therefore, in plants with long life cycles, such as dicotyledonous woody plants for example, *A. stuhlmannii* preserving embryogenic lines can be a cost-effective maintenance while those line are tested in field (Handley, 1995).

Conclusion

Sterilization protocol for *A. stuhlmannii* was successfully established. The ability to induce shoots from nodal explants and callus from leaf discs clearly indicates that this tree species is amenable to micropropagation technique. There is however need to study the effects of using other polyamines such as putrescine and spermidines in optimization of shoot multiplication and root induction for this particular tree species.

ACKNOWLEDGEMENT

The authors are grateful to ICRAF for financing this research work, Amani Nature Reserve for providing the research material and the Kenya Forest Research Institute for technical support.

REFERENCES

Abdulnour JE, Donnelley DJ, Barthakur NN (2000). The effect of boron on calcium uptake and growth in micropropagated potato plantlets. Potato Res. 43: 287-295.

Amanor K, Ghansah W, Hawthorne WD, Smith G (2003). Sustainable wild harvesting: best practices document for harvesting of *Allanblackia* Seeds from forest and farmlands with some additional notes for sustainable establishment and management of smallholder plantation and agroforestry systems that incorporate a significant *Allanblackia*component. IUCN and Unilever Report, Ghana.

Anna P, Caniato R, Cappelleti EM, Filippini R (2010). Organogenesis from shoot segments and via callus of endangered *Kosteletzkyapentacarpos (L.) Ledeb.* Plant Cell Tiss. Organ Cult. 100: 309-315. DOI: 10.1007/s11240-009-9652-5

Atangana AR, Tchoundjeu Z, Asaah EK, Simons AJ, and Khasa DP (2006). Domestication of *Allanblackia floribunda*: amenability to vegetative propagation. For. Ecol. Manag. 237: 246-251.

Attipoe L, Van Andel A, Nyame SK (2006). The *Novella* project: developing a sustainable supply chain for *Allanblackia* oil. In Ruben R., Slingerland M. and Nijhoff H. (eds) Agro-food chains and networks for development. Springer, Amsterdam, Netherlands. pp. 179-189

Bairu W, Stick WA, Staden JV (2009). Factors contributing to in vitro shoot-tip necrosis and their physiological interactions. Plant Cell Tiss. Organ Cult. 98: 239-248. DOI 10.1007/s11240-009-9560-8.

Bennett IJ, Mccomb JA, Tonkin CM, McDavid DA (1994). Alternating Cytokinins in multiplication media stimulates *in vitro* shoot growth and rooting of *Eucalyptus globulusLabill.* Ann. Botany. 74: 53-58.

Bonga JM, Klimaszewska KK, Von Aderkas P (2010). Recalcitrance in clonal propagation, in particular of conifers. Plant Cell Tiss. Organ Cult. 100: 241–254.

Bonga JM (1987). Clonal propagation of mature trees, problems and possible solutions in Bonga J.M and Durzan D.J.(Eds) Cell and Tissue culture in for. MartinusNijhoft Publishers Dordrecht. 11: 249-271.

Carlberg I, Glimelius K, Eriksson T (1983). Improved culture ability of potato protoplasts by use of activated charcoal. Plant Cell Rep. pp. 2223-2225

Carman JG, Campell WF (1990). Factors affecting somatic embryogenesis in wheat. In: Bajaj, Y.P.S., (ed) Biotechnology in agriculture and forestry,. Berlin: Springer-Verlag. p.13

Cordeiro NJ, Ndangalasi HJ (2007). Report on environmental impact assessment of Allanblackia seed collection in the East Usambara Mountains, Tanzania, Norconsult, Dar-es-salam, Tanzania.

Drew RA (1988). Rapid clonal propagation of papaya in vitro from mature field grown trees. Hort. Sci. 23: 609-611.

Egyir IS (2007). Allanblackia: standard setting and sustainable supply chain management.   Price setting and marginal cost study. Department of Agricultural Economics and Agribusiness, University of Ghana, Legon Ghana. Pp. 2-4.

El Tahir M, Mlowe D (2002). Population density of Allanblackiastuhlmannii in the disturbed, semi-disturbed and undisturbed areas of Amani Nature Reserve, East Usambara, Tanzania. TBA Tanzania project reports.

Fuller RW, Blunt JW, Boswell JL, Cardellina II JH, Boyd MR, Boyd F (2003). Guttiferone F, the first prenylatedbenzophenone from Allanblackia stuhlmannii. Planta Med. 69(9): 864-866.

Gamborg OL, Constabel F, Shyluk JP (1974). Organogenesis in callus from shoot apices of Pisum sativa. J. Plant Physiol. 30: 125-128.

Gasper TH, Coumnas M (1987). Root formation. In: Bonga J.M., and Durzan D.J.,(eds), Cell Tissue Cult. For., Martinus- Nijhoff the Hague pp.387- 412.

Glynn S, Ritzl C (2000). Predation and dispersion of Allanblackiastuhlmanii seeds in the East Usambara Mountains, Tanzania. TBA Tanzania project reports.

Gupta PK, Nadyir AL, Mascarenhas AF, Jagannthan V (1981). Tissue culture of forest trees; clonal multiplication of Tectonagrandis (L.) by tissue culture. Plant sci. 2: 195-220.

Handley LW (1995). Future uses of somatic embryogenesis in woody plant species. In: Jain S., Gupta P. and Newton R. (eds). Somatic embryogenesis in woody plants, Dordrecht; Kluwer academic Publishers. 1: 414-434.

Hepler PK (2005). Calcium: a central regulator of plant growth and development. Plant Cell. 17: 2142-2155.

Herman M (2009). The impact of the European Novel Food Regulation on trade and food innovation based on traditional plant foods from developing countries. Food Policy 43:499-507. Hilditch, T.P., 1958.Chemical composition of Natural Fats In http://www.fao.org/docrep/X5043E/x5043E0d.htm   (accessed November 1, 2004)

Hirschi KD (2004). The calcium conundrum. Both versatile nutrient and specific signal. Plant Physiol. 136: 2438-2442.

IUCN (International Union for Conservation of Nature) (2007). 2007 IUCN Red List of Threatened Species. www.iucnredlist.org [accessed August 2008]

Kataeva NV, Alexandrova IG, Butenko RG, Dragavtceva EV (1994). Effect of applied and internal hormones on vitrification and apical necrosis of different plants cultured In vitro. Plant Cell. Tissue Organ Cult. 27: 149–154.

Karkonen A, Simola KL, Kopenen T (1999). Micropropagation of several Japanese woody plants for horticultural purposes. Japanese J. Hort. 39: 142-156.

Laskshmi SG, Raghava SBV, (1993). Regeneration of plantlets from leaf disc cultures of rosewood: control of leaf abscission and shoot tip necrosis. Plant Sci. 88: 107-112

Lovett JC (1983) Allanblackiastuhlmannii and its potential as a basis of soap production in Tanzania. Umpublished. p. 25.

Luis PBC, Adriane CMGM, Silvica BRCC, Ana Christina MB (1999). Plant regeneration from seedling explants of Eucalyptus grandis-E urophylla. Plant Cell Tissue. Organ Cult. 56: 17-23.

Martinelli L, Candioli E, Costa D, Poletti V (2001). Morphogenic competence of Vitisrupestris secondary somatic embryos with a long culture history. Plant Cell Rep. 20: 279-284.

Mathayo M, Munjuga MR, Ndangalasi HJ, Cordeiro NJ (2009). Aspects on the floral and fruit biology of Allanblackiastuhlmannii (Clusiaceae), An Endemic Tanzanian tree. J. East Afri. Nat. History 98: 79-93.

Maynard CA (1988). Interaction between cloning success and forest tree improvement. In: Valentine FA, (ed) Forest and crop biotechnology, progress and prospects. Springer-Verlag, New York, pp: 335–346.

McCown BH (2000). Recalcitrance of woody and herbaceous perennials plants: dealing with genetic predeterminism. In vitro Cell Dev. Biol. Plant. 36: 149-154.

McCown BH, Sellmer JC (1987). General media and vessels suitable for woody plant culture. In: Bonga JM, Durzan L (eds) Cell and tissue culture in forestry. General Principles Biotechnol. MartinusNijhoff, Dordrecht. 1: 4–16

Mensuali-Sodi A, Panizza M, Serra G, Tognoni F (1996). Involvement of activated charcoal in the modulation of abiotic and biotic ethylene levels in tissue cultures. Sci. Hort. 54: 49-57.

Merkle SA, Dean JFD (2000). Forest tree biotechnology. Curr. Opin. Biotechnol 11:298–302.

Meshack C (2004). Indigenous knowledge of Allanblackiastuhlmanniiin east Usambaramountains, Tanzania.Technicalhand book. The Tanzania Forest Conservation Group, Dar-es-Salam, Tanzania.

Monela GC, Kajembe GC, Kaoneka ARS, Kowero G (2001). Household livelihood strategies in the miombo woodlands of Tanzania: emerging trends. Tanzania J. For. Nat. Conserv. 73: 17-33.

Mwaura L, Munjuga, M (2011). Allanblackiastuhlmannii(Engl.) Engl. In: van der Vossen, H.A.M. &Mkamilo,G.S. (Editors). PROTA 4, Wageningen, Netherlands.

Nehra NS, Kartha KK (1994). Meristem and shoot tip culture: requirements and applications. In: Plant Cell and Tissue Culture. Eds. I.K. Vasil and T.A. Thorpe. Kluwer Academic Publishers, Dordrecht, the Netherlands. Pp. 37-70.

Nyame SK (2008). Impact of Allanblackia nut harvesting on wildlife: Is the ecosystem at risk? Nat. et Faune 23: 57-58.

Osemeobo GJ (2005). Living on wild plants: evaluation of the rural household economy in Nigeria. Environ. Practice 7: 246-256

Pankaj KB, Toshiyuki M (2001). Novel micro propagationsystem: A review. Pakistan J. Biol. Sci. 4(2): 117-120.

Perez-Tornero O, Burgos L (2009). Different media requirements for micropropagation of apricot cultivars. Plant Cell Tissue Organ Cult. 63: 133–141.

Pinto G, Santos C, Neves L, Araújo C. (2002). Somatic embryogenesis and plant regeneration in Eucalyptus globulusLabill. Plant Cell Reports. 21: 208-213.

Prakash E, ShaValli Khan PS, Sairam RP, Rao KK (1999). Regeneration of plants from seed derived callus of Hybanthusenneaspermus L. Muell., a rare ethnobotanical herb. Plant Cell Rep. 18: 873-878.

Preece JE, Compton ME (1991). Problems with explants exudation in micropropagation. In: Bajaj YPS (ed) Biotechnology in agriculture and forestry. High-tech and micropropagation 1.Spinger-Vrlag, Berlin, pp.168-169.

Pye-Smith C (2009). Seeds of hope: a public-private partnership to domesticate a native tree, Allanblackia is transforming lives in rural Africa. Trees for change. The World Agroforestry Centre, Nairobi, Kenya. No. 2.

Raemakers CJJM, Jacobsen, E, Visser, RCF (1995). Secondary somatic embryogenesis and applications in plant breeding. Euphytica 81: 93-107.

Ramni J, Dawson IK, Anegbeh P, Asaah E, Atangana A, CordeironNJ, Hendrickx H, Henneh S, Kadu CC, Kattah C, Misbah M,Muchugi A, Munjuga M, Mwaura L, Ndangalasi HJ, Njau CS, Nyame SK, Ofori D, Peprah T, Russel J, Rutatina F, Sawe C, Schmidt L,Tchoundjeu Z, Simons T (2010). Allanblackia, A new tree crop in Africa for the global food industry: Market development, smallholder cultivation and Biodiversity management. Forests, Trees and livelihoods, 19: 241-268.

Saka JDK (1995). The nutritional value of edible indigenous fruits: present research status and future directions: In. J.A. maghembe,

Ntupanyama, Y. and Chirwa, P.W. (eds.), Improvement of Indigenous Fruit tree of the Miombo Woodlands of Southern Africa. ICRAF, Nairobi. pp. 50-57.

Sairam RP, Rodrigues R, Rajasekharan R (2001). Shoot organogenesis and mass propagation of Coleus forkohlii from leaf-derived callus. Plant Cell Tiss. Organ Cult. 66: 183-188.

Thorpe TA, Harry IS, Kumar PP (1990). Application of micropropagation to forestry. In: P.C. Debergh and R.H. Zimmerman, Micropropagation. Kluwer Academic Publ. Dredrecht, Netherlands.

Van Rompaey R (2005). Distribution and ecology of Allanblackia spp. (Clusiaceae) in African rain forests with special attention to the development of a wild picking system of the fruits. Report to Unilever Research Laboratories, Vlaardingen.

Werner T, Motyka V, Starnad M, Schmulling T (2001). Regulation of plant growth by cytokinin. Procc. Nat. Acad. Sci. 98: 10487-10492.

Wldi E, Schaffner W, Berger KB (1998). In vitro propagation of Petasiteshybridus (Asteraceae) from leaf and petiole explants and from inflorescence buds. Plant Cell Rep. 18: 336-340.

Yemets AI, Klimkina LAA, Tarassenko LV, Blume YB (2003). Efficient callus formation and plant regeneration of goosegrass. (Eleusineindica). Plant Cell Rep. 21: 503-510.

Zaer JB, Mapes MO (1982). Action of growth regulators. In: Bonga J.M and Durzan D.J. (eds). Plant Cell Tissue Cult. For., Martinus-NijhofDordrechet. Boston, Lancanster. pp. 231-255.

21

Genetic diversity of sesame germplasm collection (*SESAMUM INDICUM* L.): implication for conservation, improvement and use

Dagmawi Teshome Woldesenbet[1] , Kassahun Tesfaye[2] and Endashaw Bekele[2]

[1]Ethiopian Institute of Agricultural Research, Pawe Agricultural Research Center, Benishangul Gumuz, Ethiopia.
[2]Addis Ababa University, College of Natural Science, Microbial, Cellular and Molecular Biology Department, Addis Ababa, Ethiopia.

Genetic diversity assessment of genetic resources maintained at Gene-Banks has important implication for future improvement, conservation and collection activities. However, such information is not available for sesame collected by IBC, Ethiopia. Inter simple sequence repeat (ISSR) marker was used to assess the level of genetic diversity, genetic structure and genetic distance, and to indirectly estimate the level of gene flow among populations of sesame in Ethiopia. A total of 120 (82 Ethiopian and 38 exotic) sesame accessions and six ISSR primers were used. DNA was extracted using a triple CTAB extraction method from silica gel dried bulked sample of five randomly selected individual plants per accession at the stage of three to four weeks after planting. The presence of higher polymorphism was revealed among accessions collected from Ethiopia (75.85) than the exotic accessions (65.52). The average gene diversity relative to the overall population was 0.24. Samples from Welega was the most diverse, with gene diversity value of 0.26 followed by samples from Tigray (0.20) and Shewa (0.20). Samples from Gojam (0.10) and Sudan (0.12) were the least diverse. Inter-population genetic distance (D) ranged from 0.031 to 0.165 for the overall population. From the exotic accession, samples of South East Asia are distantly related to most of the Ethiopian accessions. Unweighted pair group method with arithmetic mean analysis (UPGMA) of Ethiopian sesame populations revealed two major groups and three outliers (Cultivated, Welega and Illubabore).

Key words: Bulk sampling, gene flow, genetic differentiation, genetic distance, inter simple sequence repeat (ISSR), genetic diversity and sesame landrace accessions.

INTRODUCTION

The cultivated sesame, *Sesamum indicum* L., belongs to the family Pedaliaceae. Even though the origin of sesame is still in debate, Mehra (1967) and Mahajan (2007)

stated that Ethiopian region is accepted as the origin of cultivated sesame. In addition, Bedigian (1981) argues that, owing to the wide genetic diversity in Africa, it is reasonable to assume that this subcontinent is the primary center of origin and India would then be thought of as a secondary center for sesame. It is perhaps one of the oldest crops cultivated by man, having been grown in the Near East and Africa for over 5,000 years for cooking and medicinal needs. Generally, 65% of world sesame production is used for edible oil extraction and 35% for confectionary purpose. The fatty acid composition is rather attractive, due to the high level of unsaturated fatty acids.

Sesame is grown in more than 60 countries in the world. According to FAO (2011) Ethiopia is ranked 6th in sesame production with 327,741 tons (10%) of production per year. Sesame seed is by far the leading crop in the countries oil seeds export where by more than 90% of the production is directed to export followed by Niger seed CSA (2010). The Ethiopian whitish Humera type is known for its taste (sweet) in the world market hence it is exported to the confectionary market where white seeded types are demanded by the consumers (Wijnands et al., 2009).

The main objective of sesame groups in National Agricultural Research Systems of Ethiopia is to develop cultivars concentrating on achieving a variety that combines disease resistance, white seeded and high yielding characters and works towards increased market values. The research in the country is geared towards evaluation of new collected accessions and exotic materials to develop a sesame variety with desirable characteristics with a combination of trait of interest including considerable oil content which are important for the export market (Tadele, 2005).

To strengthen the research program and broaden the genetic bases in sesame breeding, Ethiopian Institute of Agricultural Research has assembled 221 germplasm collections during 2002-2004 from different sesame growing regions in Ethiopia and preserved at Werer Agricultural Research Center. This makes the total sesame accessions preserved in Ethiopia to around 870 when it adds up with the IBC collection. However genetic variability studies, using any marker system, have not yet been conducted for the IBC collection. To meet the objectives of the national sesame improvement, information on genetic diversity and their relationship within and among collected accessions is inevitable.

Genetic markers are widely used by breeders and conservationists to study genetic diversity. A number of markers exist that can be used as tool to discern genetic variability within among populations. Each marker system has its own advantages and disadvantages in terms of ease of use and/or the degree of information provided. One of the genetic marker, molecular markers, are identifiable DNA sequence, found at specific locations of the genome, and transmitted from one generation to the next. The presence of various types of molecular markers, and differences in their principles, methodologies, and applications require careful consideration in choosing one

or more of such methods. No molecular markers are said to be ideal and fulfill all requirements needed by researchers. Thus, one can choose among the variety of molecular techniques, according to the kind of study to be undertaken. To be called a suitable molecular marker it should be highly polymorphic, co-dominant inheritance, frequent occurrence and even distribution throughout the genome, selectively neutral behavior, easy access, easy and fast assay, low cost and high throughput, high reproducibility, and transferability between laboratories, populations and/or species (Weising et al., 2005).

Inter-simple sequence repeat (ISSR) is amplification of DNA segments present at an amplifiable distance in between two identical microsatellite repeat regions oriented in opposite direction (Zietkiewicz et al., 1994). ISSR marker, in addition to its suitability to genetic diversity study, is highly polymorphic, reproducible, cost effective, requires no prior information of the sequence (Bornet et al., 2002). These facts suggest that ISSR could be an unbiased tool to evaluate the changes of diversity in agronomically important crops (Brantestam et al., 2004).

ISSR markers are now being applied for cultivar identification, assessment of genetic diversity in various plant species and in determining genetic diversity and phylogenetic relationships within and among cultivated crops (Hou et al., 2005; Joshi et al., 2000). In agronomically important crops such as sesame this marker is used to study the patterns and level of diversity. Hence, the ISSR marker assay has been chosen to study the level of diversity and patterns of distribution of sesame genetic resources in Ethiopia.

MATERIALS AND METHODS

Experiment material and DNA extraction

Sesame germplasm sample consisting of 120 accessions (82 Ethiopian and 38 exotic) were drawn from the over 636 accessions currently maintained at the gene bank at the institute of Biodiversity Conservation, Addis Ababa, Ethiopia (Table 1 and Figure 1). The sampling procedure was set in such a way that accessions from the major sesame growing regions were sufficiently represented in the sample. Germplasm passport data were used to make sure that the major regions are represented. Additional sesame samples from Africa and other parts of the world were included for comparative analysis. The samples were laid out in augmented randomized block design in Werer Agricultural Research Center, Ethiopia on station field plots. The accessions were grown in 2 row plots of 4 m length.

Young leaves (3-5 in number) were collected separately from 5 randomly selected individual plants per accession three to four weeks after planting and dried in silica gel. Approximately equal amounts of the dried leaf samples were bulked for each accession and ground with sterile pestle and mortar with addition of liquid nitrogen. Total genomic DNA were isolated from about 0.2 g of the pulverized leaf sample using modified triple Cetyl Trimethyl Ammonium Bromide (CTAB) extraction technique as describe by Borsch et al. (2003). The isolated DNA samples were visualized using 1% ethidium bromide stained agarose gel under UV light. Based on a test gel result of DNA quantity (band intensity) and

Table 1. List of Sesame accessions, altitude and location of collection used in the study (Source: IBC).

No.	Accession #	Altitude	Former Administrative Region/ Exotic
1	111501	NA	Gamo Gofa
2	208888	NA	Gamo Gofa
3	208889	1050	Gamo Gofa
4	212993	1290	Gamo Gofa
5	212994	1270	Gamo Gofa
6	212995	1290	Gamo Gofa
7	214976	NA	Gamo Gofa
8	111502	NA	Gojam
9	111503	NA	Gojam
10	111513	NA	Gojam
11	211921	1600	Gojam
12	230260	NA	Gojam
13	205173	NA	Gonder
14	235769	900	Gonder
15	235903	1150	Gonder
16	241341	600	Gonder
17	241342	600	Gonder
18	241334	750	Gonder
19	241344	980	Gonder
20	241346	1890	Gonder
21	241347	1700	Gonder
22	202512	1600	Harerge
23	208670	1830	Harerge
24	208671	1900	Harerge
25	208672	1350	Harerge
26	208673	1840	Harerge
27	228816	1500	Harerge
28	223340	NA	Harerge
29	207956	NA	Illubabor
30	207957	NA	Illubabor
31	216733	600	Illubabor
32	222876	NA	Illubabor
33	202285	1490	Shewa
34	202286	1490	Shewa
35	202374	1395	Shewa
36	203099	1240	Shewa
37	212536	1580	Shewa
38	212537	NA	Shewa
39	212540	1600	Shewa
40	234015	1090	Tigray
41	210871	NA	Eritrea
42	214202	NA	Eritrea
43	235405	1650	Tigray
44	237523	900	Tigray
45	238269	820	Tigray
46	238270	710	Tigray
47	238271	710	Tigray
48	238272	710	Tigray
49	241305	1200	Tigray
50	241311	1120	Tigray
51	111505	1440	Welega
52	202517	1420	Welega
53	202518	1360	Welega
54	207953	1300	Welega
55	207954	1200	Welega
56	208752	NA	Welega
57	237994	1290	Welega
58	215816	1435	Welega
59	212632	1460	Welo
60	202318	1525	Welo
62	202340	1555	Welo
63	202345	1565	Welo
64	202353	1640	Welo
65	202370	1590	Welo
66	215260	NA	Welo
67	203631	NA	Exotic
68	203633	NA	Exotic
69	203634	NA	Exotic
70	203637	NA	Exotic
71	203638	NA	Exotic
72	227862	NA	Exotic
73	227864	NA	Exotic
74	227865	NA	Exotic
75	227866	NA	Exotic
76	227867	NA	Exotic
77	227873	NA	Exotic
78	227874	NA	Exotic
79	227875	NA	Exotic
80	227876	NA	Exotic
81	227879	NA	Exotic
82	227903	NA	Exotic
83	227904	NA	Exotic
84	227906	NA	Exotic
85	227907	NA	Exotic
86	227908	NA	Exotic
87	227933	NA	Exotic
88	227934	NA	Exotic
89	231397	NA	Exotic
90	231399	NA	Exotic
91	227855	NA	Exotic
92	231406	NA	Exotic
93	231407	NA	Exotic
95	231409	NA	Exotic
96	231410	700	Exotic
97	205194	700	Exotic
98	205195	800	Exotic
99	205191	900	Exotic
100	205192	1100	Exotic
101	203595	760	Exotic
102	203597	NA	Exotic
103	203599	700	Exotic
104	203600	750	Exotic
105	203627	NA	Exotic
106	Abasena	1050	Released
107	E	NA	Released
108	Hir Hir or T- 85	NA	Released
109	Mehedo	NA	Released
110	Adi	1050	Released
111	Acc-BG-006	1290	Benishangulgumuz
112	Acc-BG-002	1270	Benishangulgumuz
113	Acc-BG-001	1290	Benishangulgumuz
114	Acc-BG-003	1050	Benishangulgumuz
115	Acc-BG-010	NA	Benishangulgumuz
116	Acc-BG-012(2)	1050	Benishangulgumuz
117	Acc-BG-003(2)	NA	Benishangulgumuz
118	Acc-BG-019(1)	1600	Benishangulgumuz
119	New (EW-004)	NA	Bako
120	New (EW-015)	NA	Bako

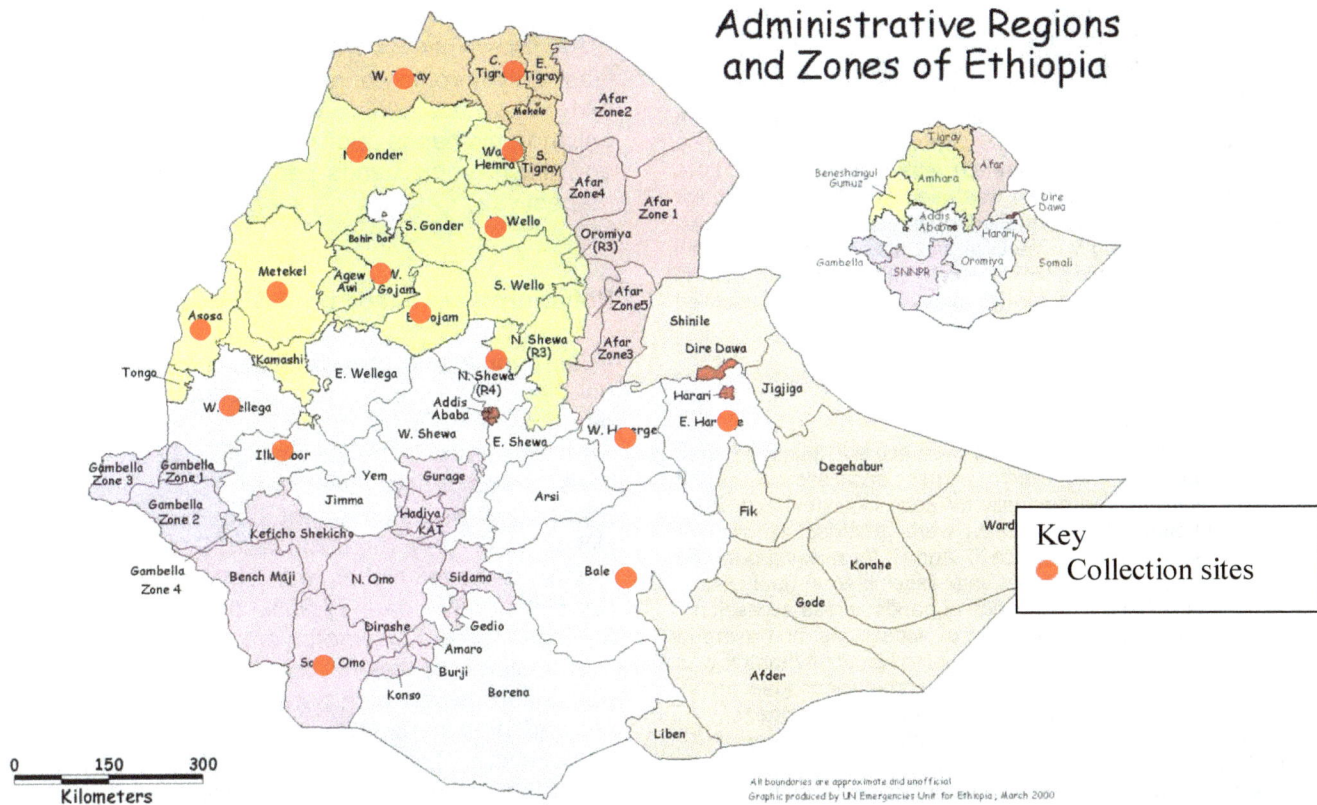

Figure 1. Former administrative region of Ethiopia showing where the sesame accessions used in this study were collected (Source: Map of the world).

Table 2. Banding patterns generated using the six primers, their annealing temperature, repeat motifs, amplification patterns and number of scored bands.

Primer	Annealing temperature (°C)	Repeat motif[1]	Amplification pattern	Number of scored bands
810-H	45	$(GA)_8T$	Good	9
818-H	45	(CA)8G	Good	9
834-H	45	$(AG)_8YT$	Good	10
844-H	45	(CT)8RC	Good	10
860-H	45	(TG)8RA	Good	9
880-H	48	(GGAGA)3	Good	11
Total				58

[1]Single-letter abbreviations for mixed base positions: R = (A, G); Y = (C, T).

quality (absence or presence of minimum smear) the second extraction of most and the first extraction of some samples were selected for further PCR amplification. The selected genomic DNA samples were diluted with sterile distilled water in 1:5 ratios.

ISSR analysis

The ISSR marker assay was conducted at Genetics Research Laboratory of the Microbial, Cellular and Molecular Biology Program Unit, Faculty of Life Sciences, Addis Ababa University, Addis Ababa. A set of primer kit was obtained from the University of British Colombia (primer kit UBC 900). Based on published work of Kim et al. (2002) and lab experience, a total of 12 primers were

used for the initial testing of primers variability and reproducibility. A total of eight sesame DNA samples from geographically distant localities were selected to screen primers and optimize the reaction conditions. Finally, a total of six polymorphic and reproducible ISSR primers were selected for the final analysis (Table 2).

The polymerase chain reaction was conducted in Biometra 2000 T3 Thermo cycler. PCR amplification was carried out in a 25 µl reaction mixture containing 1 µl template DNA, 14.0 µl H_2O, 5.0 µl dNTP (1.25 mM), 2.5 µl Taq buffer (10xThermopol reaction buffer), 2.0 µl $MgCl_2$ (2 mM), 0.3 µl primer (20 pmol/µl) and 0.2 µl Taq Polymerase (5 u/µl). The amplification program was set as 4 min preheating and initial denaturation at 94°C, followed by 40 cycles of 15 sat 94°C, 1 min primer annealing at (45°C/48°C) based on primers used, 1 min and 30 sextension at 72°C and the final extension for 7 min at 72°C. The PCR products were stored at 4°C until loading on gel for electrophoresis. The lid temperature was held at 105°C. The PCR products were stored at 4°C until loaded on gel for electrophoresis. The amplification products were differentiated by electrophoresis using an agarose gel (1.67% agarose with 100 ml 1xTBE) and 8 μl amplification product of each sample with 2 μl loading dye (6 times concentrated) was loaded on gel. DNA marker 1 k bp (molecular ladder) was used to estimate molecular weight and size of the fragments. Electrophoreses was done for 2 h at constant voltage of 100 V. The DNA was stained for 30 min with (10 mg/ml) ethidum bromide (EtBr) which was mixed with 450 ml distilled water and distained for 30 minuwith 450 ml of distilled water.

Statistical analysis

Clearly resolved, unambiguous bands were scored visually for their presence or absence for each primer and sample. ISSR profiles/bands were scored manually for each individual accession from the gel photograph. The bands were recorded as discrete characters, presence '1' or absence '0' and '?' for missing data. To analyse the scored '1' '0' fragment data assembled in data matrix. The total number of bands, distribution of bands across accessions, number of polymorphic bands in a set of accessions, and average number of bands per primer were calculated. Similarity matrix was generated based on the simple-matching coefficient (Jaccard's), using the presence/absence data for individual ISSR fragment.

Based on recorded bands different software's were used for analysis. POPGENE version1.32 software (Yeh et al., 1999) was used to calculate genetic diversity for each population as number of polymorphic loci and percent polymorphism, Shanon index, coefficient of gene differentiation (G_{ST}) (Nei, 1973), Nei's standard genetic distance (GD) (Nei, 1972) and gene flow estimates (McDermott and McDonald, 1993). Analysis of molecular variance (AMOVA) was used to calculate variation among and within population using Areliquin version 3.01 (Excoffier et al., 2006). NTSYS- pc version 2.02 (Rohlf, 2004) and Free Tree 0.9.1.50 (Pavlicek et al., 1999) software's were used to calculate Jaccard's similarity coefficient which is calculated with the formula:

$$S_{ij} = \frac{a}{a+b+c}$$

Where, 'a ' is the total number of bands shared between individuals i and j, 'b' is the total number of bands present in individual i but not in individual j and 'c' is the total number of bands present in individual j but not in individual i.

The unweighted pair group method with arithmetic mean (UPGMA) (Sneath and Sokal, 1973) was used to analyze and compare the population and generates phenogram using NTSYS-pc version 2.02 (Rohlf, 2004). The Neighbor Joining (NJ) method (Saitou and Nei1987; Studier and Keppler, 1988) was used to compare individual genotypes and evaluate patterns of genotype clustering using Free Tree 0.9.1.50 Software (Pavlicek et al., 1999).

RESULTS

Banding patterns of the ISSR primers used

Out of the 12 primers tested initially, six primers (five di-nucleotide and one penta-nucleotide) that gave relatively clear banding pattern were selected and used in this study (Table 2). The molecular weight of the bands amplified using the primers were in the range of 450 to 4000 bp. A total of 58 clear bands were scored, from 82

and 38 exotic sesame accessions. Out of the total 58 ISSR fragments, 44 were found to be polymorphic. The least polymorphic bands (four) were scored from primer 834_H and the remaining five primers all showed equal polymorphic bands (eight). The average number of bands and polymorphic bands per primer were 9.67 and 7.33, respectively.

Polymorphism based on ISSR analysis

In all populations or individuals the number of polymorphic loci ranges from four for primer-834 to eight for all other primer. Of the total 58 loci scored, 44 (75.86%) were observed to be polymorphic. From all the populations (considering Ethiopian accessions and exotic accessions as independent population) studied Welega (58.62%), Tigray (50.00%) and Gonder (44.83%) were found to have higher percent polymorphism. Gojam (25.86%) and Sudan (27.95%) on the other hand showed the least percent polymorphism.

Ethiopian accessions (considering as a single large population) when com-pared to exotic accessions they showed high percent polymorphism 75.85 and 65.52%, respectively (Table 3). Among the exotic accessions (considering samples from different country as independent population) accessions from Sudan (27.95%) were the least polymorphic and the South East Asian ones (43.1%) showed high present polymorphism. No unique bands were observed for either the accessions or the populations.

As to the primers used, none of the primers showed 100 per cent polymorphism. Primer 810, 818 and 860 showed the highest percent polymorphism (88.89%) and the least polymorphic band was scored by primer 834 which is 40.00 % and this primer also showed higher standard deviation for the genetic diversity and Shannon index which goes with its least polymorphism. The rest primers namely, 844 and 880 showed 80.00 and 72.73 per cent polymorphism, respectively (Table 4).

Genetic diversity

Grouping Ethiopian accessions into administrative region based population and exotic accessions in to country based population showed that samples from Welega were the most diverse (0.26) followed by samples from Tigray (0.20) and Shewa (0.20). Samples from Gojam (0.10) and Sudan (0.12) were the least diverse. The average gene diversity relative to the overall population was 0.24, considering accession from different administrative region of Ethiopia and exotic countries as independent population. The overall diversity index values for the total population and the Ethiopian sesame accessions were found to be 0.37 (Table 3). The extent of gene differentiation relative to the total Ethiopian population

Table 3. Number of Polymorphic Loci (NPL), percent polymorphism (PP) genetic diversity (GD) Shanon Index (I), gene differentiation (G_{ST}) and gene flow (Nm) with 120 sesame accessions and all primer.

Population	NPL	PP	GD±SD	I±SD	G_{ST}	Nm
Benishangul Gumuz	25	43.1	0.18 ± 0.22	0.26 ± 0.31		
Gamo Gofa	24	41.38	0.17 ± 0.22	0.25 ± 0.31		
Gojam	15	25.86	0.1 ± 0.18	0.15 ± 0.26		
Gonder	26	44.83	0.19 ± 0.22	0.27 ± 0.31		
Harerge	22	37.93	0.14 ± 0.2	0.21 ± 0.29		
Illubabor	22	37.93	0.17 ± 0.23	0.25 ± 0.32		
Shewa	28	48.28	0.2 ± 0.21	0.28 ± 0.31		
Tigray	29	50	0.2 ± 0.21	0.27 ± 0.3		
Welega	34	58.62	0.26 ± 0.23	0.37 ± 0.32		
Welo	23	39.66	0.17 ± 0.22	0.24 ± 0.31		
Cultivated	29	50	0.22 ± 0.23	0. 31 ± 0.32		
Ethiopian average	25.18	43.41	0.18	0.26		
Ethiopia total	44	75.86	0.24 ± 0.19	0.37 ± 0.27	0.25	1.49
Egypt	21	36.21	0.14 ± 0.21	0.21 ± 0.30		
Mexico	22	37.93	0.14 ± 0.20	0.21 ± 0.28		
Somalia	23	39.66	0.18 ± 0.23	0.26 ± 0.32		
S. E. Asia	25	43.1	0.19 ± 0.23	0.27 ± 0.32		
Sudan	16	27.95	0.12 ± 0.20	0.17 ± 0.29		
Tanzania	22	37.93	0.15 ± 0.21	0.23 ± 0.30		
Zambia	24	41.38	0.18 ± 0.22	0.25 ± 0.31		
Zimbabwe	24	41.38	0.18 ± 0.22	0.25 ± 0.31		
Exotic average	22.12	38.1925	0.16	0.23125		
Exotic Total	38	65.52	0.23 ± 0.20	0.34 ± 0.28	0.29	1.19
Average population	23.89	41.21				
Total Population	44	75.86	0.24 ± 0.19	0.37 ± 0.27	0.28	1.28

Table 4. Number of Scorable bands (NSB), number of polymorphic loci (NPL), percent polymorphism (PP) genetic diversity (GD) Shanon index (I), for each primer.

Primer	NSB	NPL	PP	GD	I
810	9	8	88.89	0.23 ± 0.18	0.36 ± 0.25
818	9	8	88.89	0.35 ± 0.17	0.51 ± 0.23
834	10	4	40.00	0.12 ± 0.18	0.19 ± 0.27
844	10	8	80.00	0.23 ± 0.17	0.37 ± 0.24
860	9	8	88.89	0.26 ± 0.2	0.40 ± 0.27
880	11	8	72.73	0.27 ± 0.21	0.39 ± 0.29
Total	58	44	75.86	0.24 ± 0.19	0.37 ± 0.27

was (G_{ST}) 0.25 and the extent of gene flow (Nm) among populations of Ethiopian sesame accessions (considering administrative region as independent population) was 1.49. For the exotic accessions the extent of gene differentiation was 0.29 and the extent of gene flow was 1.19 (Table 3).

Analysis of molecular variance

Analysis of molecular variance was carried out on the overall ISSR data score of sesame accession without and with grouping (Tables 5 and 6). AMOVA without grouping revealed that higher percentage of variation (94.09%) is attributed to the within population variation while the remaining variation is due to the among population variation (5.90%). The grouping AMOVA analysis, considering the Ethiopian accessions as one group and the exotic ones as another, also showed 5.88% among populations within group variation, 94.09% within group variation and 0.026% among group variation (Table 6).

Table 5. Analysis of Molecular Variance (AMOVA) of sesame accessions in Ethiopia without grouping.

Source of variation	Sum of squares	Variance components	Percentage variation	Fixation	P
Among populations	62.647	0.15786	5.90	0.05903	0.00
Within population	240.021	2.51660	94.09		0.00
Total	302.667	2.67446			

Table 6. Analysis of Molecular Variance (AMOVA) of sesame accessions in Ethiopia with grouping.

Source of variation	Sum of squares	Variance components	Percentage variation	Fixation	p
Among groups	3.481	0.00069	0.02577	0.05903	0.00
Among populations within groups	59.166	0.15749	5.88792		0.00
Within populations	240.021	2.5166	94.08632		0.00
Total	302.667	2.67477			

The variations was found to be highly significant at P=0.00, 1000 permutations.

Genetic distance

Inter-population genetic distance (D) ranged from 0.031 to 0.165 for the total 19 population (Table 7). Administrative region based population classification of the Ethiopian sesame accessions revealed inter population genetic distance ranging from 0.041 to 0.16. Samples from Illubabore were distantly related to samples of Wello (0.150), Shewa (0.116) Harerge (0.144) and Gamo Goffa (0.134) (Table 8). Among the pairwise population comparisons made within Ethiopian population, samples from Welega and Harerge showed the highest genetic distance (0.160) and samples from Shewa and Harerge showed the least genetic distance (0.041). In general from the exotic accessions samples of South East Asia is distantly related to most of the Ethiopian accession. Genetic distance between the other pairwise combinations of populations was very low with the least genetic distance between samples of Mexico and Tanzania (0.031).

Clustering analysis

Cluster analysis of Ethiopian sesame populations revealed two major groups and three outliers (Cultivated, Welega and Illubabore). The first major cluster again forked into two sub groups the first containing Tigray, Harerge and Shewa populations, while the second contained Gamo Goffa, and Wello populations which are characterized as the lowest sesame producers of the country.

More over the migration of the welo people to the south may have its own impact on the clustering of Gamo Goffa

and Welo (Baker, 2001). Seeds of some sesame germplasm might have been transported from Welo to Gamo Gofa along with human (Figure 2). The second major cluster comprise of Gojam, Benishangul Gumuz and Gonder populations which embodies the north western parts of the country.

DISCUSSION

Genetic diversity and application of ISSR marker

Understanding of the extent and pattern of genetic variation can be useful for several purposes. Such information can be used to design effective germplasm conservation and for setting germplasm collection mission as well as to estimate or predict the risk of genetic erosion in certain area.

From breeding point of view, knowledge of pattern of genetic variability is useful for defining heterotic patterns in hybrid breeding and for relating the observed pattern with presence of certain economically important traits. Pattern of genetic variability can be studied by morphological, isozyme or molecular markers. Among the markers ISSR markers are important to study genetic variations in plant species, as they are effective in detecting very low levels of genetic variation (Zietkiewicz et al., 1994).

Generally, ISSR primers have high resolution power in diversity analysis of different crops. This marker were observed to be very useful in detecting genetic diversity and population structure of Tef (Assefa, 2003), Coffee (Aga, 2005; Tesfaye, 2006), Lentils (Edosa et al., 2007), and rice (Gezahegne et al., 2007) collected from different parts of Ethiopia.

In addition to the advantages (inexpensive, easy to generate), ISSRs are powerful in detecting polymorphisms with high reproducibility. ISSR can detect even more polymorphism than RFLPs in maize (Kantety et

Table 7. Nei's original measures of genetic identity (above diagonal) and genetic distance (below diagonal) in 19 populations of Ethiopian and some exotic sesame accessions.

POP	BG	CU	GD	GG	GJ	HR	IL	SH	TG	WL	WO	SEA	MX	ZA	ZW	SO	SU	TZ	EG
BG	****	0.9377	0.9539	0.9073	0.9384	0.9117	0.9111	0.9302	0.9388	0.9142	0.9253	0.8799	0.9205	0.9311	0.9191	0.9078	0.9283	0.9401	0.9376
CU	0.0643	****	0.9319	0.9229	0.9161	0.9050	0.9088	0.9261	0.9381	0.9088	0.9067	0.8875	0.9131	0.9245	0.9132	0.9216	0.9068	0.9020	0.9093
GD	0.0472	0.0705	****	0.9497	0.9329	0.9336	0.9214	0.9405	0.9430	0.9067	0.9372	0.9389	0.9428	0.9421	0.8975	0.9212	0.9285	0.9389	0.9355
GG	0.0972	0.0803	0.0516	****	0.9436	0.9435	0.8746	0.9393	0.9306	0.9043	0.9482	0.9096	0.9512	0.9416	0.8843	0.9246	0.8960	0.9066	0.9165
GJ	0.0635	0.0876	0.0695	0.0581	****	0.9332	0.8939	0.9149	0.9383	0.8866	0.9215	0.8808	0.9108	0.9161	0.8942	0.9334	0.9224	0.9314	0.9171
HR	0.0924	0.0998	0.0687	0.0582	0.0691	****	0.8656	0.9594	0.9515	0.8524	0.9527	0.9014	0.9219	0.9149	0.8604	0.9259	0.9058	0.9131	0.9422
IL	0.0931	0.0957	0.0819	0.1340	0.1121	0.1443	****	0.8644	0.9132	0.8903	0.8605	0.8904	0.8682	0.9026	0.8584	0.8933	0.9153	0.9164	0.8479
SH	0.0723	0.0768	0.0613	0.0626	0.0890	0.0414	0.1457	****	0.9413	0.8983	0.9416	0.9214	0.9435	0.9277	0.8787	0.9386	0.9089	0.9112	0.9565
TG	0.0631	0.0639	0.0587	0.0720	0.0637	0.0497	0.0908	0.0605	****	0.9004	0.9187	0.9001	0.9091	0.9218	0.9205	0.9424	0.9445	0.9535	0.9445
WL	0.0898	0.0956	0.0980	0.1005	0.1203	0.1597	0.1162	0.1073	0.1050	****	0.8781	0.8743	0.8963	0.8988	0.9068	0.9051	0.8627	0.8933	0.8880
WO	0.0777	0.0980	0.0649	0.0532	0.0817	0.0485	0.1502	0.0602	0.0848	0.1300	****	0.8995	0.9428	0.9198	0.8806	0.9062	0.9024	0.9006	0.9353
SEA	0.1279	0.1194	0.0630	0.0948	0.1270	0.1038	0.1161	0.0819	0.1052	0.1344	0.1059	****	0.9177	0.8928	0.8630	0.8839	0.8927	0.8743	0.8941
MX	0.0828	0.0909	0.0589	0.0500	0.0934	0.0814	0.1413	0.0582	0.0953	0.1095	0.0590	0.0859	****	0.9692	0.8801	0.8983	0.8831	0.9117	0.9427
ZA	0.0714	0.0785	0.0596	0.0602	0.0876	0.0890	0.1025	0.0750	0.0815	0.1067	0.0836	0.1134	0.0313	****	0.8825	0.9071	0.9009	0.9188	0.9179
ZW	0.0844	0.0909	0.1081	0.1230	0.1119	0.1504	0.1526	0.1293	0.0828	0.0978	0.1271	0.1474	0.1277	0.1249	****	0.9033	0.9234	0.9012	0.9250
SO	0.0967	0.0817	0.0820	0.0784	0.0689	0.0770	0.1128	0.0633	0.0593	0.0998	0.0985	0.1234	0.1073	0.0975	0.1017	****	0.9271	0.9221	0.9253
SU	0.0744	0.0979	0.0741	0.1099	0.0808	0.0989	0.0885	0.0955	0.0571	0.1477	0.1027	0.1135	0.1243	0.1043	0.0797	0.0756	****	0.9040	0.9262
TZ	0.0617	0.1031	0.0631	0.0981	0.0711	0.0909	0.0873	0.0930	0.0476	0.1128	0.1046	0.1343	0.0925	0.0847	0.1040	0.0810	0.1009	****	0.9199
EG	0.0645	0.095	0.0666	0.0872	0.0865	0.0595	0.165	0.0444	0.0571	0.1188	0.0669	0.1119	0.059	0.0856	0.0779	0.0777	0.0766	0.0835	****

al., 1995) and more than AFLPs in rice (Blair et al., 1999).

In this study the extent and pattern of genetic variability among 82 Ethiopian and 38 exotic samples of sesame accessions were estimated using 5 di-nucleotide and one penta-nucleotide ISSR primer markers. The large number of accessions held in this study dictates the approach that can be employed. A quick, simple but reliable molecular protocol must be combined with an appropriate strategy for handling large sample sizes (Edosa et al., 2007; Edosa et al., 2010; Edosa et al., 2011 Gilbert et al., 1999). In this study, bulk sampling approach was chosen because it permits representation of the vast accession by optimum number of plants. Yang and Quiros (1993) reported that bulked samples with 10, 20, 30, 40 and 50 individuals had resulted in the same RAPD profiles as that of the individual plant constituting the bulk sample. Edosa et al. (2007) used bulked samples for diversity assessment in lentil collected from Ethiopia. The technique revealed higher genetic diversity, and, therefore, validated the usefulness of bulk sample analyses.

This result also confirms that bulked leaf samples and ISSR marker is efficient in detecting polymorphism within and among populations and accessions of sesame. The present study suggested the existence of moderate level of diversity (0.23) among sesame accessions collected from Ethiopian Institute of Biodiversity Conservation.

Thus, ISSR marker systems will provide a useful tool in the future design of collection strategies for conservation and use of sesame accessions in Ethiopia. The ISSR technique was previously performed in sesame to study the genetic relationship of sesame germplasm in Korea by Kim et al. (2002). He used 14 reliable ISSR primers and found 33% polymorphism from 79 amplification products among the 75 sesame accessions. This is low when it is compared with this results percent polymorphism (75.86%). RAPD marker was also used by Abdellatef et al. (2008) in a set of 10 sesame germplasm collected from different regions of Sudan.

A total of 64 polymorphisms (6.4 polymorphic markers per primer) out of 75 reproducible products

Table 8. Nei's original measures of genetic identity (above diagonal) and genetic distance (below diagonal) in 11 populations of Ethiopian sesame accessions.

Pop	BG	CU	GD	GG	GJ	HR	IL	SH	TG	WL	WO
Pop	****	0.9377	0.9539	0.9073	0.9384	0.9117	0.9111	0.9302	0.9412	0.9142	0.9253
BG	0.0643	****	0.9319	0.9229	0.9161	0.9050	0.9088	0.9261	0.9326	0.9088	0.9067
CU	0.0472	0.0705	****	0.9497	0.9329	0.9336	0.9214	0.9405	0.9384	0.9067	0.9372
GD	0.0972	0.0803	0.0516	****	0.9436	0.9435	0.8746	0.9393	0.9284	0.9043	0.9482
GJ	0.0635	0.0876	0.0695	0.0581	****	0.9332	0.8939	0.9149	0.9373	0.8866	0.9215
HR	0.0924	0.0998	0.0687	0.0582	0.0691	****	0.8656	0.9594	0.9536	0.8524	0.9527
IL	0.0931	0.0957	0.0819	0.1340	0.1121	0.1443	****	0.8644	0.9055	0.8903	0.8605
SH	0.0723	0.0768	0.0613	0.0626	0.0890	0.0414	0.1457	****	0.9437	0.8983	0.9416
TG	0.0606	0.0698	0.0636	0.0743	0.0647	0.0475	0.0993	0.0579	****	0.9029	0.9152
WL	0.0898	0.0956	0.0980	0.1005	0.1203	0.1597	0.1162	0.1073	0.1021	****	0.8781
WO	0.0777	0.0980	0.0649	0.0532	0.0817	0.0485	0.1502	0.0602	0.0886	0.1300	****

BG,Benishangul Gumuz; CU, cultivated; GD, Gonder; GG, Gamo Goffa; GJ, Gojam; HR, Harerge; IL, Illibabore; SH, Shewa; TG, Tigray; WL, Welega; WO, Welo; SEA, South East Asia; MX, Mexico; ZA, Zambia; ZW, Zimbabwe; SO, Somalia; SU, Sudan; TZ,Tanzania; EG,Egypt.

(7.5 fragments per primer) were obtained from the 10 primers used and low level of genetic similarity among accessions (Abdellatef et al., 2008). Seleshi (2008) studied the genetic divergence of 100 sesame accessions using 13 agro-morphological traits from EIAR collection and found wide variability for all the measured characters which agrees with this result.

The success of a crop-improvement program largely depends on the availability and knowledge of the genetic resources in a germplasm collection. In the present study all the diversity parameters confirmed that there is higher gene diversity in Ethiopian accessions than exotic accessions. Since areas of high genetic diversity contribute more accessions than those with a low diversity for further and future collection, breeding and conservation activities high priority should be given to areas with high genetic diversity. The result of this study exhibited among the ten Ethiopian areas moderate to higher genetic diversity is revealed by the accessions from the Welega, Tigray and Shewa areas. Hence, for selection based population improvement these areas have a high potential as compared to areas with lower genetic diversity like Gojam and Harerge.

Genetic distance and implication for improvement

Genetic distance is a measure of the allelic substitutions per locus that have occurred during the separate evolution of two populations or species. In this study, inter-population genetic distance for the whole population (D) ranged from 0.031 to 0.165. From the Ethiopian populations comparatively samples from Ilubabore, Harergea, Gamo Goffa, Gojam and Wello showed moderate to high genetic distance related to their respective pair wise comparison. Samples from Benishangul Gimuz, Gonder and Tigray relatively found to be more

related with each other and with other pair wise comparison showing low to very low genetic distance. Since the bulk of Ethiopian sesame production is coming from the later places this close similarity can be explained by exchange of the sesame seed among neighboring localities in the north and northwest. Although the distance between exotic accessions and local accessions varied from population to population on the average, the local accessions were separated from the exotic ones by considerable amount the south eastern Asian accessions being the most distant.

Genetic structure, patterns of distribution and implication for conservation

The genetic structure of plant populations reflects the interactions of various factors, including the long-term evolutionary history of the species (shifts in distribution, habitat fragmentation, and population isolation), genetic drift, mating system, gene flow and selection (Schaal et al., 1998). In the present study there was moderate level of genetic differentiation (G_{ST} = 0.25) among the Ethiopian sesame accessions.

Analysis of molecular variance (AMOVA) using 32 AFLP marker for sesame accessions originate from five different geographical regions representing the proposed diversity centers for sesame: India, Africa, China-Korea-Japan, Central Asia and Western Asia indicates that 5% of the variance among the patterns was due to differences among groups and 95% was due to differences within groups (Laurentin and Karlovsky 2006).

In another study, Endale et al. (2008), using SSR marker, also found 41.2% of the total genetic variation between population and 58.8 % within population for 50 sesame landraces of Ethiopia. Similarly, the result of this study is also in agreement with the above AFLP and SSR

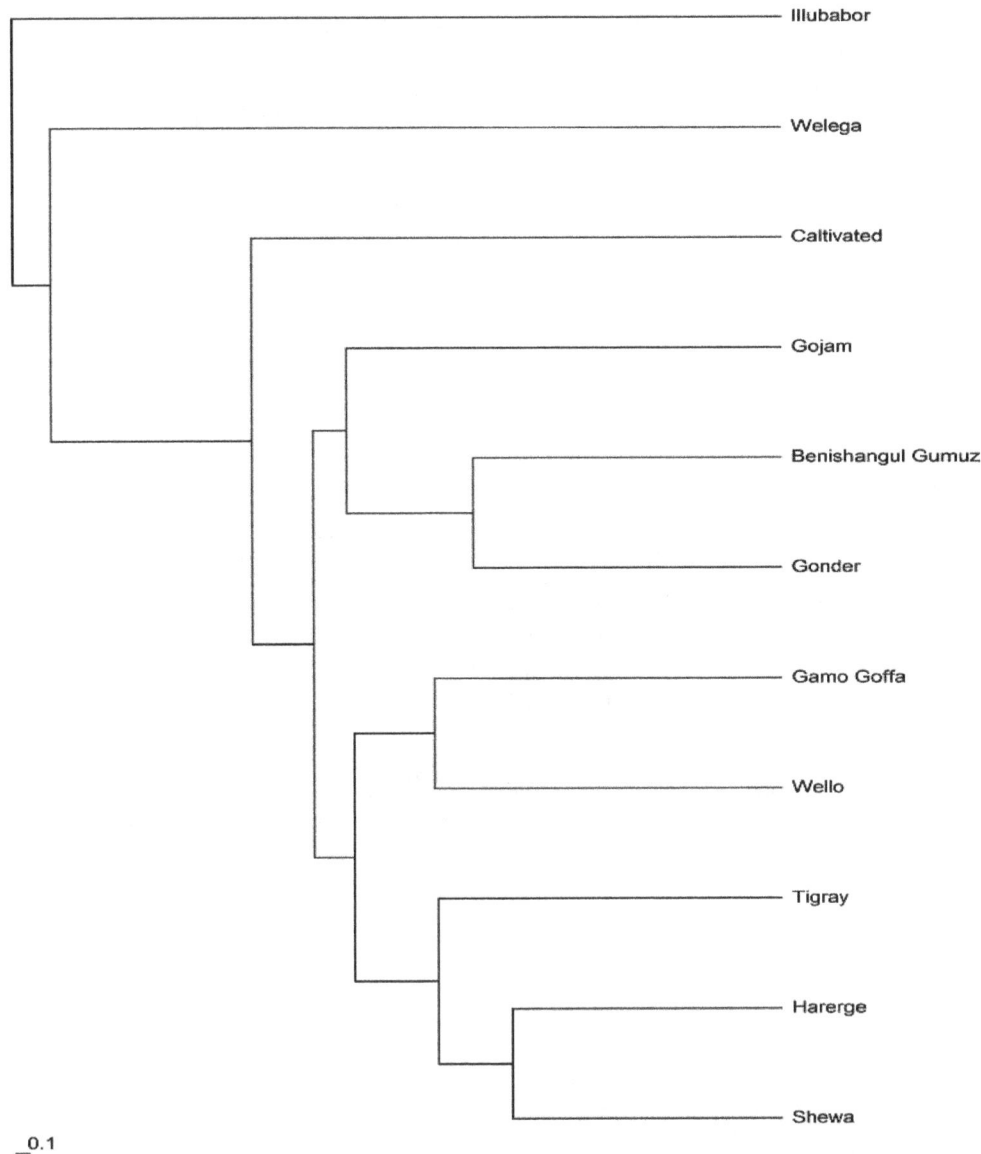

Figure 2. NJ based dendrogram for 11 Ethiopian sesame populations using 6 ISSR (5 di and1 penta nucleotide) primers.

analysis.

In this study, the AMOVA analysis showed highly significant (P=0.00) genetic differences among populations and within populations. Of the total variation, 5% attributed to among population and 95% attributed to within populations. Even though based on AMOVA analysis, a high estimate of genetic differentiation between populations of inbred species is expected this may not be a general truth if there is high gene flow repre-sented by seed movement through human involvement (Edossa et al., 2007).

Moreover, pollen movement facilitated by insect could also have a role for the observed pattern. The lower among group variance of this study can be explained in line with the above argument. Gene flow is the exchange or movement of gametes, individuals, and populations on a geographic scale (Joseph and Bruce 1993). Gene flow, in conjunction with other evolutionary forces, can result in the spread of single genes (or DNA sequences), genotypes, and even the establishment of whole populations in different regions. The movement of one individual per generation between populations is sufficient to prevent substantial differentiation between those populations (Joseph and Bruce, 1993).

The values obtained from Nm, the product of the effective size of individual populations (N) and the rate of migration among them (m); show the approximate number of individuals migrating from one population to the other, in a typical island model. Generally, if $Nm < 1$ local differentiation of populations will result, and if $Nm > 1$ there will be little differentiation among populations (Wright, 1951). The overall Nm (1.28) value of this study is considered to be higher according to Slatkin (1981, 1985), Caccone (1985) and Waples (1987) grouping. They grouped *Nm* values into three categories: high (*Nm* > 1.000), intermediate, (0.250 - 0.990), and low (0.000 - 0.249). The fact that *Nm* value of this study is higher indicates gene flow between populations is obvious which will agree with the AMOVA result showing there is low variation among population.

The G_{ST} value of this result also shows the lack of population differentiation which goes with Laurentin and Karlovsky (2006) work. They found a G_{ST} value of 0.20 (below this study 0.25) and suggested the lack of association between geographical origin and population differentiation. In their finding particularly Indian, African and Chinese-Japanese-Korean accessions are distributed throughout clusters in UPGMA analysis and the whole two-dimensional space in PCA. Toan et al. (2011) also found an intermediate estimate of gene flow (Nm) value of 0.38 on average in there study. This lack of association between geographical distribution and classification based on molecular markers in sesame was explained by the exchange of sesame seeds among widely separated locations which could be due to movement/migration of peoples from one place to another along with their own seed or the purchase of seeds from other places for cultivation. These results also became apparent from the cluster analysis.

Conclusion

This study shows that bulk sampling strategy for DNA analysis and ISSR marker are important for genetic diversity study in sesame. The study also shows, among the sesame accessions held by IBC, in comparison, Ethiopian accessions are more diverse than the exotic ones. Separate analysis of the Ethiopian accession revealed there are places with higher genetic diversity which will be valuable for the future collection, conservation and improvement strategies. Regions that showed relatively high level of genetic diversity should be the main focus for future collection and population improvement based on selection. The present study also showed weak association between genetic variation of sesame accessions and their ecological regions of origin. The most likely important factor affecting the genetic structure of sesame in this region is possibly human activities. This shows there could be a possibility of sampling plants with the same genetic constitution from different administrative regions. This problem becomes sever if an administrative region is represented by small sample as in the case of IBC.

Conflict of interest

The authors have not declared any conflict of interest.

ACKNOWLEDGEMENTS

Dagmawi T. Woldesenbet is thankful to Rural Capacity Building Project (RCBP) and the Private Public Partnership for sesame for funding the research; Institute of Biodiversity Conservation (Addis Ababa, Ethiopia) for kindly providing seeds and passport data of the landrace accessions; Werer Agricultural Research Center (Werer, Ethiopia) for provision of physical resources and other facilities; the Faculty of Life Sciences of Addis Ababa University for giving me the chance for this study. I am thankful to all the authors for their valuable assistance.

REFERENCES

Abdellatef E, Sirelkhatem R, Mohamed A, Radwan KH , Khalafalla MM (2008). Study of genetic diversity in Sudanese sesame (*Sesamum indicum* L.) germplasm using random amplified polymorphic DNA (RAPD) markers. Afr J. Biotechnol. 7(24):4423-4427.

Aga E (2005). Molecular genetic diversity study of forest coffee tree (*Coffea arabica* L.) populations in Ethiopia: Implications for conservation and breeding. Doctoral Thesis, Faculty of Landscape Planning, Horticulture and Agricultural Science, Swedish University of Agricultural Sciences (SLU).

Ali GM, Yasumoto S, Seki-Katsuta M (2007). Assessment of genetic diversity in sesame (*Sesamum indicum* L.) detected by amplified fragment length polymorphism markers. Elect. J. Biotechnol. 10:1.

Ashri A (1998). Sesame breeding. Plant Breed. Rev. 16:179-228.

Assefa K (2003). Phenotypic and molecular diversity in the Ethiopian cereal, Tef [*Eragrostis tef* (Zucc.) Trotter]. Doctoral Dissertation, Department of Crop Science, SLU. Acta Universitatis Agriculturae Sueciae. Agraria. p. 426.

Baker J (2001). Migration as a positive response to opportunity and context: the case of Welo, Ethiopia. In: Mobile Africa: changing patterns of movement in Africa and beyond. (Mirjam de, B., Rijk van, D., Foeken, D. ed.)

Barrett BA, Kidwell KK (1998). AFLP based genetic diversity assessment among wheat cultivars from the pacific northwest. Crop Sci. 38:1261-1271.

Bedigian D (1981). Origin, diversity, exploration and collection of sesame. In: Sesame: Status and Improvement, Proc. Expert Consultation, Rome, Italy, 8-12 December, 1980. FAO, Rome, Italy. pp. 164-169.

Bedigian D (2004). History and lore of sesame in Southwest Asia. Econ. Bot. 58:329-353.

Blair MW, Panaud O, McCouch SR (1999). Inter-simple sequence repeat(ISSR) amplification for analysis of microsatellite motif frequency and fingerprinting in rice (Oryza sativa L.). Theor. Appl. Genet. 98:780-792.

Bornet BC, Muller FP, Branchard M (2002). Highly informative nature of inter simple sequence repeat (ISSR) sequences amplified using triand tetra-nucleotide primers from DNA of cauliflower (*Brassica oleracea* var. 'botrytus' L.). Genome, 45:890-896.

Borsch T, Hilu KW, Quandt D, Wilde V, Neinhuis C, Barthlott W (2003). Noncoding plastid trnT-trnF sequences reveal a well resolved phylogeny of basal angiosperms. J. Evol. Biol. 16:558-576.

Brantestam AK, Bothmer RV, Dayteg C, Rashal I, Tuvesson S, Weibull J (2004). Inter simple sequence repeat analysis of genetic diversity and relationships in cultivated barley of Nordic and Baltic origin. Hereditas, 141(2):186-187.

Caccone A (1985). Gene flow in cave arthropods: A qualitative and quantitative approach. Evol. 39:1223-1235.

CSA (Central Statistic Authority) (2010). Ethiopian agricultural sample enumeration: Report on the primary results of area, production and yield of temporary crops of private peasant holdings in Meher Season, Addis Ababa, Ethiopia.

Edossa F, Kassahun T, Endashaw B (2007). Genetic diversity and population structure of Ethiopian lentil (*Lens culinaris* Medikus) landraces as revealed by ISSR marker. Afr. J. Biol. 6(12):1460-1468.

Edossa F, Kassahun T, Endashaw B (2010). A comparative study of morphological and molecular diversity in Ethiopian lentil landraces. Afr. J. Plant Sci. 4(7):241-254.

Endale D, Parzies Heiko K (2010). Genetic variability among landraces of sesame in ethiopia . Afr. Crop Sci. J. 19(1):1-13.

Excoffier L, Laval G, Schneider S (2006). Arlequin Version.3.01: An integrated software package for population genetics data analysis. Evol. Bioinform. Online 1:47-50.

FAO (2011). Food and Agricultural Organization. Rome, Italy. http://faostat.fao.org Accessed on April 2014.

Gezahegn G (2007). Relationship between wild rice species of Ethiopian rice with cultivated rice based on ISSR marker. MSc Thesis Addis Ababa University.

Gilbert JE, Lewis RV, Wilkinson MJ, Kaligari PDS (1999). Developing an appropriate strategy to assess genetic variability in plant germplasm collections. Theor. Appl. Genet. 98:1125-1131.

Godwin ID, Aitken AB, Smith LW (1997). Application of inter simple sequence repeat (ISSR) markers to plant genetics. Electrophor. 18:1524-1528.

Hou YC, Yan ZH, Wei YM, Zheng YL (2005). Genetic diversity in barley from west China based on RAPD and ISSR analysis. Barley Genet. Newslett. 35:9-12.

Joseph MM, Bruce AM (1993). Gene flow in plant pathosystems. Ann. Rev. Phytopatol. 31:353-373.

Joshi SP, Gupta VS, Aggarwal RK, Ranjekar PK, Brar DS (2000). Genetic diversity and phylogenetic relationship as revealed by inter simple sequence repeat (ISSR) polymorphism in the genus *Oryza*. Theor. Appl. Genet. 100:1311-1320.

Kantety RV, Zeng X, Bennetzen JL, Brent EZ (1995). Assessment of genetic diversity in dent and popcorn (*Zea mays* L.) inbred lines using inter-simple sequence repeat (ISSR) amplification. Mol. Breed. 1:365-373.

Kim DH, Zur G, Danin-Poleg Y, Lee SW, Shim KB, Kang CW, Kashi Y (2002). Genetic relationships of sesame germplasm collection as revealed by inter-simple sequence repeats. Plant Breed. 121:259-262.

Laurentin HE, Karlovsky P (2006). Genetic relationship and diversity in a sesame (*Sesamum indicum* L.) germplasm collection using amplified fragment length polymorphism (AFLP). BMC Genet. 7:10.

Mahajan RK, Bisht IS, Dhillon BS (2007). Establishment of a core collection of world sesame(*Sesamum indicum* l.) germplasm accessions. SABRAO J. Breed. Genet. 39(1):53-64.

Mehra KL (1967). Sesame in India. In: Oilseed Crops, Tropical Agriculture Series, (Weiss, E.A., ed.). Longman, London. pp. 282-340.

Pavlicek A, Hrda S, Flegr J (1999). Free tree free ware program for construction of phylogenetic trees on the basis of distance data and bootstrap/Jack Knife analysis of the tree robustness. Application in the RAPD analysis of genus *Frenkelia*. Folia Biol. 45:97-99.

Rohlf FJ (2004). NTSYS-pc ver 2.11T. Exter Software, Setauket, New York.

Saitou N, Nei M (1987). The neighbor-joining method: A new method for reconstructing phylogenetic trees. Mol. Biol. Evol. 4:406-425.

Seleshi A (2008). Genetic divergence and correlation study in sesame (*Sesamim indicum* L.) genotypes. MSc Thesis. Addis Ababa University.

Slatkin M (1981). Estimating levels of gene flow in natural populations. Genet. 99:323-335.

Slatkin M (1985). Rare alleles as indicators of gene flow. Evol. 39:53-65.

Sneath PHA, Sokal RR (1973). Numerical Taxonomy. Freeman, Sanfrancisco

Studier JA, Keppler KL (1988). A note on the neighbor-joining algorithm of Saitou and Nei. Mol. Biol. Evol. 5:729-731.

Tadele A (2005). Sesame (*Sesamum indicum* L.) Research in Ethiopia: a Review of Past Work and Potential and Future Prospects. In: Sesame and Safflower Newsletter.(José Fernández Martínez, IAS, Córdoba, Spain. EcoPort version by Peter Griffee, FAO).

Tesfaye K (2006). Genetic diversity of wild *Coffea arabica* populations in Ethiopia as a contribution to conservation and use planning. Ecology and development series no. 44. Doctoral Thesis. University of Bonn, Germany.

Toan DP, Mulatu G, Tri MB, Tuyen CB, Arnulf M, Anders S (2011). Comparative analysis of genetic diversity of Sesame (*Sesamum indicum* L.) from Vietnam and Cambodia using agro-morphological and molecular markers. Hereditas 148(1):28-35.

Waples RS (1987) A multispecies approach to the analysis of gene flow in marine shore fishes. Evol. 41:385-400.

Weising K, Nybom H, Wolff K, Kahl G (2005). DNA Fingerprinting in Plants: Principles, Methods and Applications. Taylor and Francis Group, USA. 444pp.

Weiss EA (1983). Oilseed Crops, Tropical Agriculture Series. Longman, London. pp. 282-340.

Wijnands JHM, Biersteker J, Van Loo EN (2009). Oilseeds business opportunities in Ethiopia 2009. Public private partnership in oil seed.

Wright S (1951). The genetical structure of populations. Ann. Eugenet. 15:323-354.

Yang RC, Boyle TJB, Ye Z, Mao JX (1999). POPGENE, the user friendly shareware for population geneticss analysis, version 1.31, Molecular Biology and Biotechnology Centre, University of Aleberta, Canada.

Yeh FC, Yang RC, Boyle TJB, Ye Z, Mao JX (1999). POPGENE, the user friendly shareware for population geneticss analysis, version 1.31, Molecular Biology and Biotechnology Centre, University of Aleberta, Canada.

Zietkiewicz E, Rafalski A, Labuda D (1994). Genome fingerprinting by simple sequence repeat(SSR)- Anchored polymerase chain reaction amplification. Genomics 20:176-183.

22

Assessment of different samples for molecular diagnosis of extra-pulmonary tuberculosis

Gasmelseed N.[1] , Aljak M. A.[1], Elmadani A. E.[1], Elgaili M. E.[2] and Saeed O. K.[3].

[1]National Cancer Institute, University of Gezira, Sudan.
[2]Department of Pathology, Faculty of Medicine, University of Gezira, Sudan.
[3]Department of Internal Medicine, Faculty of Medicine, University of Gezira, Sudan.

Extrapulmonary tuberculosis is an important clinical problem particularly in developing countries. The aim of this study was to assess two different samples (blood and fluid) for the diagnosis of extra-pulmonary tuberculosis (abdominal tuberculosis and tuberculous lymphadenitis). The study subjects were recruited from WadMedani Teaching Hospital during 2009-2013. Seventy five ascetic fluid and blood samples were collected from each suspected patient with abdominal tuberculosis and twenty five lymphatic aspirates and blood samples were collected from each suspected tuberculous lymphadenitis patient. DNA was extracted using DNP™ kit (CinnaGenInc) and polymerase chain reaction (PCR) was done using *IS6110* gene for both samples. In abdominal tuberculosis, 20/75 (27%) were positive for tuberculosis when ascetic fluid was used and 9/75 (12%) in case of blood samples. The comparison between ascetic fluid and blood samples, showed that, there was a significant difference in both results, P-value < 0.05. In tuberculous lymphadenitis, 13/25 (52%) and 3/25 (12%) were positive to tuberculosis when lymph aspirate and blood were used respectively. This study concluded that the best sample for diagnosis of abdominal TB and lymphadenitis is ascetic fluid and LN aspirate. This study recommends that ascetic fluid and lymph aspirate samples are recommended to be used in molecular diagnostic test.

Key words: Extra-pulmonary tuberculosis, molecular diagnosis, acetic fluid, lymph aspirate samples.

INTRODUCTION

Tuberculosis (TB) is among the top ten causes of global mortality, it is estimated that approximately one-third of the world's population is infected with tuberculosis bacillus, and each year eight million people develop the tuberculosis disease which annually kills 1.8 million worldwide (Ahmed et al., 2011). In Sudan, an estimated annual risk based on the data of the 1986 national prevalence survey of TB is 1.8% which gives an incidence of 90/100,000 smear positive cases, and puts Sudan among the high prevalence countries for TB in the eastern Mediterranean region (Crofton, 2000).

The World Health Organization estimated that Sudan

ranked twenty-third in the list of countries with the greatest number of estimated incident of TB cases (78 030), with an estimated rate, half again as high as the estimate for the whole world (220 vs.140 per 100 000 population) (El Sony et al., 2007). According to the Sudan annual health report 2007, the incidence of TB rate was 58 per 100,000, where Khartoum State reported the highest number of TB patients admission of 4878 cases, then Gezira State, 3105 cases, North Kordofan with 1853 cases, at the end of the list, North States became the fewest prevalence state of TB with 820 cases (Annual Health Statistical Report, 2007). In Gezira State, the incidence of TB cases from 2003 to 2008 was between 2706 - 3259, (WHO TB Program in Gezira State). The prevalence of TB in Sudan is 209 cases per 100,000 of the population and 50,000 incident cases during 2009 (Sharaf Eldin et al., 2011). Most of the cases of TB are pulmonary TB which is about 75%, the other 25% cases are extra-pulmonary TB (EPTB), (CDC, 2003).

The incidence of extra-pulmonary forms of tuberculosis varies from country to another, such that on the average between 1964 and 1989, 5 to 10% of the approximately seven million new cases each year in the developing countries were extra-pulmonary. This distribution also can be affected by origin of the individuals within a country (Talavera et al., 2001).

The diagnosis of extra-pulmonary tuberculosis is till now challenging for diagnostic routine laboratories. The aim of this study is to assess two different specimens (blood and acetic fluid for abdominal TB, blood or lymph aspirate for tuberculous lymphadenitis) using polymerase chain reaction (PCR) to diagnose abdominal TB and tuberculous lymphadenitis in patients attending Wadmedani Teaching Hospital during the period of 2009-2012.

MATERIALS AND METHODS

Study subjects

All clinically suspected patients with abdominal TB or tuberculous lymphadenitis patients attending Wadmedani Teaching Hospital during the period of study from 2009 - 2012 were recruited for this study.

Inclusion criteria of subjects

Patients with symptoms of guarding and free fluid (ascetic), abdominal pain and mass, weight loss, fever sweating, diarrhea and vomiting, and patients with past history of pulmonary TB disease or/and history of TB contact were included. Patients presented with lymphadenopathy in different body regions including cervical, inguinal and auxiliary lymph node were included in the study.

Exclusion criteria of subjects

Patients with HIV infection, Chron's disease, malignancy, cirrhosis,

ulcerative colitis, chronic diseases were excluded from this study.

Sample collection

Ascetic fluid or lymph nodes aspirate and 2 ml of venous EDTA blood from each patient were collected after a written informed consent. Ethical approval for this study was given from National Cancer Institute Research Ethical Committee (NCI-REC).

Molecular methods

DNA was extracted from lymph node aspirate and ascetic fluid blood using DNP™ kit (cinna GenInc, Cat. No.DN115C form Iran). The kit was designed to isolate double stranded DNA form human and animal sources. Lymph node aspirate and ascetic fluid samples were treated with 4% sodium hydroxide, then they were centrifuged, homogenized, supernatant discarded, and the rest used for DNA extraction as described by manufacturer.

Polymerase chain reaction (PCR)

Identification detection of *Mycobacterium tuberculosis* was done by using a specific pair of primers (the sequence of these primers, T4 and T5, are: 5'-CCT GCG AGC GTA GGC GTC GG 3' and 5' CTC GTC CAG CGC CGC TTC GG 3', respectively) designed to amplify an insertion sequence *IS6110* gene in the *M. tuberculosis* complex, the expected band size was about 123 bp. The total volume of was 25 µl for each reaction (positive and negative control was done for each PCR run). All the amplicons were checked run using agarose gel electrophoresis. The presence of 123 bp for *IS6110* gene fragments indicated a positive test for *M. tuberculosis*.

Statistical analysis

A statistical analysis was performed with SPSS statistical package version 16.0. Descriptive analysis was done and correlations between different variables were calculated using Pearson Chi-square test.

RESULTS

Hundred suspected extra-pulmonary TB cases were recruited from WadMedani Teaching Hospital during the period of September 2009 to September 2012. Seventy five out of hundred (75%) of the cases were suspected with abdominal TB while 25/100 (25%) were suspected to have lymphadinitis TB. The mean of ages of suspected abdominal subjects was 46.7 ± 18.2 years with range between (min 3 - max 80 years old). The mean age of suspected tuberculous lymphadenitis patients was 34.7 ± 20.7 years with range between (min 2 - max 80 years) as shown in Figure 1. The description of the study subjects showed that in suspected abdominal TB and TB lymphadenitis, males were 45 (60%), 10/25 (40%) respectively as shown in Table 1. The majority of the study subjects were from rural areas: 56/75 (74.7%) and 18/25 (72%) respectively as shown in Table 2. Regarding suspected abdominal TB patients, ascites was the common presenting symptom in all study subjects and

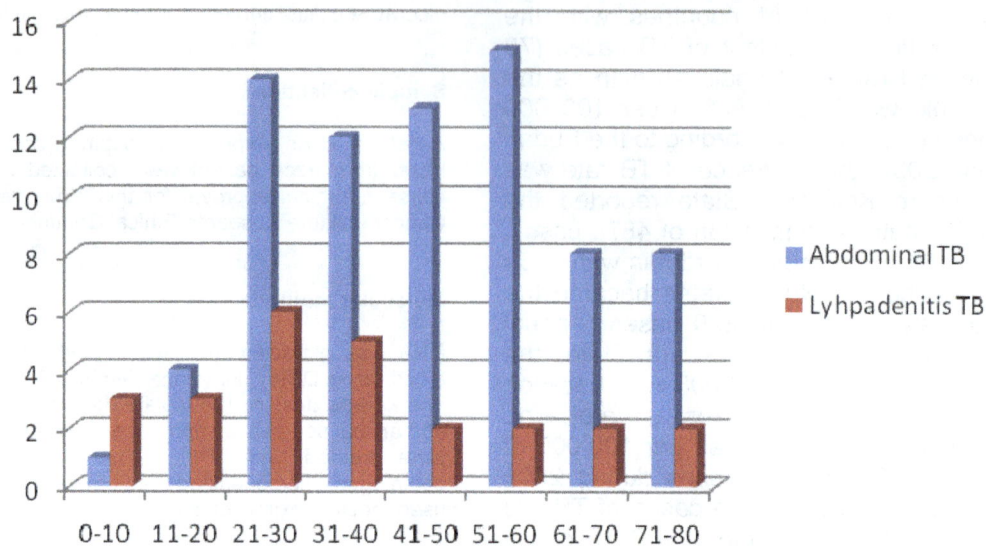

Figure 1. Description of study subjects by age groups.

Table 1. Description of study subjects by gender.

Gender	Abdominal TB (%)	Lymphadenitis TB (%)	total no. of subjects
Male	45 (60.0)	10(40.0)	55
Female	30 (40.0)	15(60.0)	45
Total	75	25	100

Table 2. Residence description of study subjects.

Sex description	Abdominal TB (%)	Lymphadenitis TB (%)	Total no. of subjects
Urban	19(25.3)	7 (28.0)	26
Rural	56(74.7)	18 (72.0)	74
Total	75	25	100

weight loss was 58/75 (77.3%), while lymphadenopathy was common presenting for TB lymphadenitis and night sweating, 10/25 (40%) were the common presenting symptoms in the suspected TB lymphadenitis study subjects.

Molecular detection of *M. tuberculosis* in abdominal TB study subjects

The *IS6110* gene is a multi-copy gene found only in *M. tuberculosis* complex. Most of the PCR studies have targeted *IS6110* gene sequence of *M. tuberculosis*

genome because of the presence of repetitive sequence of *IS6110* gene, this characteristic helps to increase the sensitivity of PCR over that obtained in the amplification of single DNA sequence. PCR was done for the ascetic fluid samples and blood for seventy five suspected abdominal TB, PCR resulted to 20/75 (26.7%) of ascetic fluid and 9/75 (12%) of blood samples were positive for *M. tuberculosis* as shown in Figure 2, PCR results indicate *M. tuberculosis* DNA with length 123 bp as in Appendix 1.

The comparison between ascetic fluid samples blood samples results, showed that there was a significant difference, between the two samples using person Chi-

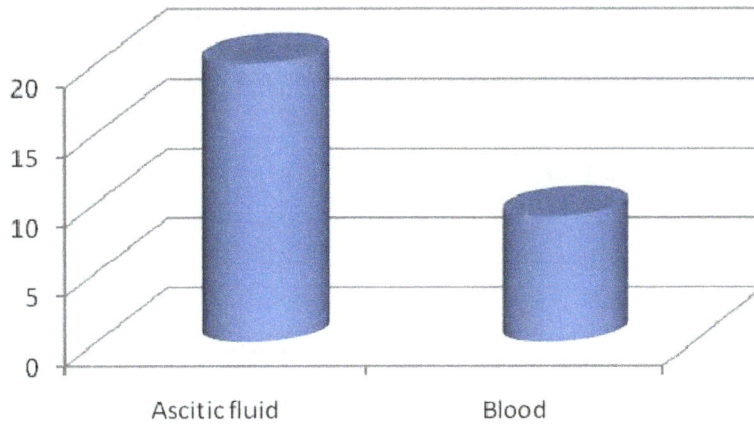

Figure 2. The positive PCR results in ascetic fluid and blood samples of study subjects. N=75.

Table 3. Comparison between ascetic fluid and blood.

		PCR results for blood		Total
		Positive	Negative	
PCR results for fluid	Positive	9	11	20
	Negative	0	55	55
Total		9	66	75

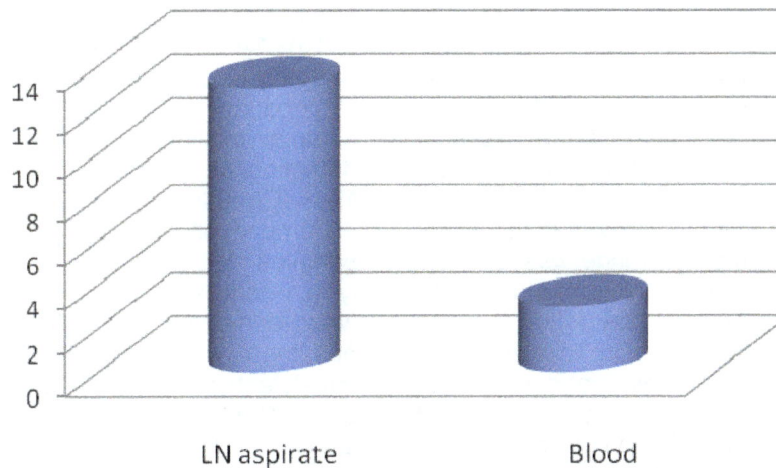

Figure 3. The positive PCR results in LN aspirate and blood samples by *IS6110* in suspected LN TB patients. N=25.

Square test = 28.1, P- value = 0.000 as shown in Table 3.

Detection of *M. tuberculosis* in tuberculous lymphadenitis study subjects

PCR was done using *IS6110* gene for 25 lymph node aspirate samples, 13/25 (52%) of the study subjects were positive for *M .tuberculosis* indicating bands with 123 bp. For the blood samples 3/25 (12%) of the study subjects were positive for PCR as shown in Figure 3. In comparison between the lymph node aspirate samples and blood samples, there was a significant difference, Person Chi-square test = 28.1, P- value = 0.000 as

Table 4. Comparison between LN and blood samples results.

		PCR results for blood		Total
		Positive	Negative	
PCR results for LN aspiration	Positive	3	10	13
	Negative	0	12	12
Total		3	22	25

shown in Table 4.

DISCUSSION

Extra-pulmonary tuberculosis (EPTB) is considered as an important clinical problem, it can occur in isolation or along with a pulmonary focus as in disseminated tuberculosis. It has been observed that EPTB constituted about 15 to 20% of all cases of TB (Sharma and Mohan, 2004). Abdominal TB involve the gastrointestinal tract, peritoneum, lymph nodes or solid viscera, constitutes up to 12% of extra-pulmonary TB and 1-3% of the total TB cases. The disease can mimic many conditions, including inflammatory bowel disease, malignancy and other infectious diseases, thus diagnosis is therefore often delayed (Uzunkoy et al., 2004). In this study, in 100 suspected EPTB, 75 were abdominal TB, 20/75 (26.7%) of them were positive. Tuberculosis of the lymph node (tuberculous lymphadenitis) is the most common form of extrapulmonary tuberculosis. In developed countries, tuberculosis is implicated in as few as 1.6% of patients with lymphadenopathy (Sarwar et al., 2004). But in developing countries almost two third of the cases of lymphadenopathy are due to tuberculosis (Sarwar et al., 2004).

The mean age of the abdominal TB cases in this study was 47.8 ±18.2 years, while the age range was found to be between 21-70 years which was similar to study done by Uzunkoy et al. (2004) in Turkey; they found that the mean age was 39 years (range 18-65). In Taiwan, Huan-Lin et al. (2009) found that patients with abdominal TB, had age range from 22 to 88 years, with a mean age of 50 ± 18[9]. In the tuberculous lymphadenitis patients, the mean ages was 36.4±16.6 years and the age range was 21-40 years, this was different from study done by in Sudan that showed found that a mean age of thirty patients was 26.9±11.2 years (Sharaf-eldin et al., 2002).

Extra-pulmonary tuberculosis, which remained without bacteriologic or histological confirmation, and the diagnosis based on recognition of signs and symptoms is raised from involvement of particular organ or system. Molecular technique using nucleic acid amplification methods to detect M. tuberculosis in clinical specimens are increasingly in use as a tool for TB diagnosis. IS6110 gene is used as a target by the majority of the

investigators performing PCR based diagnosis of TB (Kulkarni et al., 2006). The principal reason for using IS6110 gene is because it is considered to be a good target for amplification as this insertion sequence is found in almost all members in high copy number in most strains of the M. tuberculosis complex, which was thought to confer higher sensitivity (Kulkarni et al., 2006; Sharma et al., 2013).

The aim of this study was to compare between different samples molecular diagnosis of extra pulmonary TB by using IS6110 gene. In comparison between ascetic fluid and blood samples, the study showed that there was a significant difference (P<0.05) between the two different samples, this means that ascetic fluid had a high positivity than blood and it may be due to the presence of the microorganism at the location of the lesion while low level of was found in the blood. This is consistent with a study done in Pakistan using ascetic fluid DNA sample and IS6110 specific primer sequence for M. tuberculosis, 50/55 (90.9%) patients were positive (Phulpoto et al., 2010). Several studies had the same results with high positivity from ascetic fluid (Uzunkoy et al., 2004; Protopapas et al., 2003). Thus, ascetic fluid is a suitable sample for diagnosis of abdominal TB. Therefore, detecting M. tuberculosis from blood samples was less likely (Chhuttani et al., 1986)

Comparison between lymph node aspiration and blood samples for patient with lymphadenitis, showed that lymph node aspirate samples 13/25 (52%) were positive when compared with only 3/25 (12%) in blood samples, and there was a significant difference (P<0.05), this is consistent with a study conducted in lymphodenopathy clinic, Central Police Hospital, Burri, Khartoum, Sudan, showing that circulating M. tuberculosis DNA was detected in the aspirate fluid of all the 28 patients (100%) with multiple lymph nodes (Sharaf-eldin, 2002). Another study detected M. tuberculosis from blood, in 191 suspected patients with extrapulmonary TB, they found a low rate of detection (Kolk et al., 1998). Same result was also found in a study done on sputum-smear positive patients with M. tuberculosis (Aguado et al., 1996).

Conclusion

For PCR diagnosis of abdominal TB and tuberculous

lymphadenitis, the best samples are ascetic fluid and lymph nodes aspirate, respectively.

Recommendation

Ascetic fluid and lymph aspirate are recommended to be used as sample for diagnosis of extra-pulmonary tuberculosis.

Conflict of Interests

The author(s) have not declared any conflict of interests.

ACKNOWLEDGEMENTS

Thanks to Wadmedani Teaching Hospital staff for assisting in the collection of samples. Great thanks to Ministry of Higher Education and Research in Sudan for financing this research project.

REFERENCES

Aguado JM, Rebollo MJ, Palenque E, Folgueria L (1996). Blood-based PCR assay to detect pulmonary tuberculosis. Lancet. 347:1836-1837.

Ahmed HE, Nassar AS, Ginawi I (2011). Screening for tuberculosis and its histological pattern in patients with enlarged lymph node. Pathol. Res. Int. 41:35-76.

Chhuttani PN, Dilawari JB (1986). Tuberculosis of the stomach. In:Shah SJ, ed. API Text book of Medicine. Bombay. Assoc. Phys. India. p. 570.

Crofton J (2000). Reforms to the health sector must retain vertical programmes like those for tuberculosis. BMJ. 320:1726.

El Sony A, Slama K, Salieh M, Elhaj H, Adam K, Hassan A, Enarson DA (2007). Feasibility of brief tobacco cessation advice for tuberculosis patients: a study from Sudan. Int. J. Tuberc. Lung. Dis. 11(2):150-155.

Kolk AH, Kox LF, van Leeuwen J, Kuijper S, Jansen HM (1998). Clinical utility of the polymerase chain reaction in the diagnosis of extrapulmonary tuberculosis. Eur. Respir. J. 11:1222-1226.

Kulkarni S, Vyas S, Supe A, Kadival G (2006). Use of polymerase chain reaction in the diagnosis of abdominal tuberculosis. J. Gastroenterol. Hepatol. pp. 819-823.

Phulpoto JA (2013). Diagnosis of TuberculousAscities; Role of polymerase chain reaction and adenosine deaminase activity. Professional Med. J. 20(3):381-384. Protopapas A, Milingos SD (2003). Miliary tuberculous. J. Am. Assoc. Gynecol. Laparosc. 10(2):139.

Sarwar A, Haque AU, Aftab S, Mustafa M, Moatasim A, Siddique S and Sani A (2004). Spectrum of Morphological Changes in Tuberculous Lymphadenitis. Int. J. Pathol. 2(2):85-89.

Sharaf Eldin GS, Fadl-Elmula, I, Ali MS, Ali AB, Abdel Latif GAS, Mallard K, Bottomley Ch, McNerney R (2011). Tuberculosis in Sudan: a study of Mycobacterium tuberculosis strain genotypeand susceptibility to anti-tuberculosis drugs. Bio. Med. Central Infect. Dis. 11:219.

Sharaf-Eldin GS, Saeed NS, Hamid ME, Jordaan AM, Van der GDS, Warren R M, Van Helden P D, Victor TC (2002). Molecular Analysis of Clinical Isolates of Mycobacterium tuberculosis Collected from Patients with Persistent Disease in the Khartoum Region of Sudan. J. Infect. 44:244-251.

Sharma SK, Mohan A (2004). Extrapulmonary tuberculosis. Indian J. Med. Res. 120:316-353.

Sharma K, Sinha SK, Sharma A, Nada R, Prasad KK, Goyal K, Rana SS, Bhasin DK, Sharma M (2013). Multiplex PCR for rapid diagnosis of gastrointestinal tuberculosis. J. Glob. Infect. Dis. 5(2):49-53.

Talavera W, Miranda R, Lessnau K, Klapholz L (2001). Extrapulmonary tuberculosis In L N Friedman (ed.), Tuberculosis: current concepts and treatment. CRC Press, Inc. Boca Raton Fla. 2nd ed pp.139-190.

Uzunkoy A, Harma M, Harma M (2004). Diagnosis of abdominal tuberculosis: Experience from 11 cases and review of the literature. World J. Gastroenterol. 10(24):3647-3549.

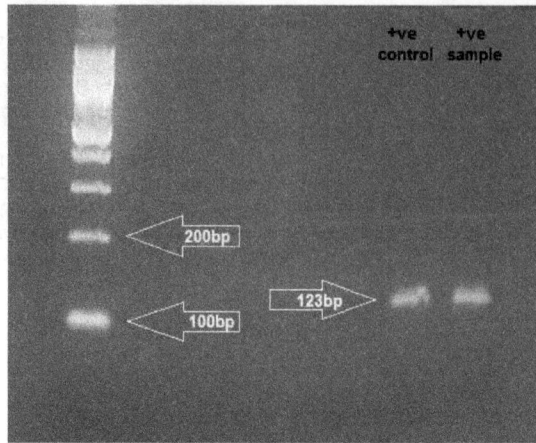

Appendix 1. A band of 123 bp for positive sample using *IS6110*.

23

Genetic variation of 17 Y-chromosomal short tandem repeats (STRs) loci from unrelated individuals in Iraq

Imad H.[1] , Cheah Q.[1], Mohammad J.[2] and Aamera O.[3]

[1]Department of Molecular Biology, Putra University, Kuala Lumpur City, Malaysia.
[2]Department of Molecular Biology, Babylon University, Hilla City, Iraq.
[3]Institute of Medico-legal in Baghdad, Ministry of Health of Iraq, Iraq.

Analysis of 17 Y chromosomal short tandem repeats (STRs) loci in a population sample of 400 unrelated male living in the middle and south of Iraq was done to evaluate allele frequencies and gene diversity for each Y-STR locus of the Y filer™ PCR amplification kit. The seventeen loci include DYS635, DYS437, DYS448, DYS456, DYS458, YGATA H4, DYS389I, DYS389II, DYS19, DYS391, DYS438, DYS390, DYS439, DYS392, DYS393, DYS385a and DYS385b. Total DNA from blood cells was extracted using the FTA™ paper DNA extraction. A total of 361 unique haplotypes were identified among the 400 individuals studied. The DYS385b had the highest diversity (GD = 0.8392), while loci DYS392 had the lowest (D = 0.2695).

Key words: Allele frequency, gene diversity, Iraq, Y filer™.

INTRODUCTION

Microsatellites (Ellegren et al., 2004) are DNA regions with repeat units that are 2 to 7 bp in length or most generally short tandem repeats (STRs) or simple sequence repeats (SSRs). The classification of the DNA sequences is determined by the length of the core repeat unit and the number of adjacent repeat units. It may contain several hundred to thousands (Butler et al., 2012; Andrea et al., 2008) of them. Length and sequence polymorphisms (Silvia et al., 2009) may be found within the DNA.

DNA can be used to study human evolution. Besides, information from DNA typing are important for medico-legal matters with polymorphisms leading to more biological studies (Walkinshaw et al., 1996). Since the STR markers are important for human identification purposes (Carolina et al., 2010), the number of repeats can be highly variable among individuals and can be used for identification purposes.

Chromosome Y microsatellites or STRs seem to be ideal markers to delineate differences between human populations for several reasons: (a) They are transmitted in uniparental (paternal) fashion without recombination, (b) They are very sensitive for genetic drift, and (c) They allow a simple highly informative haplotype construction. Also, for forensic applications this ability to differentiate distinct Y chromosomes makes Y-STR's an advantageous addition to the well characterized autosomal STRs. For a number of forensic applications, Y-STRs could be superior to autosomal STRs. Especially in rape cases where (1) the differential lysis was unsuccessful, (2) the number of sperm cells is very low, (3) to vasectomy epithelial cells instead of sperm cells from the ejaculate of the perpetrator have to be analysed, and (4) the perpetrator, due to a familial relationship shares many autosomal bands with the victim, Y-STR's could provide crucial evidence. Also, in the case of male-male rape or rape cases with multiple perpetrators, Y-STRs could lead to essential qualitative evidence. In all such cases Y-STRs facilitates a simple and reliable exclusion of suspects (Park et al., 2007; Parson et al.,

2003; Butler et al., 2002).

Multiplex polymerase chain reaction (PCR) is defined as the simultaneous amplification of multiple regions of DNA templates by adding more than one primer pair to the amplification reaction mixture. Since its first description in 1988, PCR multiplexing has been applied in many areas of DNA testing including the analysis of deletions, mutations and STRs (Chamberlain et al., 1988; Henegariu et al., 1994; Kimpton et al., 1996). Furthermore, the wide availability of genetic information due to the publishing of the sequence of the human genome makes the demand for multiplex PCR even greater (Venter et al., 2000; Lander et al., 2001). For example, more than 1.4 million single nucleotide polymorphisms (SNPs) have been identified in the human genome (Venter et al., 2001; Lander et al., 2001). Multiplex PCR primer sets have been used for linkage studies to track genetic diseases (Evans et al., 2001; Kwak et al., 2005). Eukaryotic genomes are full of repeated DNA sequences.

The use of the Y chromosome for forensic purposes was until recently restricted by a lack of polymorphic markers (Kayser et al., 1997). The Y chromosome is less variable than the other chromosomes. The majority of Y STRs including the ones presented here are located within its non-recombining region and are passed without recombination. Thus, results from individual markers cannot be combined using the product rule. Many markers are thus needed to obtain a high degree of discrimination between unrelated males (Kuppareddi et al., 2010; Hanson and Ballantyne, 2007).

The first aim of this study was to determine the genetic structure of Iraq population. The second one was to evaluate the importance of these loci for forensic genetic purposes.

MATERIALS AND METHODS

Population

Four hundred unrelated healthy males from the middle and south of Iraq provinces (Figure 1) were used.

DNA extraction

Total DNA from blood cells was extracted using the FTA™ paper DNA extraction (Dobbs et al., 2002).

PCR

PCR is the process used to amplify a specific region of DNA. It is possible to create multiple copies from the small amount of template DNA using this process. 12Plex amplification was performed by the commercial kit Y filer™ PCR amplification kit (Applied Biosystems, Foster City, CA) that amplifies 17 Y-STR loci (DYS635, DYS437, DYS448, DYS456, DYS458, YGATA H4, DYS389I, DYS389II, DYS19, DYS391, DYS438, DYS390, DYS439, DYS392, DYS393, DYS385a and DYS385b) and a segment of the

amelogenin gene, according to manufacturer's instructions but in a total reaction volume of 25 ml.

Typing

The ABI Prism1 3730xl Genetic Analyzer 16-capillary array system (Applied Biosystems, Foster City, CA, USA) following manufacturer's protocols, with POP-7™ Polymer and Data Collection Software, GeneMapper version 3.5 software (Applied Biosystems) were used. The allele designations were determined by comparison of the PCR products with those of allelic ladders provided with the kit. Nomenclature of loci and alleles is according to the International Society of Forensic Genetics (ISFG) guidelines reported in Gill et al. (2001). By comparison of the size of a sample's alleles to size of alleles in allelic ladders for the same loci being tested in the sample, the STR genotyping was conducted.

Quality control

Allelic ladders, male DNA (positive internal control), female DNA (negative control) and the amelogenin (internal control), provided by Applied Biosystems, were used in each reaction with the Y filer™ kit.

Statistical data analysis

Allele frequencies were calculated by direct counting. Allele diversity was calculated as per Nei (1987):

$$D = \frac{n}{n-1} \left(1 - \sum_{i=1}^{n} p_i^2 \right)$$

Where, n is the sample size and pi is the frequency of the ith allele.

RESULTS AND DISCUSSION

We identified 317 different haplotypes in our study sample (Table 2). 216 of which (82.3%) were unique, 15 were found twice (4.7%) and 13 were found in three individuals (4.1%). The most frequent haplotype was haplotype number 77 (Table 1). Haplotype 77 seems to be specific to Iraq. This is to be corroborated by future investigations. The DYS385b and DYS458 had the highest diversity (GD = 0.8392 and 0.806, respectively), while loci DYS392 and DYS439 had the lowest (D = 0.2695 and 0.2991, respectively).

Data comparison between our samples and a previously published sample from the Iraqi population (Nadia et al., 2011) was performed for markers which are common to both studies using the exact test for population differentiation implemented in GENEPOP (Raymond and Rousset, 1995).

In another study on 17 Y-STR Y-chromosomal short tandem repeat loci from the Cukurova region of Turkey, the DYS391 recorded lowest gene diversity in this region to be 0.51 and the highest 0.95 for DYS385a/b and no significant differences were found when comparing this

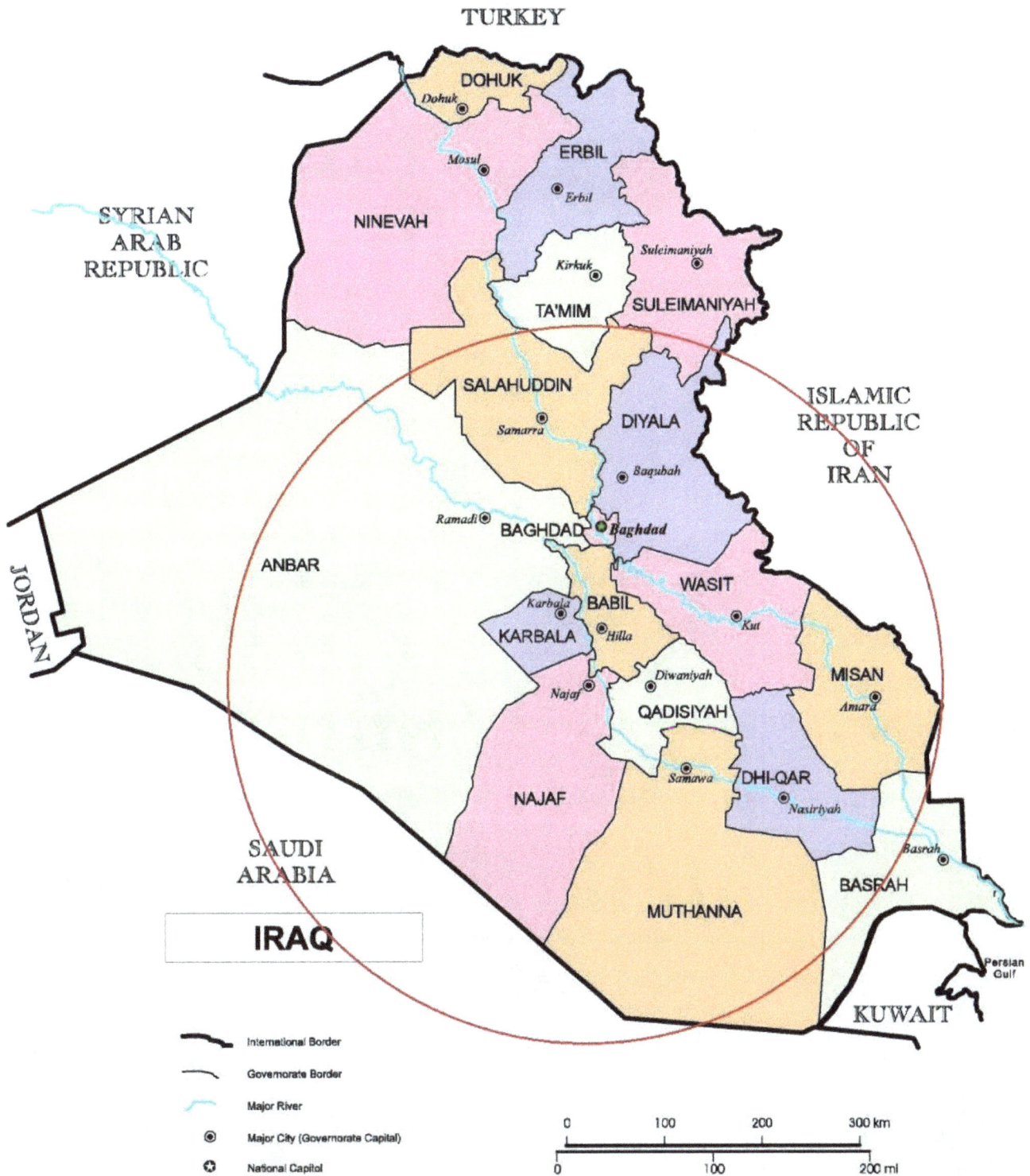

Figure 1. An Iraq administrative map showing the population and the central governorates where our samples came from (areas included in the red circle) (Source: UNOSAT, www.unosat.org, United Nations).

data with haplotype data of other Turkish populations (Ayse et al., 2011). In Northern Greece, the haplotype diversity was 0.9992 in seventeen Y STR loci typed in a population sample of unrelated male individuals. Haplotypes are presented for the following loci: DYS456, DYS389I, DYS390, DYS389II, DYS458, DYS19, DYS385a/b, DYS393, DYS391, DYS439, DYS635, DYS392, Y GATA H4, DYS437, DYS438 and DYS448

Table 1. Allele frequencies of 17 Y-STR loci in 400 Iraq males.

Alleles	DYS635	DYS437	DYS448	DYS456	DYS458	YGATA H4	DYS19	DYS385a	DYS385b	DYS389I	DYS389II	DYS390	DYS391	DYS392	DYS393	DYS438	DYS439
8	-	-	-	-	-	-	-	-	-	-	-	-	-	-	-	-	-
9	-	-	-	-	-	-	-	-	-	-	-	-	0.0563	-	-	0.1063	0.0094
10	-	-	-	-	-	0.1030	-	-	-	-	-	-	0.2188	0.0438	-	**0.6594**	0.0594
11	-	-	-	-	-	0.2030	-	0.1625	-	-	-	-	**0.4969**	**0.8500**	-	0.1281	0.3281
12	-	-	-	-	-	**0.4981**	-	0.0438	-	0.2750	-	-	0.1250	0.0688	0.1844	0.1063	**0.4094**
13	-	-	-	0.0982	-	0.0901	0.3594	0.5219	0.0625	0.3406	-	-	0.1031	0.0375	**0.6594**	-	0.1250
14	-	**0.4881**	-	0.0880	0.0531	0.1000	**0.4219**	0.1719	**0.2500**	**0.3844**	-	-	-	-	0.1313	-	0.0688
15	-	0.2500	-	0.5031	0.3300	-	0.1625	0.0094	0.1063	-	-	-	-	-	-	-	-
16	-	0.2601	-	0.2380	0.1882	-	0.0343	0.0188	0.1438	-	-	-	-	-	-	-	-
17	-	-	-	0.0450	0.1001	-	0.0218	0.0563	0.0438	-	-	-	-	-	-	-	-
18	-	-	0.1300	-	0.0730	-	-	-	0.1688	-	-	-	-	-	-	-	-
19	-	-	**0.5032**	-	0.1030	-	-	0.0156	0.1719	-	-	-	-	-	-	-	-
20	0.0781	-	0.2430	-	0.1431	-	-	-	0.0531	-	-	-	-	-	-	-	-
21	0.1650	-	0.1003	-	-	-	-	-	-	-	-	0.0500	-	-	-	-	-
22	0.0530	-	0.1000	-	-	-	-	-	-	-	-	0.0594	-	-	-	-	-
23	0.2880	-	-	-	-	-	-	-	-	-	-	**0.4406**	-	-	-	-	-
24	**0.3781**	-	-	-	-	-	-	-	-	-	-	0.3688	-	-	-	-	-
25	0.0301	-	-	-	-	-	-	-	-	-	-	0.0813	-	-	-	-	-
28	-	-	-	-	-	-	-	-	-	-	0.0906	-	-	-	-	-	-
29	-	-	-	-	-	-	-	-	-	-	**0.4125**	-	-	-	-	-	-
30	-	-	-	-	-	-	-	-	-	-	0.2938	-	-	-	-	-	-
31	-	-	-	-	-	-	-	-	-	-	0.1188	-	-	-	-	-	-
32	-	-	-	-	-	-	-	-	-	-	0.0500	-	-	-	-	-	-
33	-	-	-	-	-	-	-	-	-	-	0.0344	-	-	-	-	-	-
GD*	0.737	0.631	0.651	0.671	0.806	0.682	0.6650	0.6547	0.8332	0.6606	0.7140	0.6573	0.5802	0.2695	0.5140	0.5262	0.2991

GD*: Genetic diversity; In **bold** are the most common allele for each locus.

This database study provides additional informa-tion for the application of Y chromosomal STRs to forensic identification efforts in Greece (Leda et al., 2008).

This paper follows the guidelines for publication of population data requested by Lincoln et al. (2000).

Conclusions

Power of discrimination values for all tested loci

means that those loci can be used as a DNA-based database. Different alleles were observed across the population. The lowest gene diversity was DYS392 and DYS439. The high gene diversity was DYS385b and DYS458. Based on statistical parameters, the population of Iraq may

Table 2. Haplotypes for the 17 Y-STR loci observed 400 Iraq males.

Haplotype	DYS635	DYS437	DYS448	DYS456	DYS458	YGATA H4	DYS19	DYS385a	DYS385b	DYS389I	DYS389II	DYS390	DYS391	DYS392	DYS393	DYS438	DYS439	N	F
H1	24	15	20	15	14	12	13	11	16	12	30	23	10	11	13	10	13	1	0.0025
H2	24	15	18	15	14	13	13	11	16	12	30	23	8	11	13	9	13	1	0.0025
H3	24	15	20	15	15	12	13	11	16	12	31	23	8	11	12	9	10	1	0.0025
H4	24	15	19	15	15	11	13	11	14	12	31	23	8	11	13	11	10	1	0.0025
H5	24	15	19	16	15	10	13	13	14	12	31	23	10	11	13	10	10	1	0.0025
H6	24	15	19	16	17	12	13	13	14	12	28	23	10	11	13	10	13	1	0.0025
H7	24	15	19	14	17	12	13	13	14	12	28	23	9	11	14	12	13	1	0.0025
H8	24	15	19	16	17	11	13	13	14	12	28	23	10	11	13	12	10	1	0.0025
H9	24	15	20	16	17	14	13	13	14	12	29	23	10	11	13	12	10	5	0.0125
H10	24	15	19	16	15	14	13	13	14	12	29	23	8	11	14	12	11	1	0.0025
H11	24	15	19	13	15	11	13	13	14	12	29	23	9	11	14	10	11	1	0.0025
H12	24	15	21	13	15	10	13	13	14	14	29	23	10	11	12	11	11	1	0.0025
H13	24	15	21	15	15	13	13	13	14	14	29	23	10	11	13	10	11	1	0.0025
H14	24	15	22	15	15	12	13	13	14	13	29	23	10	11	14	10	10	1	0.0025
H15	24	15	19	15	20	12	13	13	18	13	29	23	10	11	14	12	10	1	0.0025
H16	24	15	19	15	16	12	13	13	18	13	28	23	8	11	13	12	13	1	0.0025
H17	24	15	19	15	18	12	13	13	18	13	28	23	8	11	13	12	10	1	0.0025
H18	24	15	19	14	18	13	13	13	16	13	28	23	10	11	13	12	10	1	0.0025
H19	24	15	19	16	18	12	13	14	16	13	31	23	10	11	13	10	10	1	0.0025
H20	24	15	21	16	15	12	13	14	16	12	31	23	10	11	13	10	10	1	0.0025
H21	24	14	21	16	16	12	13	11	16	12	31	23	10	11	12	10	10	1	0.0025
H22	24	14	19	15	15	12	13	11	13	12	28	23	10	11	12	10	10	1	0.0025
H23	24	14	19	15	20	12	13	11	13	12	28	23	10	11	14	9	10	1	0.0025
H24	24	14	19	13	15	12	13	11	13	12	30	23	10	11	14	10	10	1	0.0025
H25	24	14	20	15	15	12	13	11	18	12	30	23	10	11	13	10	10	3	0.0075
H26	24	16	18	15	19	12	13	11	14	12	30	25	10	11	13	10	10	3	0.0075
H27	24	16	18	16	15	12	13	11	14	12	32	23	10	11	13	10	10	1	0.0025
H28	24	14	19	13	15	12	13	11	18	12	33	23	10	11	13	10	10	1	0.0025
H29	24	14	19	15	20	12	13	13	18	14	30	21	10	11	13	10	10	1	0.0025
H30	24	14	20	15	19	12	13	13	18	14	29	24	10	11	13	10	10	1	0.0025
H31	24	14	20	16	15	12	13	13	18	13	29	24	10	11	12	10	11	1	0.0025
H32	24	14	20	13	15	11	13	13	18	14	29	25	10	11	13	10	11	1	0.0025
H33	24	16	20	15	15	10	13	13	18	14	29	21	10	11	13	10	11	1	0.0025
H34	24	15	20	15	16	12	13	13	18	14	29	23	10	11	13	10	11	1	0.0025
H35	24	15	21	15	16	10	13	13	18	13	29	23	10	11	13	10	11	1	0.0025
H36	24	15	21	14	16	13	13	13	18	12	29	23	10	11	14	10	11	1	0.0025
H37	24	15	19	15	16	11	13	13	18	14	29	24	10	11	13	10	11	1	0.0025
H38	24	14	20	15	15	14	13	13	16	13	30	23	10	11	13	10	11	1	0.0025
H39	24	14	19	15	15	10	13	13	19	13	28	23	10	11	13	10	11	1	0.0025
H40	24	14	19	14	15	12	13	13	19	13	30	24	10	11	13	10	11	1	0.0025
H41	24	14	19	13	15	12	13	13	19	13	32	21	10	11	13	10	11	1	0.0025
H42	24	14	19	16	15	12	13	13	19	13	32	25	10	11	12	10	11	1	0.0025
H43	24	14	19	15	16	12	13	13	14	13	28	25	10	11	12	10	11	1	0.0025
H44	24	14	19	15	20	12	13	13	14	13	29	25	10	11	12	10	11	1	0.0025
H45	24	14	19	15	19	12	13	12	14	13	32	25	10	11	13	10	12	1	0.0025
H46	24	14	21	15	16	12	13	12	14	13	30	25	10	11	13	10	12	1	0.0025
H47	24	14	18	15	16	11	13	12	14	13	28	23	10	11	12	10	12	1	0.0025

Table 2. Contd.

H48	23	14	19	15	16	13	13	12	14	13	28	23	10	11	14	10	12	1	0.0025
H49	23	14	19	13	15	12	13	12	14	12	28	24	10	11	13	10	12	1	0.0025
H50	23	14	19	16	15	14	13	12	14	12	30	24	10	11	12	10	10	1	0.0025
H51	23	14	21	16	15	11	13	11	14	14	30	24	10	11	12	9	10	1	0.0025
H52	23	14	20	15	15	11	13	11	14	14	30	25	10	11	12	10	13	2	0.0050
H53	22	14	20	15	16	11	13	11	14	13	30	23	10	11	12	10	14	1	0.0025
H54	22	14	18	15	20	11	13	11	14	12	30	23	10	11	13	10	10	1	0.0025
H55	23	14	20	15	19	11	13	11	14	14	30	24	11	11	13	10	13	1	0.0025
H56	23	14	20	15	18	12	13	11	19	13	30	24	11	11	13	10	12	1	0.0025
H57	23	14	20	14	18	11	13	11	16	12	30	24	11	11	14	11	10	1	0.0025
H58	23	14	20	15	17	12	13	11	16	14	30	24	11	11	13	9	10	1	0.0025
H59	23	14	19	17	17	11	13	11	15	13	32	22	8	11	14	10	10	1	0.0025
H60	23	16	19	15	17	14	13	11	15	13	32	24	10	11	13	10	10	1	0.0025
H61	23	14	19	16	17	13	13	11	15	13	30	25	10	11	13	12	10	1	0.0025
H62	23	14	18	16	17	11	13	11	15	12	30	25	10	11	13	11	10	1	0.0025
H63	23	15	19	13	17	11	13	11	15	14	30	25	9	11	14	11	9	1	0.0025
H64	23	15	19	14	17	12	13	11	15	14	32	25	10	11	14	11	10	1	0.0025
H65	23	15	19	17	20	12	13	11	15	14	29	25	10	11	13	10	9	3	0.0075
H66	23	14	20	17	19	12	13	11	15	14	30	23	8	11	13	10	9	1	0.0025
H67	23	14	20	15	19	11	13	11	15	14	30	23	10	11	12	10	14	1	0.0025
H68	23	14	21	16	19	13	13	11	15	14	30	24	9	11	13	10	13	1	0.0025
H69	23	14	21	16	20	12	13	11	18	13	30	24	9	11	13	9	10	1	0.0025
H70	23	14	19	16	18	12	13	11	18	14	29	24	9	11	13	12	10	1	0.0025
H71	23	14	19	15	20	12	13	11	18	12	29	24	9	11	13	11	10	1	0.0025
H72	23	14	20	16	17	12	13	11	18	12	29	23	12	11	13	11	10	1	0.0025
H73	23	14	21	16	17	10	13	11	19	12	29	23	10	11	13	11	10	1	0.0025
H74	23	14	19	16	15	11	13	11	19	12	29	22	10	11	13	11	11	1	0.0025
H75	23	14	19	14	15	14	13	11	19	12	29	23	9	11	13	11	11	1	0.0025
H76	23	14	19	15	15	12	13	11	14	12	29	23	9	11	13	10	11	1	0.0025
H77	23	14	18	15	15	13	13	11	14	12	29	24	11	11	13	10	11	7	0.0175
H78	23	16	18	15	15	12	13	11	14	13	29	24	10	11	13	12	11	1	0.0025
H79	23	15	18	15	20	12	13	11	14	12	30	24	10	11	13	10	11	1	0.0025
H80	23	16	18	13	20	11	13	11	14	14	30	24	9	11	13	10	14	1	0.0025
H81	23	16	19	13	19	11	13	11	14	14	30	23	11	11	13	10	14	1	0.0025
H82	23	16	19	15	19	12	13	16	13	14	30	25	10	11	13	10	10	1	0.0025
H83	23	16	19	15	16	12	13	16	13	12	30	25	10	11	13	9	10	1	0.0025
H84	23	14	19	15	15	12	13	16	13	12	30	25	10	11	13	10	10	1	0.0025
H85	23	16	20	15	15	11	13	14	13	12	30	24	9	11	13	10	10	1	0.0025
H86	23	16	20	16	16	13	13	14	17	12	30	24	11	11	13	10	14	1	0.0025
H87	23	14	21	15	16	11	13	14	17	12	29	24	12	11	13	10	14	1	0.0025
H88	23	15	21	15	19	12	13	14	20	12	29	24	11	11	13	11	12	1	0.0025
H89	23	14	21	15	20	12	13	14	20	12	29	25	9	11	13	11	13	1	0.0025
H90	20	16	21	15	18	12	13	14	20	12	29	24	10	11	13	9	13	3	0.0075
H91	20	16	21	15	17	12	13	14	18	12	29	24	10	11	13	10	13	3	0.0075
H92	20	16	21	13	17	11	13	14	19	12	29	24	12	11	13	12	13	1	0.0025
H93	20	16	19	16	17	12	13	14	19	12	29	24	10	11	13	10	13	1	0.0025
H94	20	15	19	15	15	12	13	14	16	12	29	23	10	11	13	11	13	1	0.0025
H95	20	14	19	15	20	12	13	14	16	12	29	23	9	11	13	10	13	1	0.0025
H96	20	14	19	17	15	12	13	14	16	12	29	23	8	11	13	10	14	1	0.0025
H97	20	14	19	14	15	14	13	14	16	12	29	23	12	11	13	10	10	1	0.0025
H98	24	14	19	13	15	13	13	14	16	14	29	21	12	11	13	11	10	1	0.0025
H99	24	14	19	16	15	10	13	14	16	14	29	22	8	11	13	11	11	1	0.0025

Table 2. Contd.

H100	24	14	19	17	15	12	13	13	16	14	29	22	10	11	13	11	13	1	0.0025
H101	21	14	19	17	16	12	13	13	16	14	29	25	10	11	13	11	12	1	0.0025
H102	21	14	19	15	16	11	13	13	16	14	32	23	10	11	13	10	12	1	0.0025
H103	21	14	20	15	16	12	14	13	16	14	28	22	10	13	13	10	12	1	0.0025
H104	21	14	20	15	19	12	14	13	16	14	33	23	10	13	13	10	11	1	0.0025
H105	21	14	18	14	19	10	14	13	16	14	28	25	10	11	13	10	11	1	0.0025
H106	21	14	18	15	19	11	14	13	19	14	28	25	10	11	13	10	14	1	0.0025
H107	21	14	21	15	19	10	14	13	15	14	28	23	10	11	13	10	14	1	0.0025
H108	21	14	22	15	19	12	14	13	19	14	28	23	10	11	13	10	14	1	0.0025
H109	21	14	22	15	20	12	14	13	20	14	32	23	10	12	13	10	11	5	0.0125
H110	21	14	19	15	15	12	14	13	20	13	32	23	10	11	13	10	11	1	0.0025
H111	21	14	20	15	16	12	14	14	20	13	29	23	10	11	13	10	11	1	0.0025
H112	21	14	22	15	16	11	14	14	13	13	29	23	10	11	13	10	11	1	0.0025
H113	21	14	19	15	15	11	14	14	13	13	29	25	10	13	13	10	11	1	0.0025
H114	22	14	19	16	15	10	14	14	13	13	31	23	10	13	13	10	11	1	0.0025
H115	23	14	22	16	15	13	14	14	13	12	31	24	10	13	13	10	11	1	0.0025
H116	22	14	18	16	15	12	14	14	13	13	31	25	10	12	13	10	11	1	0.0025
H117	21	14	19	16	20	12	14	14	19	14	30	23	10	11	13	12	10	1	0.0025
H118	21	14	19	15	16	12	14	14	18	12	30	23	10	11	13	9	10	1	0.0025
H119	21	16	19	15	16	11	14	14	16	12	30	23	10	11	13	9	13	1	0.0025
H120	21	15	19	15	18	12	14	14	18	12	30	23	10	12	13	10	11	2	0.0050
H121	21	14	19	16	18	14	14	14	18	12	30	23	9	11	14	10	11	3	0.0075
H122	21	14	19	13	17	14	14	14	18	12	30	23	11	11	12	10	11	1	0.0025
H123	21	16	19	15	15	14	14	14	18	14	30	23	11	11	14	10	11	1	0.0025
H124	21	15	19	15	15	12	14	14	18	13	30	23	11	12	14	10	11	1	0.0025
H125	21	16	19	15	20	12	14	14	18	13	30	23	10	14	14	10	13	1	0.0025
H126	21	16	19	16	20	12	14	14	14	13	28	23	10	11	13	10	10	1	0.0025
H127	21	14	19	16	19	12	14	14	14	13	29	23	10	14	13	9	10	1	0.0025
H128	24	14	22	16	19	11	14	14	14	13	30	24	12	11	13	9	10	1	0.0025
H129	24	14	20	16	19	13	14	14	14	14	28	24	12	11	12	9	10	1	0.0025
H130	24	14	20	14	15	13	14	14	14	14	28	25	12	13	13	9	12	1	0.0025
H131	24	14	20	13	15	12	14	14	14	12	28	24	11	13	13	9	13	1	0.0025
H132	24	14	20	13	16	12	14	14	14	14	31	24	11	14	13	12	10	1	0.0025
H133	24	15	20	15	14	12	14	14	14	14	31	22	11	11	14	10	10	1	0.0025
H134	24	16	20	15	17	12	14	14	14	14	28	22	9	11	12	10	10	2	0.0050
H135	22	16	20	15	18	12	14	14	14	13	28	23	9	11	12	12	11	2	0.0050
H136	24	16	18	15	17	11	14	14	14	13	29	24	9	11	12	11	11	1	0.0025
H137	24	16	19	15	15	11	14	16	14	13	29	24	10	11	12	10	11	1	0.0025
H138	24	15	19	16	15	13	14	16	14	13	29	24	9	11	12	10	10	1	0.0025
H139	24	14	19	16	16	12	14	16	13	13	29	24	9	11	14	10	10	1	0.0025
H140	24	16	19	16	14	12	14	16	13	13	29	24	11	11	13	10	10	1	0.0025
H141	21	16	19	13	14	12	14	15	17	13	29	24	12	11	13	10	11	1	0.0025
H142	21	15	19	16	15	12	14	14	17	12	30	24	12	12	13	10	11	1	0.0025
H143	23	15	22	16	19	12	14	14	19	12	30	24	12	11	13	10	11	1	0.0025
H144	23	14	18	16	19	11	14	14	19	12	30	23	12	11	14	10	11	5	0.0125
H145	23	16	18	16	15	11	14	14	19	14	30	24	11	11	13	10	11	1	0.0025
H146	23	16	21	16	15	13	14	14	19	14	30	22	10	13	14	10	11	1	0.0025
H147	23	16	19	15	15	12	14	14	19	14	28	23	10	11	12	12	10	1	0.0025
H148	23	16	22	15	15	10	14	12	19	14	28	22	10	11	12	9	10	4	0.0100
H149	23	14	22	14	15	10	14	12	16	14	29	23	10	11	13	9	11	1	0.0025
H150	23	14	19	15	16	10	14	17	16	14	29	21	10	11	13	11	11	1	0.0025
H151	23	14	20	15	16	12	14	17	16	14	29	24	10	12	12	11	11	2	0.0050

Table 2. Contd.

H152	23	14	20	13	16	12	14	17	16	13	29	24	10	11	14	11	11	2	0.0050
H153	23	16	20	14	17	12	14	17	19	13	29	24	10	11	13	11	11	1	0.0025
H154	23	16	21	15	14	12	14	13	19	13	29	24	10	13	13	10	10	1	0.0025
H155	23	15	20	15	14	11	14	13	19	13	28	24	10	11	12	10	10	1	0.0025
H156	23	14	20	15	16	11	14	13	15	13	29	25	10	11	12	10	11	1	0.0025
H157	23	15	20	16	17	11	14	13	15	13	33	25	10	13	13	10	11	1	0.0025
H158	20	15	20	13	15	11	14	13	15	13	33	25	10	11	12	10	11	1	0.0025
H159	20	15	18	15	15	13	14	13	15	14	33	24	11	11	13	10	11	1	0.0025
H160	20	15	22	15	15	12	14	13	15	14	33	21	11	14	13	10	11	1	0.0025
H161	20	16	21	15	15	12	14	12	15	12	33	23	11	11	13	9	11	1	0.0025
H162	20	16	19	15	15	10	14	13	15	12	33	23	11	11	13	12	11	1	0.0025
H163	20	14	19	15	16	12	14	13	18	12	28	23	11	11	13	12	14	1	0.0025
H164	20	14	19	16	16	10	14	13	20	12	32	23	11	11	13	10	10	1	0.0025
H165	21	14	19	16	20	12	14	13	18	12	32	23	11	11	13	10	12	1	0.0025
H166	21	16	20	15	15	12	14	12	14	12	28	23	11	11	13	9	12	1	0.0025
H167	21	16	20	14	15	11	14	13	14	12	30	23	10	11	13	10	10	1	0.0025
H168	24	16	20	14	16	11	14	13	14		30	23	9	12	13	10	10	1	0.0025
H169	24	15	20	15	16	11	14	13	14	13	30	23	12	12	12	10	10	1	0.0025
H170	24	14	20	15	20	11	14	13	14	13	30	21	11	13	12	10	10	1	0.0025
H171	24	14	20	16	16	11	14	13	14	13	30	21	12	11	12	10	13	3	0.0075
H172	24	14	19	16	15	12	14	13	19	13	30	23	12	11	12	11	11	2	0.0050
H173	24	16	19	13	15	12	14	13	19	14	30	23	10	11	12	10	11	1	0.0025
H174	24	15	19	15	15	10	14	19	17	14	30	23	9	11	12	10	11	1	0.0025
H175	25	15	19	15	17	12	14	13	17	14	31	23	11	11	13	10	10	1	0.0025
H176	25	15	19	15	20	12	14	12	19	14	33	23	11	11	13	10	11	4	0.0100
H177	25	14	19	15	19	12	14	13	19	14	29	23	9	11	13	10	11	1	0.0025
H178	25	14	20	15	17	12	14	13	20	14	30	23	9	11	13	12	10	1	0.0025
H179	23	14	20	15	14	12	14	13	20	14	31	24	10	11	13	10	13	1	0.0025
H180	23	15	20	15	15	10	14	13	18	14	31	24	10	11	13	10	14	1	0.0025
H181	23	15	19	15	15	12	14	13	15	14	31	24	10	11	13	9	11	1	0.0025
H182	23	14	19	15	16	12	14	16	18	14	31	24	12	11	13	9	11	1	0.0025
H183	23	16	19	15	20	11	14	17	18	14	28	24	9	11	13	10	11	1	0.0025
H184	23	16	22	15	20	13	14	17	18	14	29	24	11	11	13	10	10	1	0.0025
H185	24	16	21	15	20	14	14	13	18	14	29	24	10	11	13	10	10	1	0.0025
H186	24	14	21	15	18	12	14	13	16	14	32	24	10	11	13	10	10	1	0.0025
H187	24	14	19	15	19	14	14	13	16	14	29	24	10	11	13	10	13	1	0.0025
H188	24	14	19	15	16	12	14	13	19	14	29	24	10	11	13	10	11	1	0.0025
H189	24	14	19	15	16	11	14	13	19	14	30	24	9	11	13	10	10	5	0.0125
H190	24	14	19	15	16	13	14	13	14	14	30	24	9	11	13	11	12	1	0.0025
H191	24	14	20	15	16	10	14	16	14	14	29	24	9	11	13	11	11	1	0.0025
H192	24	14	20	15	15	12	14	13	14	14	29	24	9	11	13	9	10	1	0.0025
H193	24	16	18	15	15	10	14	13	14	14	33	24	9	11	13	9	10	1	0.0025
H194	24	15	18	15	19	12	14	17	14	14	29	24	11	11	13	9	10	1	0.0025
H195	24	14	18	15	20	12	14	13	14	14	29	24	12	11	13	9	10	1	0.0025
H196	24	16	18	15	20	14	14	13	14	13	28	23	8	11	13	10	10	1	0.0025
H197	24	16	20	15	20	12	14	13	14	13	28	23	10	11	13	10	10	1	0.0025
H198	25	16	20	14	20	11	14	13	14	13	32	22	9	11	13	10	10	1	0.0025
H199	21	16	20	16	15	12	14	13	14	13	30	22	9	11	13	9	10	1	0.0025
H200	21	14	20	16	16	12	14	13	14	13	30	24	11	11	13	9	11	3	0.0075
H201	21	14	22	16	15	12	14	13	17	13	30	24	11	11	13	11	11	3	0.0075
H201	21	15	22	13	15	12	14	13	15	13	29	24	10	11	13	11	11	1	0.0025
H202	21	16	22	13	14	12	14	13	15	13	29	23	9	11	13	11	11	1	0.0025

Table 2. Contd.

H203	22	16	18	15	14	10	14	13	18	13	29	23	9	11	13	10	13	1	0.0025
H204	24	16	19	15	14	12	14	13	18	13	29	21	11	11	13	10	13	1	0.0025
H205	24	15	19	16	16	12	14	19	13	13	29	24	11	11	13	11	13	1	0.0025
H206	24	14	19	14	16	11	14	13	13	14	30	24	11	11	13	12	13	1	0.0025
H207	24	15	19	15	16	11	14	13	13	14	30	22	9	11	13	12	13	1	0.0025
H208	24	15	20	15	20	11	14	13	19	12	30	22	9	11	12	10	13	1	0.0025
H209	24	15	21	15	15	14	14	13	19	12	30	24	10	11	12	10	10	2	0.0050
H210	24	16	21	15	18	14	14	13	16	12	28	24	10	11	12	10	10	2	0.0050
H211	24	16	18	15	18	12	14	13	19	12	30	24	10	11	12	12	12	2	0.0050
H212	24	14	18	16	16	12	14	13	19	14	29	22	10	11	12	9	12	1	0.0025
H213	24	14	19	16	16	13	14	13	19	14	30	23	10	11	12	11	11	1	0.0025
H214	24	14	20	16	16	14	14	13	17	13	30	23	10	11	12	12	12	1	0.0025
H215	24	16	19	16	19	12	14	13	19	13	30	23	9	11	12	12	12	1	0.0025
H216	24	16	19	15	20	12	14	17	19	13	29	23	11	11	12	12	10	1	0.0025
H217	24	15	19	13	15	12	14	13	15	13	29	23	8	11	12	12	10	1	0.0025
H218	24	15	22	15	15	11	14	13	15	13	31	23	9	11	12	11	10	1	0.0025
H219	24	15	18	15	15	10	14	13	15	14	29	23	9	11	12	11	13	1	0.0025
H220	24	15	21	15	17	12	15	17	18	14	29	23	9	11	12	11	10	1	0.0025
H221	24	15	19	15	16	12	15	17	18	14	29	24	10	11	12	11	14	1	0.0025
H222	24	14	19	13	16	14	15	13	18	14	29	24	10	11	12	11	11	1	0.0025
H223	24	14	20	14	16	11	15	13	18	14	31	23	12	11	14	10	11	1	0.0025
H224	22	16	20	13	15	11	15	13	18	14	31	21	12	11	13	10	11	6	0.0150
H225	21	15	20	15	14	11	15	13	18	14	31	24	9	11	14	10	11	1	0.0025
H226	21	15	19	15	14	10	15	13	14	14	29	24	9	11	13	9	14	1	0.0025
H227	21	15	18	15	14	14	15	19	19	14	29	24	9	11	13	10	13	4	0.0100
H228	21	16	19	16	15	12	15	13	19	12	29	24	10	11	13	10	10	1	0.0025
H229	23	14	19	14	15	10	15	13	16	12	30	22	11	11	13	10	10	1	0.0025
H230	23	15	19	16	20	12	15	13	16	12	30	23	10	11	13	10	10	1	0.0025
H231	23	15	22	16	17	12	15	17	20	12	28	23	10	11	13	9	10	3	0.0075
H232	21	16	20	15	17	12	15	17	16	13	29	23	9	11	12	12	11	2	0.0050
H233	21	16	19	16	16	12	15	13	16	13	29	23	9	14	12	12	10	1	0.0025
H234	25	14	19	15	16	11	15	13	16	12	29	23	8	14	12	10	11	1	0.0025
H235	25	16	19	14	16	11	15	13	16	12	29	24	8	13	12	10	11	1	0.0025
H236	25	16	22	14	15	14	15	13	16	12	29	23	9	11	13	10	13	1	0.0025
H237	25	14	19	13	15	13	15	13	16	13	29	23	9	13	13	9	14	1	0.0025
H238	21	14	18	15	15	13	15	13	16	12	29	23	9	13	13	11	10	1	0.0025
H239	21	15	18	15	20	12	15	13	13	14	29	23	10	11	13	11	10	3	0.0075
H240	22	16	18	16	20	12	15	12	13	13	29	23	10	11	14	11	10	1	0.0025
H241	22	16	19	16	19	12	15	12	13	14	29	23	12	11	13	10	12	3	0.0075
H242	22	14	19	16	15	13	15	12	20	14	29	23	12	11	13	10	11	1	0.0025
H243	24	14	19	14	15	11	15	13	19	14	29	23	12	13	12	10	10	1	0.0025
H244	24	15	19	15	15	11	15	13	19	14	31	23	12	13	14	10	10	1	0.0025
H245	24	14	20	15	19	11	15	13	15	14	31	23	9	11	13	10	10	1	0.0025
H246	20	16	20	15	18	14	15	13	18	14	31	21	9	12	13	10	10	1	0.0025
H247	20	16	21	15	18	12	15	13	18	14	31	23	9	11	12	10	14	1	0.0025
H248	20	14	21	13	18	12	15	13	18	14	31	24	9	12	13	10	10	1	0.0025
H249	20	14	18	15	15	12	15	13	18	12	29	24	9	12	13	10	11	1	0.0025
H250	20	14	22	15	20	11	15	13	18	13	28	24	9	11	13	10	10	1	0.0025
H251	20	14	22	16	15	13	15	15	18	14	28	24	9	11	13	10	13	1	0.0025
H252	20	15	22	14	15	11	15	13	18	14	30	24	9	11	14	10	13	1	0.0025
H253	20	15	19	15	16	11	15	13	14	14	31	24	9	12	13	10	10	1	0.0025
H254	20	14	19	15	19	11	15	13	14	13	31	22	9	12	13	10	10	1	0.0025

Table 2. Contd.

H255	20	16	19	15	18	10	15	13	17	13	31	22	11	11	13	10	10	1	0.0025
H256	20	14	19	16	18	10	15	15	20	13	31	23	8	11	13	10	10	1	0.0025
H257	24	14	22	16	20	12	15	13	19	13	29	23	10	11	13	10	10	1	0.0025
H258	24	16	22	14	20	12	15	13	19	13	29	24	9	11	13	10	10	2	0.0050
H259	24	16	19	14	16	12	15	13	17	13	29	24	9	11	13	10	10	2	0.0050
H260	24	16	20	13	16	12	15	13	19	13	31	24	9	11	13	10	10	1	0.0025
H261	24	16	21	15	19	12	15	13	15	13	29	24	12	11	13	10	10	1	0.0025
H262	24	15	21	15	20	12	15	19	15	13	30	24	10	11	13	10	10	1	0.0025
H263	24	14	21	15	15	12	16	13	15	13	29	24	10	11	14	10	10	1	0.0025
H264	24	14	19	15	18	12	16	13	15	12	30	24	10	11	12	10	11	1	0.0025
H265	23	14	19	16	15	10	16	13	15	14	30	24	10	11	13	10	11	1	0.0025
H266	23	15	19	14	15	10	16	13	15	14	30	23	10	11	12	10	11	1	0.0025
H267	23	16	19	14	16	12	16	17	15	14	28	23	10	11	13	10	11	1	0.0025
H268	23	16	20	15	16	12	16	17	15	14	28	21	10	11	12	10	13	1	0.0025
H269	23	14	20	14	19	12	16	17	19	14	29	24	10	11	13	10	10	1	0.0025
H270	23	16	21	14	20	13	16	17	19	14	29	22	10	11	13	10	10	1	0.0025
H271	23	14	21	16	20	13	16	13	16	14	29	24	10	11	13	10	10	3	0.0075
H272	23	14	18	16	16	13	16	13	16	13	29	24	10	11	14	10	10	1	0.0025
H273	23	15	18	15	15	11	16	13	14	13	30	24	10	11	13	10	10	1	0.0025
H274	23	14	19	15	18	11	17	13	18	13	30	24	10	11	13	10	10	1	0.0025
H275	23	16	19	13	15	11	17	13	20	13	29	24	10	14	13	10	14	1	0.0025
H276	23	15	19	16	15	12	17	19	14	13	29	24	10	14	13	10	14	2	0.0050
H277	23	14	19	17	15	12	17	13	17	13	31	21	9	13	13	10	11	1	0.0025
H278	23	15	19	17	18	12	17	14	17	13	30	24	9	12	13	9	10	1	0.0025
H279	23	14	19	15	15	11	13	14	17	13	32	23	9	11	13	11	13	1	0.0025
H280	24	14	19	16	15	10	13	14	19	13	30	22	9	11	14	10	12	1	0.0025
H281	24	14	20	16	20	11	13	12	19	15	30	25	9	11	12	12	13	1	0.0025
H282	20	14	22	15	20	12	13	17	13	15	30	23	10	12	14	10	12	1	0.0025
H283	20	14	18	15	16	12	13	17	16	13	30	21	8	14	12	9	13	1	0.0025
H284	20	16	18	17	16	14	13	13	13	15	32	25	11	12	14	9	10	1	0.0025
H285	20	16	19	16	19	14	14	13	20	13	31	21	12	13	13	9	11	1	0.0025
H286	24	16	20	17	15	14	14	11	20	15	28	22	8	12	12	11	10	1	0.0025
H287	24	15	20	15	15	11	16	17	20	13	30	25	10	12	14	9	10	1	0.0025
H288	23	15	20	14	15	12	16	11	16	13	29	25	12	11	12	10	11	1	0.0025
H289	21	15	20	15	14	12	16	13	16	12	28	21	11	14	13	9	10	1	0.0025
H290	21	15	20	15	20	12	17	13	16	13	28	24	12	11	13	12	13	1	0.0025
H291	22	15	19	15	19	12	13	13	18	12	30	24	8	13	13	9	13	1	0.0025
H292	22	14	19	15	16	12	13	17	16	12	28	21	9	12	13	9	10	1	0.0025
H293	24	14	19	13	15	12	13	17	15	14	29	25	10	12	14	9	12	4	0.0100
H294	24	16	19	14	15	10	13	12	18	12	31	22	9	11	14	9	10	1	0.0025
H295	23	16	19	16	15	14	13	13	13	14	30	23	11	11	12	10	10	1	0.0025
H296	23	14	19	16	19	10	16	13	14	14	33	23	10	11	13	12	10	4	0.0100
H297	20	14	22	16	20	10	14	13	18	14	28	23	8	12	13	11	10	1	0.0025
H298	23	14	18	17	16	10	14	19	18	12	31	23	9	13	13	10	10	1	0.0025
H299	23	15	18	14	16	13	14	16	18	13	29	23	9	13	12	10	13	1	0.0025
H300	23	15	20	17	16	12	14	17	16	13	30	24	9	14	14	10	13	1	0.0025
H301	24	14	20	14	16	11	14	12	16	14	30	22	9	12	14	10	11	1	0.0025
H302	21	14	19	17	16	14	14	11	16	13	28	21	10	11	14	10	11	1	0.0025
H303	24	14	18	13	17	12	14	13	13	12	29	24	10	12	12	12	11	1	0.0025
H304	24	14	22	15	17	14	15	13	15	12	32	24	10	13	12	11	12	1	0.0025
H305	24	16	22	15	19	12	13	13	16	12	28	24	12	11	12	9	10	1	0.0025
H306	21	16	19	17	14	14	13	16	16	13	31	24	11	11	14	9	10	1	0.0025

Table 2. Contd.

																		N	F
H307	22	16	19	15	14	12	15	13	20	13	33	24	12	13	13	10	13	1	0.0025
H308	24	16	18	15	15	12	13	11	17	13	28	24	8	12	13	12	13	1	0.0025
H309	24	16	20	16	15	10	13	16	13	15	29	24	11	14	13	9	13	1	0.0025
H310	24	14	20	15	15	12	15	13	15	15	29	21	8	11	13	9	11	1	0.0025
H311	23	14	20	15	17	11	13	13	13	13	33	23	11	13	14	9	11	1	0.0025
H312	23	15	20	17	17	11	14	13	15	13	30	22	10	11	14	10	10	1	0.0025
H313	23	14	22	15	16	11	14	11	15	12	29	22	12	14	13	11	10	1	0.0025
H314	20	14	22	14	16	14	14	16	16	14	28	21	12	11	12	12	10	1	0.0025
H315	21	14	18	15	20	10	14	19	20	12	32	23	11	11	12	11	10	2	0.0050
H316	21	16	19	17	19	13	14	17	20	15	29	24	10	11	14	9	13	1	0.0025
H317	21	15	20	15	14	14	14	12	20	13	28	24	8	11	13	10	11	1	0.0025

N: Number of males observed for each haplotype. F: frequency of each haplotype in the sample of 105 males.

use this 17 STR loci as a vital tool for forensic identification and paternity testing.

REFERENCES

Andrea V, Nicoletta C, Fausta G, Anna P, Elena M, Francesco DF (2008). Population Data for 15 Autosomal STRs Loci and 12 Y Chromosome STRs Loci in a Population Sample from the Sardinia Island (Italy). Leg. Med. 11(1):37-40.

Ayse S, Husniye C, Behnan A, Yasar S (2011). Haplotype frequencies of 17 Y-chromosomal short tandem repeat loci from the Cukurova region of Turkey. Croat Med. J. 52(6):703-708.

Butler JM, Hill CR (2012). Biology and genetics of new autosomal STR loci useful for forensic DNA analysis. Forensic Sci. Rev. 24(1):15-26.

Butler JM, Schoske R, Vallone PM, Kline MC, Redd AJ, Hammer MF (2002). A Novel Multiplex for Simultaneous Amplification of 20 Y-Chromosome STR Markers, Forensic Sci. Int., 129(1):10-24.

Carolina N, Miriam B, Cecilia S, Yolanda C, Jianye G, Bruce B, et al. (2010). Reconstructing the population history of Nicaragua by means of mtDNA, Y-chromosome STRs and autosomal STR markers. Am. J. Phys. Anthropol. 143(4):591-600.

Chamberlain JS, Gibbs RA, Rainer JE, Nguyen PN, Casey CT (1988). Deletion screening of the Duchenne muscular dystrophy locus via multiplex DNA amplification. Nucleic Acids Res. 16:11141-11156.

Dobbs LJ, Madigan MN, Carter AB, Earls L (2002). Use of FTA gene guard filter paper for the storage and transportation of tumor cells for molecular testing, Arch. Pathol Lab Med. 126(1):56-63.

Ellegren H (2004). Microsatellites: simple sequences with complex evolution. Nat. Rev. Genet. 5:435-445.

Evans JC, Frayling TM, Cassell PG, Saker PJ, Hitman GA, Walker M, Levy JC, O'Rahily S, Rao PV, Bennett AJ, Jones EC, Menzel S, Prestwich P, Simecek N, Wishart M., Dhillon R., Fletcher C., Milward A., Demaine A., Wilkin T., Horikawa Y., Cox NJ, Bell GI, Ellard S, McCarthy MI, Hattersley AT (2001). Studies of association between the gene for calpain-10 and type 2 diabetes mellitus in the United Kingdom. Am. J. Hum. Genet. 69:544-552.

Gill P, Brenner C, Brinkmann B, Budowle B, Carracedo A, Jobling MA et al. (2001). DNA Commission of the International Society of Forensic Genetics: recommendations on forensic analysis using Y-chromosome STRs, Forensic Sci. Int. 124:5-10.

Hanson EK, Ballantyne J (2007). An ultra-high discrimination Y chromosome short tandem repeat multiplex DNA typing system, PLoS ONE 2(8):e688.

Henegariu O, Hirschman P, Killian K, Kirsch C, Lengauer R, Maiwald K (1994). Rapid screening of the Y chromosome in idiopathic sterile men, diagnostic for deletions in AZF, a genetic Y factor expressed during spermatogenesis. Andrologia 26:97-106.

Kayser M, Caglia A, Corach D, Fretwell N, Gehrig C, Graziosi G et al. (1997). Evaluation of Y-chromosomal STRs: a multicenter study. Int. J. Leg. Med. 110:125-133.

Kimpton CP, Oldroyd NJ, Watson SK, Frazier RRE, Johnson PE, Millican ES, Urquhart A, Sparkes BL, Gill P (1996) Validation of highly discriminating multiplex short tandem repeat amplification systems for individual identification. Electrophoresis 17:1283-1293.

Kuppareddi B, Suhasini G, Vijaya M, Kanthimathi S, Nicole M, Martin T et al. (2010). Y chromosome STR allelic and haplotype diversity in five ethic Tamil Populations from Tamil Nadu, India. Leg. Med. 12:265-269.

Kwak KD, Jin HJ, Shin DJ, Kim JM, Roewer L, Krawczak M et al. (2005). Y-Chromosomal STR Haplotypes and Their Applications in Forensic and Population Studies in East Asia. Int. J. Leg. Med. 119(4):195-201.

Lander ES, Linton LM, Birren B, Nusbaum C, Zody MC, Baldwin J, Devon K, Dewar K, Doyle M, Fitz HW et al. (2001) Initial sequencing and analysis of the human genome . Nat. 409:8-921.

Leda K, Jessica L, Jodi A (2008). Population genetics of Y-chromosome STRs in a population of Northern Greeks, Forensic Sci. Int. 175(2-3):250-255.

M Nei (1987). Molecular Evolutionary Genetics, Columbia University Press, New York.

Nadia Al-Zahery, Maria P, Vincenza B, Viola G, Mohammed A, Baharak H (2011). In search of the genetic footprints of Sumerians: a survey of Y-chromosome and mtDNA variation in the Marsh Arabs of Iraq. BMC Evol. Biol. 11:288.

Park MJ, Lee HY, Chang U, Kang SC, Shin KJ (2007) .Y-STR analysis of degraded DNA using reduced-size amplicons. Int. J. Leg. Med. 121(2):152-157.

Parson W, Niederstätter H, Brandstätter A, Berger B (2003). Improved specificity of Y-STR typing in DNA mixture samples. Int. J. Leg. Med. 117(2):109-114.

Raymond M, Rousset F (1995). GENEPOP (version 1.2). Population genetics software for exact tests and ecumenicism. J. Heredity 86:248-249.

Silvia B, Marta CD, Andrea Z, Dario B, Tatiana G (2009). Integration of genomic and gene expression data of childhood ALL without known aberrations identifies subgroups with specific genetic hallmarks. Genes Chromosomes Cancer 48:22-38.

Venter JC, Adams MD, Myers EW, Li PW, Mural RJ, Sutton GG, Smith HO, Yandell M et al. (2001). The sequence of the human genome. Science 291(5507):1304-1351.

Walkinshaw M, Strickland L, Hamilton H, Denning K, Gayley T (1996). DNA Profiling in two Alaskan Native Populations Using HLA-DQA1, PM, and D1S80 Loci. J. Forensic Sci. 41:47.

24

Diversity analysis of sugarcane genotypes by microsatellite (SSR) markers

Smiullah[1], Farooq Ahmed Khan[1], Aqeel Afzal[1], Abdullah[1], Ambreen Ijaz[2] and Usman Ijaz[1]

[1]Department of Plant Breeding and Genetics, University of Agriculture, Faisalabad, Pakistan.
[2]Department of Bioinformatics and Biotechnology, GC University, Faisalabad, Pakistan.

Thirty (30) simple sequence repeat (SSR) primer pairs chosen randomly from the SSR primer collection were used to detect polymorphism in 17 sugarcane accessions. A total of 62 DNA fragments were generated by the 30 primers with an average of about 2.14 bands per primer. Bands that a primer yielded in the study ranged from 1 to 4. The genetic distances for SSR data using 17 sugarcane accessions, was constructed based on Nei (1978) and relationships between accessions were portrayed graphically in the form of a dendrogram. The value of genetic similarity ranging from 62.90 to 90.30% was observed among the 17 sugarcane accessions. The highest genetic similarity of 90.03% was seen among genotypes S-2003-US-118 and S-2003-US-312. From the present study, it may be concluded that SSRs markers are best tool for investigation of genetic diversity in sugarcane.

Key words: Simple sequence repeat (SSR), polymorphism, genetic diversity.

INTRODUCTION

Sugarcane (*Saccharum* spp. hybrids) is a genetically complex crop of major economic importance in tropical and sub-tropical countries (Khan et al., 2004). It is mainly used for sugar production but recently gained increased attention because of its employment generation potential and recent emphasis on production of bio-fuels. The importance of sugarcane has increased in recent years because cane is an important industrial raw material for sugar and allied industries producing alcohol, acetic acid, butanol, paper, plywood, industrial enzymes and animal feed (Arencibia, 1998). Considering the current needs of cane industry it is imperative to breed high sugar producing varieties that also have other desired agronomic traits.

Saccharum is a complex genus characterized by high ploidy levels and composed of at least six distinct species - *Saccharum officinarum, Saccharum barberi, Saccharum sinensi, Saccharum spontaneum, Saccharum robustum* and *Saccharum edule* (Daniels and Roach, 1987). Sugar recovery can be increased from current average of 8.32 to 10-11% with the development of improved cane varieties. For development of improved varieties, genotypic studies of sugarcane are required. Described as an allopolyploid, modern cultivated sugar-cane have approximately 80-140 chromosomes with 8-18 copies of a basic set ($x = 8$ or $x = 10$ haploid chromo-some number) (Ming et al., 2001). Continuous selection for the same traits may narrow genetic diversity to the extent that it may be difficult to predict diversity based on pedigree history alone. With the advent of molecular markers, it is now possible to make direct comparison of genetic diversity at the DNA level without some of the over simplifying assumptions associated with calculating genetic diversity based on pedigree history (McIntyre et al., 2001). Rapid advances in the field of molecular biology and its allied sciences made the use of molecular markers a routine practice providing plant breeders a precise tool in analyzing genetic diversity for plant improvement (Andersen and Lubberstedt, 2003).

The molecular markers are of many types e.g. RFLPs,

Table 1. Description of seventeen genotypes used in genetic diversity study.

Genotype	Source of collection
CPF-247	AARI , Faisalabad
SPF-245	AARI , Faisalabad
S-2003-US-618	AARI , Faisalabad
S-2003-US-628	AARI , Faisalabad
S-2002-US-247	AARI , Faisalabad
HSF-240	AARI , Faisalabad
CPF-237	AARI , Faisalabad
CPF-234	AARI , Faisalabad
S-2003-US-718	AARI , Faisalabad
S-2003-US-778	AARI , Faisalabad
S-2003-US-165	AARI , Faisalabad
S-2003US-312	AARI , Faisalabad
HSF-242	AARI , Faisalabad
CP-77-400	AARI , Faisalabad
CP-72-2086	AARI , Faisalabad
SPF-246	AARI , Faisalabad
SPF-213	AARI , Faisalabad

TRAPs, RAPDs, SNPs, simple sequence repeats (SSRs) and AFLPs. In the present study, microsatellite or SSR marker was used to analyze genetic diversity of different sugarcane genotypes. Microsatellites or simple sequence repeats (SSRs), are stretches of DNA, consisting of tandemly repeated short units of 1-6 base pairs in length. They are ubiquitous in eukaryotic genomes and can be analyzed through PCR technology. The sequences flanking specific microsatellite loci in a genome are believed to be conserved within a particular species, across species within a genus and rarely even across related genera. Simple sequence repeats (SSR) markers reveal polymorphisms due to variation in the lengths of microsatellites at specific individual loci. Microsatellites are born from regions in which variants of simple repetitive DNA sequence motifs are already over represented (Tautz et al., 1989). It is now well established that the predominant mutation mechanism in microsatellite tracts is 'slipped-strand mispairing'. This process has been well described by Eisen (1999). When slipped-strand mispairing occurs within a microsatellite array during DNA synthesis, it can result in the gain or loss of one, or more, repeat units depending on whether the newly synthesized DNA chain loops out or the template chain loops out, respectively. The relative propensity for either chain to loop out seems to depend in part on the sequences making up the array, and in part on whether the event occurs on the leading (continuous DNA synthesis) or lagging (discontinuous DNA synthesis) strand. SSR allelic differences are, therefore, the results of variable numbers of repeat units within the microsatellite structure; they are therefore, multiallelic and co-dominant in

nature, thus proving to be very informative. Among the range of DNA-based molecular marker techniques, a promising polymerase chain reaction (PCR)-based technique used extensively for genetic mapping (McIntyre et al., 2001), as well as fingerprinting of sugarcane clones (Piperidis et al., 2000; Pan et al., 2002), is microsatellites or SSRs. SSR genetic markers are the best tool to demonstrate the genetic diversity in sugarcane (Smiullah et al., 2012).

The present study was undertaken to investigate the genetic diversity and establish the relationship between different sugarcane genotypes in Pakistan, using SSR markers. Obtaining accurate estimates of the genetic diversity among germplasm sources may increase the efficiency of plant breeding. Knowledge of genetic diversity and relationships among breeding genome, their polymorphic nature, codominance and materials has a significant impact on crop improvement.

MATERIALS AND METHODS

The genetic diversity studies were done as a collaborative research, in Department of Plant Breeding and Genetics, University of Agriculture, Faisalabad and Agriculture Biotechnology Research Institute (ABRI), Ayub Agricultural Research Institute (AARI), Faisalabad during 2010-2012. The plant material used for the study of genetic diversity was comprised of seventeen sugarcane accessions (Table 1). These accessions were collected from the germplasm source in the Sugarcane Section of Ayub Agricultural Research Institute, Faisalabad. The genetic material includes commercial cultivars and elite lines.

PCR amplification

Fresh young leaves were collected from the field experiment for isolation of the DNA. Total genomic DNA of the plants was extracted by using modified (CTAB) method (Hoisington et al., 1994; Doyle and Doyle, 1990). DNA concentration was determined, using a Nano Drop spectrophotometer (ND1000). Primer selection was based on previous investigation on SSR analysis, carried out with sugarcane genotypes and somaclones in this laboratory. Primer pairs obtained from Gene link company (USA) were used in PCR reaction for each genotype. For SSR analysis, concentration of genomic DNA, I0 × PCR buffer with $(NH_4)_2SO_4$, $MgCl_2$, dNTPs primers and taq DNA polymerase were optimized.

A reaction mixture of 20 µl was used to amplify genomic DNA in a thermal cycler (Eppendorf DNA Thermal Cycler 9600). To confirm that the observed bands were amplified genomic DNA and not the primer artifacts, genomic DNA was omitted from control reaction. A negative control was also run to confirm if the master/reaction mixture is correctly prepared or not. The PCR products were electrophoresed at 90 V, in 2% agarose gel for approximately 2 h, using 0.5 × tris-boric acids EDTA (TBE) buffer, along with a DNA molecular size marker.

The gel contained 0.5 µg/ml ethidium bromide to stain the DNA and photographed under UV light using gel documentation system. Reactions were duplicated to check the consistency of the amplified products. Only easily resolved bright DNA bands were scored as presence (1) and absence of bands (0). Coefficient of similarity among somaclones was calculated according to Nei and Li (1978). Similarity coefficient was utilized to generate a dendrogram by means of unweighted pair.

Data analysis

The data on bands generated by the 30 primers were selected for analysis of genetic diversity (Table 2). The bands were counted by starting from the top and ending with the bottom of the lanes. All segregating bands that were well resolved and unambiguous were scored for the presence (1) or absence (0) in the 17 genotypes. The data of the primers were used to estimate the dissimilarity on the basis of number of unshared amplified products and a dissimilarity matrix was generated using Nei's similarity indices (Nei, 1978). In addition, population relationships were inferred using the un-weighted pair group of arithmetic means (UPGMA) clustering method using the Popgen software (version 3.5).

RESULTS AND DISCUSSION

In recent years, the popularity of SSR-based markers has increased considerably. The main reasons which make microsatellites an especially attractive tool for a number of applications are: their high levels of allelic variation and their co-dominant character, which means that they deliver more information per unit assay than any other marker systems, thus reducing costs; microsatellites are assayed using PCR, so only small amounts of tissue are required.

Thirty (30) SSR primer pairs chosen randomly from the SSR primer collection were used to detect polymorphism in 17 sugarcane accessions. The PCR product was observed by running on agarose gel to study poly-morphism, most of the primers were polymorphic except five primers which were monomorphic and produced only one fragment per primer (Figure 1). All the primers were found to give reproducible bands. A total of 62 DNA fragments were generated by the 30 primers with an average of about 2.14 bands per primer. Bands that a primer yielded in the study ranged from 1 to 4. Generally, the size and the number of bands produced were dependent upon the nucleotide sequence of the primer pair, size of the primer used and the source of the template DNA. In this study the primer used were of the size ranging from 300-420 bp. Reactions were duplicated form to check the consistency of the amplified products. Only easily resolved bright DNA bands were scored.

Cluster analysis

Pattern of polymorphism by SSRs

About 85.25% polymorphism was estimated as 55 out of 62 fragments were polymorphic with 30 primers used among the 17 sugarcane accessions. The rest of the 7 bands were monomorphic in the 17 accessions. In the present study, the 17 sugarcane accessions appeared to show variability with the 30 primers used. Although none of the primers individually was as informative as to differentiate all the accessions; highly polymorphic profiles were obtained with of the primers SMs35.

(Sugarcane Microsatellite primer no.35) while five primer pairs such as SMs46, SMs47, SMs48 SMs49 and SMs50 were found to be monomorphic. Therefore, it may be concluded from the present results that SSRs can be used for identification of genetic diversity and the relationship between the members of the complex species. Jannoo et al. (2001) studied diversity in 96 sugarcane genotypes with just two primer pairs and reported a high level of heterozygosity. Cordeiro et al. (2001) applied 21 primer sets to five sugarcane geno-types, and among them, 17 pairs were polymorphic, but the level of polymorphism (PIC value) in the cultivars detected by these SSRs was low (0.23). The level of polymorphism indicates that distinction between any two varieties is possible with appropriate SSR primer pair. This supports the use of SSR markers, as an excellent tool, for diversity analysis and loci mapping.

Genetic distances/similarities between the accessions

The genetic distance for SSR data using 17 sugarcane accessions, was constructed based on Nei (1978) and relationships between accessions were portrayed graphi-cally in the form of a dendrogram in Figure 2, the value of genetic similarity ranging from 62.90 to 90.30% was observed among the 17 sugarcane accessions. The lowest genetic distance of 62.90% was seen among genotypes S-2003-US-118 and S-2003-US 312. These two genotypes differed from each other only in 5 bands with 14 different primers. The most dissimilar of all the accessions was S-2003-US-118 and SPF-213 with genetic distance of 90.30%. Genomic SSRs have been shown to produce a greater number of alleles and higher PIC values than those from EST derived SSRs in sugarcane (Pinto et al., 2006).

In several other studies, elite sugarcane (Saccharum hybrids) germplasm showed genetic diversity as well (Selvi et al., 2003; Cordeiro et al., 2003). Selvi et al. (2003) revealed a broad range (0.324-0.8335) of pair-wise similarity values when tested on 30 or 40 commercial sugarcane cultivars.

Clustering pattern

The cluster analysis based on similarity values has classified all the sugarcane accession in two of the four major groups (I, II, III and IV). The first major group consisted of two accessions CPF-247 and S-2003-US-165 forming the most distinct cluster I. Second major group was further grouped into IIA, IIB and IIC. Group IIA consisted of three accessions namely SPF-245, S-2003-US-618 and HSF-242. Group IIB consists of four genotypes viz. HSF-240, CP-72-2086, S-2003-US-778 and SPF-213. Group IIC contained two accessions CPF-

Table 2. Name of the primers used for detection of polymorphism in sugarcane genotypes.

Primer no.	Band size	Primer sequence (F/R)	Annealing temperature
SMs1	600-2000	GGTTTGTTACTCTACTCCCGT GGTTTGTTACTCTACTCCCGT	55
SMs2	550-900	CATCTGCTCCCTCTTCCT TGAGCAAAGAAAGAGAAGTAGTC	55
SMs3	400-550	CATCTGCTCCCTCTTCCT CTCTGGCGGCTTGGTCCTG	52
SMs5	400-800	CTCTGCGGCTTGGTCCTG CATCCTCCAAGCATCTGT	54
SMs6	500-600	GACTCCTGTCACCGTCTTC ATACTTCAACCGTCTCCTCC	55
SMs7	400-500	CTAAGCAAGAACACAGGAAAG AGCAACAGCAGAGAGCAG	54
SMs8	400-550	CTGACTAAGGAGGAAGTGGAG GACGACGATAGATGAAACA	55
SMs9	400-500	GAGCCGCAAGGAAGCGAC CATACAAGCAGCAAGGATAG	50
SMs10	500-700	CTCTCTTCTCGTCTCCTCATT GTCCTTCTTCTTCTCGTGGT	55
SMs11	400-500	ACACGCATCGCAAGAAGG AAGAACACTCAACAGAAGCAC	55
SMs12	400 - 600	AAATGTCTTCGCACTAACC AAGGAGATGCTGATGGAGA	55
SMs16	400 - 500	CCCAGAGGACAAGGAACT GTAATGGAAGGAAGCAACTGA	50
SMs17	400-450	GGCTCCTCCTACTCGTTC GAGCCTTTGGATGTGGTC	55
SMs18	400-600	CTACACATCTCCATTCCACAG TTTAGGGTTCGTTAGGGTAAG	55
SMs19	300-500	GGCTCCTCCTACTCGTTC GAGCCTTTGGATGTGGTC	53
SMs20	350-500	CTACACATCTCCATTCCACAG TTTAGGGTTCGTTAGGGTAAG	50
SMs21	400-600	GGCTCCTCCTACTCGTTC GAGCCTTTGGATGTGGTC	50
SMs22	300-400	CTACACATCTCCATTCCACAG TTTAGGGTTCGTTAGGGTAAG	55
SMs23	350-600	GGCTCCTCCTACTCGTTC GAGCCTTTGGATGTGGTC	50
SMs24	400-450	CTACACATCTCCATTCCACAG TTTAGGGTTCGTTAGGGTAAG	53
SMs31	550-650	TTCTCGCCCTCCCGCTAC TTCTCTCCTCCTCCTCTTTC	55
SMs35	400-850	TTCTCGCCCTCCCGCTAC TTCTCTCCTCCTCCTCTTTC	53
SMs42	400-500	GTTTCTCCACCTCCAACTC ACAGACACAGGCGGGCGA	55
SMs43	400-500	CCCAGTGCTTCCTCTCTC TAGCACTCCATTCAGCAAA	55
SMs45	400	CTTCCCTCCCTCTCCTCT AGCCTTCTACTAAACTATCTGCT	55
SMs46	400	GTGAGTGAGACCAGACCAG CCGTGCTGTAGTTGTTGTAG	50
SMs47	400	ATACGCTACTCTGAATCCCAC CAATCACTATGTAAGGCAACA	50
SMs48	400	ACTCCTCTTCCTCTTCCTCTT GTTGTTCCCGTTCCCGCC	53
SMs49	250 - 400	ACTCGGTCATCTCATCACTC GTTCTTCGGGTCATCTGG	55
SMs50	400-500	ACGGTGAGCGAGGACTAC CTTGGGTGGCATCAGGAA	55

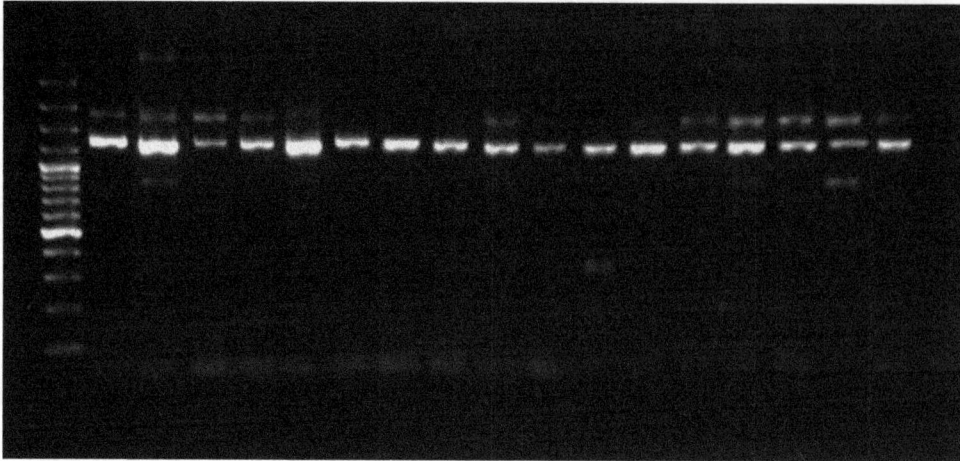

Figure 1. Result of electrophoresis of SSR product of 17 genotypes using sugarcane microsatellite primer no.18.

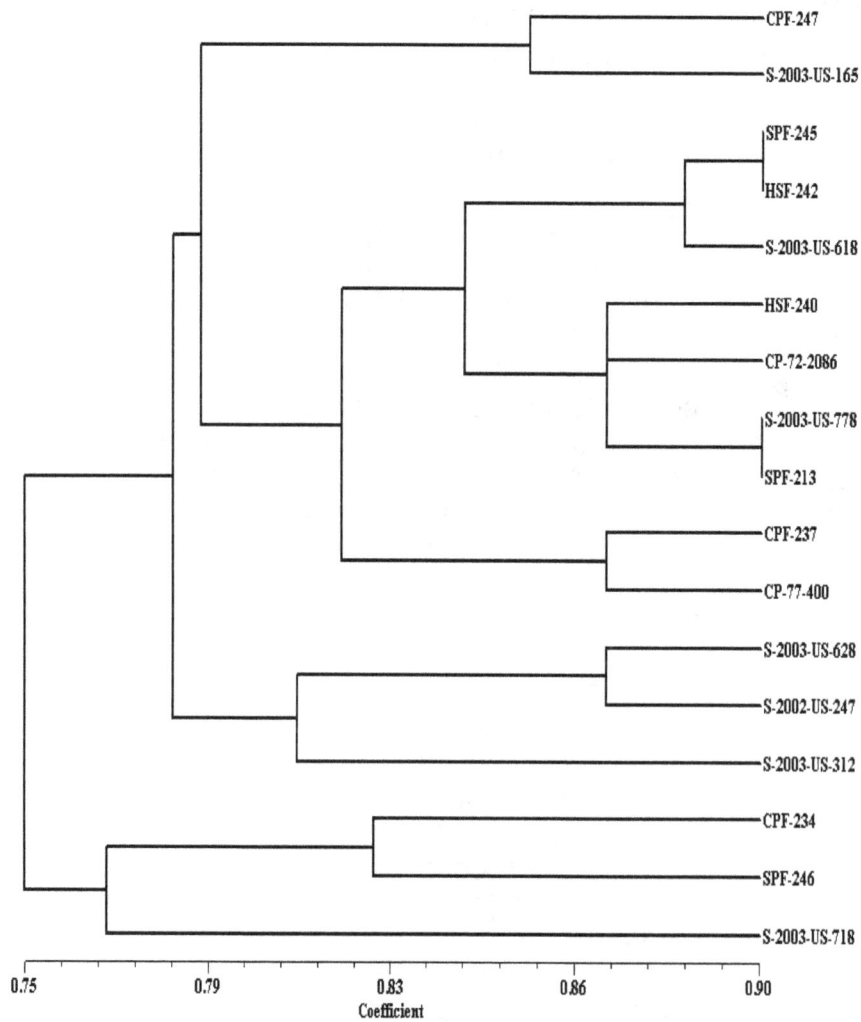

Figure 2. Dendrogram of 17 sugarcane accessions developed from SSRs data using unweighted pair group of arithmetic means (UPGMA) based on Nei's (1978) genetic distance.

237 and CP-77-400. Group III comprised of three genotypes viz. S-2003-US-628, S-2003-US-312 and S-2002-US-247. Group IV consisted of three accessions viz. CPF-234, SPF-246 and S-2003-US-718. Genotypes included in same cluster are more similar to each other but these are less similar to the genotypes in other clusters.

Conclusions

The analysis of variations in SSR fragments provides an effective tool for examining diversity to improve plant breeding strategies. Identifying useful SSRs is critical but in sugarcane this can be a lengthy and difficult process due to their abundance and the complexity of the sugarcane genome. Less information is available on the genetic diversity within and between *Saccharrum* cultivars which has been based mainly on morphological characteristic. Thus, it can be concluded that estimates of genetic similarity based on molecular markers may provide more accurate information to plant breeder. This data will support the exploitation of sugarcane germplasm on molecular basis. SSR markers used in the study may also be used by researcher for genetic mapping and gene tagging in sugarcane. Locus mapping ability of these SSR markers will provide more information than those available through diversity. These markers may be used for construction of genetic map in sugarcane. Future breeding efforts involving crosses between and within the groups identified in this study may provide useful strategies for combining beneficial genes and alleles in new sugarcane varieties while maintaining genetic diversity.

REFERENCES

Andersen JR, Lubberstedt T (2003). Functional markers in plants. Trends Plant Sci. 8:554-560.

Arencibia A (1998). Gene transfer in sugarcane. In: Biotechnology of Food Crops in Developing Countries. (Eds.): T. Hohn and K.M. Leisinger. Springer-Verlag, New York. pp. 79-104.

Cordeiro GM, Casu R, McIntyre CL, Manners JM (2001). Microsatellite markers from sugarcane (Saccharum spp.) ESTs cross transferable to Erianthus and Sorghum. Plant Sci. 160:1115-1123.

Cordeiro GM, Pan YB, Henry RJ (2003). Sugarcane microsatellites for the assessment of genetic diversity in sugarcane germplasm. Plant Sci.165:181-189.

Daniels J, Roach BT (1987). Taxonomy and Evolution. In: Sugarcane Improvement through Breeding (Heinz DJ, ed.). Elsevier Press, Amsterdam. 7-84.

Doyle JJ, Doyle JL (1990). Isolation of plant DNA from fresh tissue. Focus. 12:13-15.

Eisen JA (1999). Mechanistic basis for microsatellilte instability. In: Goldstein DB, Schlotterer, C (eds) Microsatellites: evolution and applications. Oxford University Press, Oxford. pp. 34-48.

Hoisington D, Khairallah M, González D (1994). Laboratory Protocols: CIMMYT Applied Molecular Genetics Laboratory. CIMMYT, Mexico. 165-173.

Jannoo N, Forget L, Dookun A (2001). Contribution of Microsatellites to Sugarcane Breeding Program in Mauritius. International Society of Sugar Cane Technologists, Proceedings of the XXIV Congress, Brisbane. pp. 637-639.

Khan IA, Khatri A, Nizamani GS, Siddiqui MA, Khanzada MH, Dahar NA, Seema N, Naqvi MH (2004). *In vitro* studies in sugarcane. Pak. J. Biotech. 1:6-10.

McIntyre L, Aitken K, Berding N, Casu R, Drenth J, Jackson P, Jordan D, Piperidis G, Reffay N, Smith G, Tao Y, Whan V (2001). Identification of DNA Markers Linked to Agronomic Traits in Sugarcane in Australia. In: International Society of Sugar Cane Technologists. Proceedings of the XXIV Congress, Brisbane, Australia, 17-21 September. Australian Society of Sugar Cane Technologists, Brisbane. 560-562.

Nei M (1978). Estimation of average heterozygosity and genetic distance from a small number of individual. Genetics 89:583-590.

Pan YB, Cordeiro GM, Henry RJ, Schnell RJ (2002). Microsatellite Fingerprints of Louisiana Sugarcane Varieties and Breeding Lines. In: Plant, Animal and Microbe Genomes X Conference, 12-16 January, San Diego.

Pinto LR, Oliveira KM, Marconi T, Garcia AAF, Ulian EC, de Souza AP (2006). Characterization of novel sugarcane expressed sequence tag microsatellites and their comparison with genomic SSRs. Plant Breed. 125:378-378.

Piperidis G, Christopher MJ, Carroll BJ, Berding N, D'Hont A (2000). Molecular contribution to selection of intergeneric hybrids between sugarcane and the wild species *Erianthus arundinaceus*. Genome 43:1033-1037.

Selvi A, Nair NV, Balasundaram N, Mohapatra T (2003). Evaluation of maize microsatellite markers for genetic diversity analysis and fingerprinting in sugarcane. Genome 46:394-403.

Smiullah, Farooq AK, Abdullah, Aqeel A, Muhammad AJ, Zafar I, Rameez I, Javed IW (2012). In vitro Regeneration, detection of Somaclonal Variation and screening for SCMV in Sugarcane (*Saccharum* spp.) Somaclones. Afr. J. Biotech., 11(48):10841-10850.

Tautz D (1989). Hypervariablity of simple sequences as a general source for polymorphic DNA markers. Nucl. Acids Res. 17:6443-6471.

25

Induction and regeneration of somatic embryos from *Vitex doniana* (Lamiaceae) leaf explants

Colombe Dadjo[1], Jane Kahia[2], Catherine Muthuri[3], Lucien Diby[2], Christophe Kouame[2], Peter Njenga[4] and Modeste Kouassi[5]

[1]Institute of Basic Sciences, Technology and Innovation, Pan African University, P. O. Box 62000-00200, Nairobi, Kenya.
[2]World Agroforestry Centre (ICRAF), Cote d' Ivoire Country Program Cocody Mermoz, Abidjan, Côte d'Ivoire.
[3]ICRAF World Agroforestry Centre (ICRAF), P. O. Box 30677-00100, Nairobi, Kenya.
[4]Jomo Kenyatta University of Agriculture and Technology, P. O. Box 62000-00200, Nairobi, Kenya.
[5]Centre National de Recherche Agronomique (CNRA), Laboratoire Central de Biotechnologies (LCB), 01 BP 1740 Abidjan 01, Côte d'Ivoire.

The present study was conducted with the aim of evaluating some of the factors that influence induction and regeneration of somatic embryos in *Vitex doniana* since there are no available reports on tissue culture of this tree species. Leaves from plants growing under temporary shed were cultured on Murashige and Skoog supplemented with silver nitrate and four amino acids (proline, tryptophan, lysine and leucine) at varying concentrations, 0.11 mg/l thidiazuron, 2% sucrose and 100 mg/l myo-inositol in separate experiments. The explants cultured on media supplemented with tryptophan at 30.6 mg/l produced the optimal (6.5) number of embryos per explant. This number was fivefold more than the number obtained in the control. On the other hand, it was observed that the explants on media supplemented with silver nitrate at 8.45 mg/l gave the same mean (6.5) number of embryos per explant. These first ever results on the induction of somatic embryo in *V. doniana* could be used for mass propagation and to select useful traits of this tree species at the cellular level. However, further work needs to be done on the conversion of the regenerated embryos.

Key words: Amino acids, leaf explant, silver nitrate, somatic embryogenesis, *Vitex doniana*.

INTRODUCTION

Many neglected and underutilized wild species (NUS) are nutritionally rich (Ghane et al., 2010; Johns and Eyzaguirre, 2006). Therefore, their erosion can have immediate consequences on the nutritional status and food security of the poor and their enhanced use can bring about better nutrition and fight hunger. Even though the link between agrobiodiversity and diet diversity is not automatic (Burchi et al., 2011), it is agreeable that the diminution of agrobiodiversity, to some extent, places considerable strain on the ease with which households are able to enjoy diversified, balanced diets. Awareness of the importance and value of crop wild relatives and of the need to conserve them has increased one of such high value plant that is in dire need for conservation, *Vitex doniana* Sweet (Black plum). It had been considered to belong to Verbenaceae family by different authors but in recent works, it has been transferred into Lamiaceae based on different evidences (Wagstaff et al., 1998). It is the most abundant and widespread *Vitex* species in Africa (Orwa et al., 2009). Black plum (*V. doniana)* is an indigenous species important for the livelihoods of rural populations in Benin Republic in particular and West Africa in general (Codjia et al., 2003). The fruits and leaves are the edible part of the trees. They are either eaten raw or after processing. The plant is also widely used in traditional system for medicinal purposes. The leaves, fruits, roots, barks and seed of the plant have been used as medication for liver disease, anodyne, stiffness, leprosy, backache, hemiplegia, conjunctivity, rash, measles, rachitis, febrifuge, as tonic galactagogue to aid milk production in lactating mothers, sedative, digestive regulator, treatment of eye troubles and kidney troubles. It has also been used for treatment of conditions such as infertility, anemia, jaundice, dysentery, gonorrhea, headaches, diabetes, chickenpox, rash and fever (Louppe et al., 2008; Orwa et al., 2009). Despite the widely known nutritional, medicinal and economic uses of *V. doniana* products, the species is still under-utilized, unimproved. To date, the species has been chosen as a model species to be domesticated in Benin (Codjia et al., 2003; Dadjo et al., 2012).The strong anthropic pressure affecting this species has caused its numbers to fall increasingly in its natural environment (Achigan-Dako et al., 2010). The planting of seedlings is negligible and the seeds of this tree have a very weak germinating capacity (Thies 1995). Sanoussi et al. (2012) reported that macropropagation rate of this tree species by stem cuttings is slow. Therefore, there is need to seek for alternative propagation methods. Tissue culture methods are considered as the most promising means to protect and propagate tree species of economic interest. *In vitro* propagation through somatic embryogenesis is the most feasible alternative to the other *in vitro* methods (Zimmerman, 1993, Saiprasad, 2001). Somatic embryos are widely considered to be of single cell origin; hence this is advantageous for transformation studies. Moreover, the process of somatic embryogenesis offers a mean to propagate large numbers of transgenic plants over a short period of time.

The success in tissue culture depends on the effectiveness of the sterilization methods used on the explants prior to culture initiation (Yildiz and Celal, 2002). Sterilization is the process of eliminating contamination from explants before establishment of *in vitro* cultures. Various sterilization agents such as calcium hypochlorite

and sodium hypochlorite are commonly used to decontaminate the tissues.

Ethylene is known to reduce somatic embryogenic competence in many plants and the use of silver nitrate (AgNO$_3$), an ethylene action inhibitor (Beyer, 1976), has been shown to influence *in vitro* somatic embryogenesis (Kong et al., 2012). It has also been reported that amino acids play a key role in plant growth and development because they are good source of nitrogen (Kirby et al., 1987; Shanjani, 2003). Amino acids such as glutamine, proline and tryptophan, have been identified as enhancers of somatic embryogenesis in some species (Deo et al., 2010). Their efficacy in embryogenesis has been attributed to their contribution to various cellular processes such as improving cell signaling processes in various signal transduction pathway (Lakshmanan and Taji, 2000) and as precursor molecules for certain growth regulators. So far, there are no reports on tissue culture of *V. doniana*. During the current study, we hypothesized that the various concentrations of AgNO$_3$ and some amino acids could enhance induction and regeneration of somatic embryos in *V. doniana*. Therefore, the aim of this study was to evaluate the effects of various concentrations of AgNO$_3$ and some amino acids on induction and regeneration of somatic embryos in *V. doniana*.

MATERIALS AND METHODS

V. doniana seedlings (2-4 years old) originally wildings collected from the field in Glo in the south part and Cove in the central part of Benin were transported to Abidjan, Côte d' Ivoire where they were maintained in a temporary shed and watered daily for one month before harvesting the leaves.

Explant preparation and surface sterilization

Healthy looking young leaves (2nd pair) were collected and cleaned with cotton wool soaked with liquid soap. Thereafter, they were immersed in 0.5% (w/v) fungicide (Ridomil) containing two drops of Tween-20 (wetting agent) for an hour. The leaf explants were then transferred to the lamina flow cabinet for surface sterilization. Pre sterilization was carried using 70% ethanol solution for thirty seconds and then rinsing twice with sterile distilled water. The explants were subjected to further sterilization using varying (1, 1.5 and 2%) concentrations of calcium hypochlorite (CaOCl$_2$) containing 2-3 drops of Tween 20 and varying time duration. In an attempt to increase the number of clean explants, double sterilization (two steps) was conducted. This involved sterilizing the explants using 2% calcium hypochlorite for 30 min rinsing twice with sterile distilled water followed by quick dip in 70% ethanol. They were then subjected to a second step, by sterilizing them with 2% calcium hypochlorite for 15 min and finally rinsing with sterile distilled water four times.

Media preparation and culture conditions

Silver nitrate (at concentrations 8.45; 16.9 and 25.35 mg/l) and four amino acids namely proline (5.75; 11.5 and 17.25 mg/l), tryptophan (10.2; 20.4 and 30.6 mg/l), lysine (7.3; 14.6 and 21.9 mg/l) and leucine (6.55; 13.1 and 19.65 mg/l) were added to half strength

Table 1. Effect of CaOCl$_2$ on elimination of surface contamination from *V. doniana* leaves explants.

Concentrations of calcium hypochlorite (in % w/v)	Exposure time (min)	Percentage of clean explants (%)
1	25	0
1	30	0
1	35	0
1	40	0
1.5	20	0
1.5	25	0
1.5	30	0
1.5	35	20
2	15	40
2	20	70
2	25	70
2	30	70

Table 2. Effect of different AgNO$_3$ concentrations on induction and regeneration of somatic embryos in *V. doniana*.

AgNO$_3$ conc. (mg/l)	Embryogenic cultures (%)	Mean number of embryos/explant ± SE
0	91	1.2 ± 0.13b*
8.45	90	6.5 ± 0.41a
16.9	67	1.5 ± 0.27b
25.35	25	1.75 ± 0.25b
P-value		0.000

*Means followed by the same letter are not significantly different at p > 0.01.

Murashige and Skoog (MS) (1962) medium supplemented with 100 mg/l myo-inositol, 0.11 mg/l TDZ and 2% sucrose in separate experiments. The pH of the medium was adjusted between 5.7 and 5.8 prior to the addition of the solidifying agent, and autoclaved at 121°C for 15 min. The MS medium without AgNO$_3$ and amino acids is referred to as control. All cultures were incubated in a dark room maintained at 25 ± 2°C.

Experimental design, data collection and analysis

Completely randomized design was used for all the experiments. Each single explant was considered as experimental unit. Twenty (20) replicates per treatment were used at the outset of experiments and each treatment repeated at least twice. The data were subjected to one-way analysis of variance (ANOVA) and the significant differences between treatment means were assessed by using Minitab version 14 Software. The results of the sterilization experiment are expressed as percentage (%) while for regeneration of somatic embryos data are presented as means ± standard error (SE).

RESULTS

Surface sterilization

The highest percentage (70%) of clean leaf explants

were obtained when 2% calcium hypochlorite was used for 20, 25 and 30 min, respectively (Table 1). However, this percentage decreased further and after two weeks only 50% clean explants were observed. The evaluation of double sterilization treatment resulted in increase in percent clean explants from 50 to 91% after one month and this procedure was used for all the subsequent experiments.

Effect of AgNO$_3$ on induction and regeneration of somatic embryos

The addition of AgNO$_3$ to the culture medium had a significant effect on the mean number of embryos per explant (Table 2). The percent embryogenic cultures decreased with increasing AgNO$_3$ concentration. The highest (91 and 90 %) frequency of embryogenic cultures were observed in the control and in the media supplemented with 8.45 mg/l AgNO$_3$. The later AgNO$_3$ concentration produced the highest (6.5) mean number of embryos. On the other hand, the lowest (1.2) mean number of somatic embryos was produced in the control. Significant differences were detected between the number of somatic embryos per explant at 8.45 mg/l

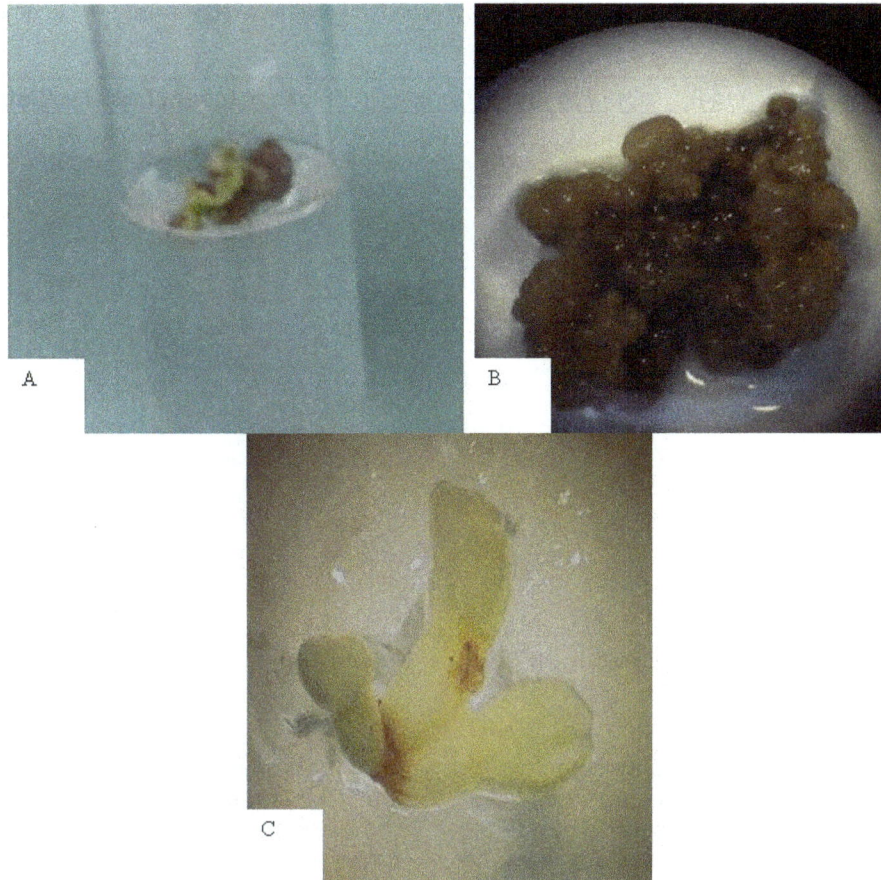

Figure 1. A) Callus, B) globular embryos, C) cotyledonary embryos.

$AgNO_3$ and the other concentrations of $AgNO_3$ at p < 0.01. Callus was observed from the cut edges of the leaf and globular embryos were obtained four weeks after culture (Figure 1A and B). Cotyledonary stage of embryos was observed two months after culture as shown in Figure 1C.

Effect of some amino acids induction and regeneration of somatic embryos

Among the four amino acids evaluated, the highest leucine concentration (19.65 mg/l) was found to inhibit embryo formation (Table 3). However, when leaf explants were cultured on media supplemented with the lower concentrations (6.55 and 13.1mg/l), the mean number of embryos obtained were double those obtained in the control. It was observed that increasing the concentration of proline from 5.75 to 17.25 mg/l decreased significantly the mean number of embryos. Leaf explants cultured on medium supplemented with 30.6 mg/l tryptophan produced the highest (6.5) mean number of somatic embryos which was fivefold more than the embryos obtained in the control.

DISCUSSION

The current study was conducted with the aim of evaluating some of the factors that influence induction and regeneration of somatic embryos since there were no available reports on tissue culture of *V. doniana*. Before any explants is placed into culture, it is essential to destroy all microorganisms and the success in tissue culture depends on the effectiveness of the sterilization methods used on the explants prior to culture initiation (Yildiz and Celal, 2002). The outer surface of plants growing under natural or greenhouse conditions is normally infected with spores and other microbial cells. The use of field grown plants as a direct source of explant material for obtaining 'clean' explant, presents a major challenge. Previous attempts to initiate clean explants from field grown coffee, especially those from canopy close to the ground, resulted in 100% contamination (Kahia, 1999). In an attempt to obtain clean *in vitro* cultures, sources of contamination other than surface contaminants need to be considered. Even if the surface of the explant is effectively sterilized, the contaminants could emanate from the inner tissues when the plant materials are dissected into small explants after

Table 3. Effect of different amino acid concentrations on induction and regeneration of somatic embryos in *V. doniana.*

Amino acid	conc. (mg/l)	Embryogenic cultures (%)	Mean number of embryos/explant ± SE
Control	0	91	$1.2 \pm 0.13a*$
	6.55	25	$2.67 \pm 1.2b$
Leucine	13.1	50	$2.4 \pm 0.51b$
	19.65	0	0^x
	7.3	30	$1.33 \pm 0.21a$
Lysine	14.6	38	$1.5 \pm 0.34a$
	21.9	50	$1.6 \pm 0.4a$
	5.75	10	$3.5 \pm 0.5b$
Proline	11.5	10	$2.0 \pm 0.0a$
	17.25	6	$1.5 \pm 0.5a$
	10.2	42	$2.33 \pm 0.21b$
Tryptophan	20.4	17	$1.5 \pm 0.5a$
	30.6	60	$6.5 \pm 0.66 c$
P-value			0.000

x. The zero response is not included in the statistical analysis. Means followed by the same letter are not significantly different at $p > 0.01$.

the surface sterilization. Systemic contaminants, for example, are not eliminated by surface sterilization (Webster et al., 2003). For this reason, a systemic fungicide such as Ridomil used in this study can be helpful to control the incidence of internal fungal infection in explants. The disinfectant widely used for surface sterilization of explants in tissue culture is sodium hypochlorite which dates back to the mid-18th century (Miche and Balandreau, 2001). It is usually purchased as household laundry bleach and as such it is readily available and can be diluted to proper concentrations. A balance between concentration and time must be determined empirically for each type of explant because of phytotoxicity. Previous reports indicate that bleach (1.4% NaOCl solution for 1 min) was found to be effective in sterilizing greenhouse-derived leaf explants in *Aquilaria crasna* and *Aquilaria sinensis* (Okudera and Ito, 2009). On the other hand, effective sterilization was achieved by using 50% bleach for 20 min on shoot tips of *Aquilaria hirta*, from greenhouse (Hassan et al., 2011) while leaf explants of *Allanblackia stuhlmannii* were best surface sterilized using 8% sodium hypochlorite (Neondo et al., 2011). The other sterilant used for decontaminating explants is calcium hypochlorite. It is known to be less injurious to plant tissues and is generally used at a concentration of 3.25% (CSS 451 2009). During the present study, calcium hypochlorite was used as attempts to use the commercial bleach even at very low concentrations was found to be phytotoxic and led to the death of explants. The lower concentrations of calcium hypochlorite were not effective in decontaminating the explants. These results concur with those of Mihaljević et al. (2013) who reported high percent sterilization (80%) of

sour cherry nodes when a high (3%) concentration of $CaOCl_2$ was used. A two-step sterilization was adopted in the work being reported in order to increase the percent clean explants and using this procedure, 91% clean explants was recorded. Similar results were reported by Nieves and Evalour (2011) who recorded 90% clean explants from cotyledon of *Moringa oleifera* using two steps comprising of 5% $Ca(OCl)_2$.

There are conflicting reports on the use of silver nitrate in induction and regeneration of somatic embryos. When added at low concentrations of 5 to 50 µM (0.845 to 8.45 mg/l), it was found to inhibit somatic embryo formation in *Coffea canephora* leaf explants (Hatanaka et al., 1995). On the other hand, media supplemented with $AgNO_3$ has been shown to improve somatic embryogenesis in species such as *Triticum durum* (Poaceae) (Fernandez et al., 1999), *Coffea canephora* (Rubiaceae) (Fuentes et al., (2000), *Spinacia oleracea* (Ishizaki et al., 2000), *Carthamus tinctorius* (Asteraceae) (Mandal et al., 2001), *Paspalum scrobiculatum* (Poaceae) (Vikrant, 2002), *Bactris gasipaes* (Arecaceae) (Steinmacher et al., 2007), *Paspalum scrobiculatum* (Poaceae) and *Eleusine coracana* (Poaceae) (Kothari-Chajer et al., 2008), *Hedychiurn bousigonianum* (Gingeberacea) (Sakhanokho et al., 2009), *Gossypium nelsonii* (Malvaceae), *Gossypium australe* (Yan et al., 2010) and *Pinus taeda* (Pinaceae) (Pullman et al., 2003). Kong and Yeung (1995) reported that 100 µM $AgNO_3$ (16.9 mg/l $AgNO_3$) stimulated embryo formation in white spruce. The exact mechanism by which $AgNO_3$ affects somatic embryogenesis is not completely understood (Kong et al., 2012). During the current study, incorporating 8.45 mg/l silver nitrate in the media led to a fivefold increase in the number of embryos as compared to the control. However, it was observed that increasing the concentration from 8.45 to 25.35 mg/l decreased both the percentage of embryogenic cultures and the mean number of embryos. These results are contrary to those of Kong et al. (2012) who reported that the number of somatic embryos per explant in Manchurian ash increased with increasing $AgNO_3$ concentration. These workers reported that the lowest number of embryos per explant (1.0) was induced at 2.5 mg/l $AgNO_3$ while the higher concentration of 10 mg/l resulted to a threefold increase in the number of somatic embryos with mean (3.86) embryos per explant. Comparable results have also been reported by Ishizaki et al. (2000) in *Spinacia oleracea* where addition of 10 mM $AgNO_3$ (1.69 g/l $AgNO_3$) to the medium resulted in formation of about three times more embryos as compared to the controls. Fuentes et al. (2000) found that the addition of $AgNO_3$ caused only small modifications in the ionic equilibrium of the medium and concluded the effects of the compound on somatic embryogenesis were not attributable to any substantial changes in available nutrients.

Requirement of exogenous supply of amino acids for *in vitro* somatic embryogenesis has been reported in a number of plant species (Basu et al., 1989; Claparols et al., 1993). For instance, glutamine was found to be beneficial for embryo development in date palm (El-Shiaty et al., 2004). These workers reported that MS medium supplemented with 100 mg/l glutamine gave the highest (3.33) mean number of embryos. In another report, incorporating proline in sugarcane cultures enhanced somatic embryogenesis (Gill et al., 2004). In *Peucedanum oreoselinum,* embryo formation and maturation was enhanced by addition of proline in MS medium (Coste et al., 2011). On the other hand, in wheat culture, the efficiency of the amino acids was found to be genotype-based (Duran et al., 2013). Sarker et al. (2007) reported that there were significant responses when L-asparagine at 150 mg/l concentration was used in four popular Bangladeshi wheat cultivars *viz* Kanchan, Shourav, Gourav and Satabdi. In the work being reported, it was found that supplementing MS media with amino acid promoted the induction and regeneration of somatic embryos. These results concur with those of Shahsavari (2011) who reported that when 100 to 300 µM (20.4 to 61.2 mg/l) tryptophan was added to callus induction medium of rice, there was a great enhancement of the frequency of embryogenic cultures and regeneration of somatic embryos. The lowest concentration of proline evaluated in the current study increased almost three folds the mean number of embryos as compared to the control. However, increasing the proline concentration resulted in decrease in the frequency of embryogenic cultures and the mean number of embryos. These results are contrary to those reported by Chowdhry et al. (2003) who observed that increasing the concentration of proline enhanced embryogenesis in rice. Differential responses

of different amino acids indicate the requirement of specific amino acids for somatic embryo regeneration in *V. doniana.*

Conclusion

This study clearly demonstrates that this tree species is amenable to somatic embryogenesis. The results of the current study will be a valuable tool to complement the production of planting materials and thus help in exploiting the medicinal and nutrition value of this tree for the rural poor in Benin and Africa in general. *Ex situ* and domestication programs of *V. doniana* could benefit from the findings of the current study. There is however need to do more work on the conversion of somatic embryos.

Conflict of interests

The authors hereby declare that there is no conflict of interest.

ACKNOWLEDGMENTS

The authors thank the World Agroforestry Centre (ICRAF), Côte d' Ivoire country Program for allowing them to use their facilities to carry out the work. The help rendered by the staff of ICRAF Somatic Embryogenesis Laboratory in Côte d' Ivoire is highly appreciated

REFERENCES

Achigan-Dako EG, Pasquini MW, Assogba-Komlan F, N'danikou S, Yédomonhan H, Dansi A, Ambrose-Oji B (2010). Traditional vegetables in Benin: diversity, distribution, ecology, agronomy, and utilisation. Institut National des Recherches Agricoles du Bénin, Benin.

Basu A, Sethi U, Guha-Mukherjee S (1989). Regulation of cell proliferation and morphogenesis by amino acids in Brassica cultures and its correlation with threonine deaminase. Plant Cell Rep. 8:333-335.

Beyer EM (1976). A potent inhibitor of ethylene action in plants. Plant Physiol. 58:268-271.

Burchi F, Fanzo J, Frison E (2011). The Role of Food and Nutrition System Approaches in Tackling Hidden Hunger. Int. J. Environ. Res. Public Health 8:358-373.

Chowdhry CN, Tyagi AK, Maheshwari N, Maheshwari SC (1993). Effect of L-proline and L-tryptophan on somatic embryogenesis and plantlet regeneration of rice (*Oryza sativa* L. cv. Pusa 169). Plant Cell Tiss. Org. Cult. 32:357-361.

Claparols I, Santos M.A, Torne JM (1993). Influence of some exogenous amino acids on the production of maize embryogenic callus and on endogenous amino acid content. Plant Cell Tiss. Org. Cult. 34 (1):1-11.

Codjia JTC, Assogbadjo AE, Ekue MRM (2003). Diversité et valorisation au niveau local des resource vegetales forestieres alimentaires du Benin. Cahiers Agric. 12:321-331.

Coste A, Oltean B, Halmagyi A, Deliu C (2011). Direct somatic embryogenesis and plant regeneration in *Peucedanum oreoselinum* (L.) Moench. Rom. Biotech. Lett. 16(4):6451-6459.

CSS 451 (2009). Biotechnology Applications for Plant Breeding and Genetic. January 13, Course and laboratory orientation; sterile techniques and tissue culture http://www.msu.edu/course/css/451/

Dadjo C, Assogbadjo AE, Fandohan B, Glèlè Kakaï R, Chakeredza S, Houehanou TD, Van Damme P, Sinsin B (2012). Uses and management of black plum (*Vitex doniana* Sweet) in Southern Benin. Fruits, 67(4):239-248.

Deo PC, Tyagi AP, Taylor M, Harding RM, Becker DK (2010). Factors affecting somatic embryogenesis and transformation in modern plant breeding. South Pac. J. Nat. Appl. Sci. 28(1):27-40.

Duran RE, Coskun Y, Demırcı T (2013). Comparison of amino acids for their efficiency on regeneration in wheat embryo culture. Asian J. Plant Sci. Res. 3(1):115-119.

El-Shiaty OH, El-Sharabasy SF, El-Kareim AHA (2004). Effect of some amino acids and biotin on callus and proliferation of date palm (*Phoenix dactylifera* L.) Sewy cultivar. Arabian J. Biotechnol. 7:265-272.

Fernandez S, Michaux-Ferriere N, Coumans M (1999). The embryogenic response of immature embryo cultures of durum wheat (*Triticum durum* Desf): histology and improvement by AgNO₃. Plant Growth Regul. 28:147-155.

Fuentes SRL, Calheiros MBP, Manetti-Filho J, Vieira LGE (2000). The effects of silver nitrate and different carbohydrate sources on somatic embryogenesis in *Coffea canephora*. Plant Cell Tiss. Org. Cult. 60:5-13.

Ghane SG, Lokhande VH, Ahire ML, Nikam TD (2010). *Indigofera glandulosa* Wendl. (Barbada) a potential source of nutritious food: underutilized and neglected legume in India. Genet. Resour. Crop Evol. 57(1):147-153.

Gill NK, Gill R, Gosal SS (2004). Factors enhancing somatic embryogenesis and plant regeneration in surgarcane (*Saccharum officinarum* L.). Indian J. Biotechnol. 3:119-123.

Hassan NH, Ali NAM, Zainudin F, Ismail H (2011). Effect of 6-benzylaminopurine (BAP) in different basal media on shoot multiplication of *Aquilaria hirta* and detection of essential oils in the *in vitro* shoots. Afr. J. Biotechnol. 10(51):10500-10503.

Hatanaka T, Sawabe E, Azuma T, Uchida N, Yasuda T (1995). The role of ethylene in somatic embryogenesis from leaf discs of Coffea canephora. Plant Sci. 107:199-204.

Ishizaki T, Komai F, Msuda K (2000). Exogenous Ethylene Enhances Formation of Embryogenic Callus and Inhibits Embryogenesis In Cultures of Explants of Spinach roots. J. Am. Soc. Hort. Sci. 125(1):21-24.

Johns T, Eyzaguirre PB (2006). Symposium on Wild-gathered plants: basic nutrition, health and survival" Linking biodiversity, diet and health in policy and practice. Proceed. Nutr. Soc. 65: 182-189.

Kahia WJ (1999). *In vitro* propagation of the new Coffea arabica cultivar- Ruiru 11. PhD thesis. University of London.

Kirby EG, Leustek T, Lee MS (1987). Nitrogen nutrition. In: Cell and Tissue Culture in Forestry. Volume 1. Edited by Bonga JM, DJ Durzan.. Martinus Nijhoff Publishers, Dordrecht, Boston, Lancaster; 237.

Kong D, Shen H, Li N (2012). Influence of AgNO₃ on somatic embryo induction and development in Manchurian ash (*Fraxinus mandshurica* Rupr.). Afr. J. Biotechnol. 11(1):120-125.

Kong LS, Yeung EC (1995). Effects of silver nitrate and polyethylene glycol on white spruce (*Picea glauca*) somatic embryo development: enhancing cotyledonary embryo formation and endogenous ABA content. Physiol. Plant. 93:298-304.

Kothari-Chajer A, Sharma M, Kachhwaha S, Kothari S (2008). Micronutrient optimization results into highly improved in vitro plant regeneration in kodo (*Paspalum scrobiculatum* L.) and finger millets (*Eleusine coracana* (L.) Gaertn.). Plant Cell Tiss. Org. Cult. 94:105-112.

Lakshmanan P, Taji A (2000). Somatic embryogenesis in leguminous plants. Plant Biol. 2:136-148.

Louppe D, Oteng-Amoako AA, Brink M (2008). Timbers 1. Plant Resources of Tropical Africa. Wageningen, the Netherlands. PROTA Foundation. Backhuys Publishers. CTA.

Mandal AKA, Dutta Gupta S, Chatterji AK (2001). Factors affecting somatic embryogenesis from cotyledonary explants of safflower. Biol. Plantarum. 44:503-507.

Miche L, Balandreau J (2001). Effects of rice seeds sterilization with Hypochlorite on inoculated *Burkholderia vietnamiensis*. Appl. Environ. Microbiol. 67(7): 3046-3052.

Mihaljević I, Dugalić K, Tomaš V, Viljevac M, Pranjić A, Čmelik Z, Puškar B, Jurković Z. (2013). *In vitro* sterilization procedures for micropropagation of 'OBLAČINSKA' sour cherry. J. Agric. Sci. 58 (20): 117-126.

Murashige T, Skoog F (1962). A revised medium for rapid growth and bioassay with tobacco tissue cultures. Plant Physiol. 15:473-497.

Neondo J, Machua J, Muigai A, Nyende AB, Munjuga M, Jamnadass R, Muchugi A (2011). Micropropagation of *Allanblackia stuhlmanii*: Amenability to tissue culture technique. Int. J. Biotechnol. Mol. Biol. Res. 2(11):185-194

Nieves MC, Evalour TA (2011). Callus Induction in Cotyledons of *Moringa oleifera* Lam. Phillip. Agric. Sci. 94(3):239-247.

Okudera Y, Ito M (2009). Production of agarwood fragrant constituents in *Aquilaria* calli and cell suspension cultures. Plant Biotechnol. 26:307-315.

Orwa C, Mutua A, Kindt R, Jamnadass R, Anthony S (2009). Agroforestree Database: a tree reference and selection guide version 4.0
http://www.worldagroforestry.org/sites/treedbs/treedatabases.asp)

Pullman GS, Namjoshi K, Zhang Y (2003). Somatic embryogenesis in loblolly pine (Pinus taeda L.): Improving culture initiation with abscisic acid, silver nitrate, and cytokinin adjustments. Plant Cell Rep. 22:85-95.

Saiprasad GVS (2001). Artificial seeds and their applications. Resonance, 6:39-47.

Sakhanokho HF, Rajasekaran K, Kelley RY (2009). Somatic embryogenesis in *Hedychium bousigonianum*. Hort. Sci. 44:1487-1490.

Sanoussi A, Ahoton LE, Odjo Th (2012). Propagation of Black Plum (*Vitex doniana* Sweet) Using Stem and Root Cuttings in the Ecological Conditions of South Benin. Tropicultura, 2:107-112.

Sarker KK, Kabir AH, Sharmin SA, Nasrin Z, Alam MF (2007). Improved somatic embryogenesis using L-Asparagine in wheat (*Triticum aestivum* L.). Sjemenarstvo, 24(3):187-196

Shahsavari E (2011). Contribution of sorbitol on regeneration of embryogenic calli in upland rice. Int. J. Agric. Biol. 13 838–840

Shanjani PS (2003). Nitrogen effect on callus induction and plant regeneration of *juniperus excelsa*. Int. J. Agric. Biol. 5(4):419-422.

Steinmacher DA, Cangahuala-Inocente GC, Clement CR, Guerra MP (2007). Somatic embryogenesis from peach palm zygotic embryos. In Vitro Cell Dev. Biol. Plant. 43:124-132.

Thies E (1995). Principaux ligneux agro forestiers de la Guinée, Zone de transition. Guinée Bissau, Guinée, Côte d'Ivoire, Ghana, Togo, Bénin, Nigéria, Cameroun. Schriftenreihe der, GTZ No 253.

Vikrant RA (2002). Somatic embryogenesis from immature and mature embryos of a minor millet Paspalum scrobiculatum L. Plant Cell Tiss. Org. Cult. 69:71-77.

Wagstaff SJ, Hickerson L, Spangler R, Reeves PA, Olmstead RG (1998). 'Phylogeny In labiatae S. L., Inferred From Cpdna Sequences'. Plant Syst. Evol. 209(3-4):265-274. doi:10.1007/bf00985232.

Webster S, Mitchell SA, Ahmad MH (2003). A novel surface sterilization method for reducing fungal and bacterial contamination of field grown medicinal explants intended for *in vitro* culture. Proceedings of 17th SRC conference entitled 'Science and Technology for Economic Development: Technology Driven Agriculture and Agro-Processing' SRC, Jamaica.

Yan SF, Zhang Q, Wang JE, Sun YQ, Daud MK, Zhu SJ (2010). Somatic embryogenesis and plant regeneration in two wild cotton species belong to G genome. *In Vitro* Cell Dev. Biol. Plant, 46:298-305.

Yildiz M, Celal E (2002). The effect of Sodium hypochlorite solution on *in vitro* seedling growth and shoot regeneration of flax (*Linus usitatissimum*), Springer-Verlag.

Zimmerman JL (1993). Somatic embryogenesis: a model for early development in higher plants. Plant Cell, 5(10):1411-1423.

Permissions

List of Contributors

Oluwasayo Kehinde Moyib
Department of Petroleum and Chemical Sciences, Tai Solarin University of Education, Lagos-Benin Express Road, P. M. B. 2118, Ijebu-Ode, Ogun State, Nigeria
Cassava Breeding Unit, International Institute of Tropical Agriculture (IITA) P. M. B. 5320, Oyo Road, Ibadan, Oyo State, Nigeria
Department of Biochemistry, University of Ibadan, Ibadan, Oyo State, Nigeria

Jonathan Mkumbira
Cassava Breeding Unit, International Institute of Tropical Agriculture (IITA) P. M. B. 5320, Oyo Road, Ibadan, Oyo State, Nigeria

Oyeronke Adunni Odunola
Department of Biochemistry, University of Ibadan, Ibadan, Oyo State, Nigeria

Alfred Godwin Dixon
Cassava Breeding Unit, International Institute of Tropical Agriculture (IITA) P. M. B. 5320, Oyo Road, Ibadan, Oyo State, Nigeria
Sierra Leone Agricultural Research Institute, Tower Hill, P. M. B. 1313 Freetown, Sierra Leone

Syed Naseer Shah
Genetics and Plant Propagation Division, Tropical Forest Research Institute, Mandla Road, Jabalpur 482 021, India

Amjad M. Husaini
Centre for Plant Biotechnology, Division of Biotechnology, SKUAST-K, Shalimar, Srinagar-191121, India

Fatima Shirin
Genetics and Plant Propagation Division, Tropical Forest Research Institute, Mandla Road, Jabalpur 482 021, India

Mariana Carmen Chifiriuc
Department of Microbiology, MICROGEN (Center for Research in Genetics, Microbiology and Biotechnology), Faculty
of Biology, University of Bucharest, Ale. Portocalilor 1-3, Sector 5, 77206-Bucharest, Romania

Coralia Bleotu
Institute of Virology Stefan S. Nicolau, 285 Mihai Bravu Ave. 030304, Bucharest, Romania

Diana-Roxana Pelinescu
Department of Microbiology, MICROGEN (Center for Research in Genetics, Microbiology and Biotechnology), Faculty of Biology, University of Bucharest, Ale. Portocalilor 1-3, Sector 5, 77206-Bucharest, Romania

Veronica Lazar
Department of Microbiology, MICROGEN (Center for Research in Genetics, Microbiology and Biotechnology), Faculty of Biology, University of Bucharest, Ale. Portocalilor 1-3, Sector 5, 77206-Bucharest, Romania

Lia- Mara Ditu
Department of Microbiology, MICROGEN (Center for Research in Genetics, Microbiology and Biotechnology), Faculty of Biology, University of Bucharest, Ale. Portocalilor 1-3, Sector 5, 77206-Bucharest, Romania

Tatiana Vassu
Department of Microbiology, MICROGEN (Center for Research in Genetics, Microbiology and Biotechnology), Faculty
of Biology, University of Bucharest, Ale. Portocalilor 1-3, Sector 5, 77206-Bucharest, Romania

Ileana Stoica
Department of Microbiology, MICROGEN (Center for Research in Genetics, Microbiology and Biotechnology), Faculty of Biology, University of Bucharest, Ale. Portocalilor 1-3, Sector 5, 77206-Bucharest, Romania

Olguta Dracea
Cantacuzino Institute, Sp. Independentei 103, Bucharest, Romania

Ionela Avram
Department of Microbiology, MICROGEN (Center for Research in Genetics, Microbiology and Biotechnology), Faculty of Biology, University of Bucharest, Ale. Portocalilor 1-3, Sector 5, 77206-Bucharest, Romania

Elena Sasarman
Department of Microbiology, MICROGEN (Center for Research in Genetics, Microbiology and Biotechnology), Faculty of Biology, University of Bucharest, Ale. Portocalilor 1-3, Sector 5, 77206-Bucharest, Romania

Nderitu Peris Wangar
Moi University, P.O. Box 1125-30100, Eldoret, Kenya

Kinyua Miriam Gacheri,
Moi University, P.O. Box 1125-30100, Eldoret, Kenya

Mutui Mwendwa Theophilus
Moi University, P.O. Box 1125-30100, Eldoret, Kenya

Ngode Lucas
Moi University, P.O. Box 1125-30100, Eldoret, Kenya

Clayton Miguel Costa
Department of Surgery, Federal University of Santa Catarina, UFSC, Brazil

Geraldo Bernardes
Department of Surgery, Federal University of Santa Catarina, UFSC, Brazil

Sandro Melim Sgrott
Department of Surgery, Federal University of Santa Catarina, UFSC, Brazil

Jorge Bins Ely
Department of Surgery, Federal University of Santa Catarina, UFSC, Brazil

Luismar Marques Porto
Graduate Program in Chemical Engineering, Federal University of Santa Catarina, UFSC, Brazil

Armando José d'Acampora
Department of Surgery, Federal University of Santa Catarina, UFSC, Brazil

Ammad Ahmad Farooqi
Institute of Molecular Biology and Biotechnology (IMBB), University of Lahore, Pakistan

Qaisar mansoor
Institute of Biomedical and Genetic Engineering (IBGE), Islamabad, Pakistan

Aamir Rana
Institute of Molecular Biology and Biotechnology (IMBB), University of Lahore, Pakistan

Zeeshan Javed
Institute of Molecular Biology and Biotechnology (IMBB), University of Lahore, Pakistan

Shahzad Bhatti
Institute of Molecular Biology and Biotechnology (IMBB), University of Lahore, Pakistan

Alka Grover
Division of Crop Improvement, Central Potato Research Institute, Shimla 171001, H.P. India

Abhinav Grover
Department of Biochemical Engineering and Biotechnology, Indian Institute of Technology Delhi, Hauz Khas, New Delhi 110016, India

S. K. Chakrabart
Division of Crop Improvement, Central Potato Research Institute, Shimla 171001, H.P. India

Wamik Azmi
Department of Biotechnology, Himachal Pradesh University, Shimla,171005, H.P. India

Durai Sundar
Department of Biochemical Engineering and Biotechnology, Indian Institute of Technology Delhi, Hauz Khas, New Delhi 110016, India

S. M. P. Khurana
Amity Institute of Biotechnology, Amity University, Haryana, 122413, India

Chinnaswamy Ramesha
Silkworm Breeding and Molecular Genetics Laboratory, Andhra Pradesh State Sericulture Research and Development Institute (APSSRDI), Kirikera-515 211, Hindupur, AP, India

Savarapu Sugnana Kumari
Biology Division, Indian Institute of Chemical Technology, Tarnaka, Hyderabad - 500 007, AP, India

Chebba Moremnagari Anuradha
Department of Biotechnology, College of Engineering and Technology, Sri Krishnadevaraya University, Anantapur-515 003, AP, India

Hothur Lakshmi
Silkworm Genetics and Breeding Laboratory, Central Sericultural Research and Training, Institute (CSR and TI), Central Silk Board, Berhmpore - 742 101, West Bengal, India

Chitta Suresh Kumar
Bioinformatics Centre, Department of Biochemistry, Sri Krishnadevaraya University, Anantapur- 515 003, AP, India

Nibras S. Al-Ammar
Department of Microbiolody, College of Medicine, University of Basrah, Basrah, Iraq

Ihsan Al-Saimary
Department of Microbiolody, College of Medicine, University of Basrah, Basrah, Iraq

Saad Sh. Hamadi
Department of Medicine, College of Medicine, University of Basrah, Basrah, Iraq

Ma Luo
Department of Medicine, College of Medicine, University of Basrah, Basrah, Iraq

T. Saravanan
Molecular Virology Laboratory, Indian Veterinary Research Institute, Bangalore Campus, Hebbal, Bangalore - 560 024, Karnataka, India

C. Ashok Kumar
Molecular Virology Laboratory, Indian Veterinary Research Institute, Bangalore Campus, Hebbal, Bangalore - 560 024, Karnataka, India

G. R. Reddy
Molecular Virology Laboratory, Indian Veterinary Research Institute, Bangalore Campus, Hebbal, Bangalore - 560 024, Karnataka, India

H. J. Dechamma
Molecular Virology Laboratory, Indian Veterinary Research Institute, Bangalore Campus, Hebbal, Bangalore - 560 024, Karnataka, India

G. Nagarajan
Molecular Virology Laboratory, Indian Veterinary Research Institute, Bangalore Campus, Hebbal, Bangalore - 560 024, Karnataka, India

P. Ravikumar
Molecular Virology Laboratory, Indian Veterinary Research Institute, Bangalore Campus, Hebbal, Bangalore - 560 024, Karnataka, India

G. Srinivas
Molecular Virology Laboratory, Indian Veterinary Research Institute, Bangalore Campus, Hebbal, Bangalore - 560 024, Karnataka, India

V. V. S. Suryanarayana
Molecular Virology Laboratory, Indian Veterinary Research Institute, Bangalore Campus, Hebbal, Bangalore - 560 024, Karnataka, India

Meera Sumanth
Department of Pharmacology, Visveswarapura Institute of Pharmaceutical Sciences, 22nd Main, 24th Cross, B. S. K II stage, Bangalore-560070, Karnataka, India

G. Nagarjuna Chowdary
Department of Pharmacology, Visveswarapura Institute of Pharmaceutical Sciences, 22nd Main, 24th Cross, B. S. K II stage, Bangalore-560070, Karnataka, India

Srijata Mitra
Department of Biotechnology, The University of Burdwan Golapbag Burdwan, West Bengal, India

Pranab Roy
Department of Biotechnology, The University of Burdwan Golapbag Burdwan, West Bengal, India

Sucik Maylinda
Faculty of Animal Husbandry, Brawijaya University, Malang – Indonesia

Aamir Rana
Institute of Molecular Biology and Biotechnology, University of Lahore, Pakistan   National Institute for Genomics and Advanced Biotechnology, NARC, Islamabad, Pakistan

Shahzad Bhatti
Institute of Molecular Biology and Biotechnology, University of Lahore, Pakistan

Ghulam Muhammad Ali
National Institute for Genomics and Advanced Biotechnology, NARC, Islamabad, Pakistan

Shoukat Ali
Plant Biotechnology Program, NARC, Islamabad, Pakistan

Nazia Rehman
Pakistan Agricultural Research Council (PARC), Institute of Advanced Studies in Agriculture, Islamabad, Pakistan

Sabir Hussain Shah
Pakistan Agricultural Research Council (PARC), Institute of Advanced Studies in Agriculture, Islamabad, Pakistan

Ammad Ahmad Farooqi
Institute of Molecular Biology and Biotechnology, University of Lahore, Pakistan

Ourid Ibtissam
Centre de Recherche Forestière, Rabat, Marocco
Faculté des Sciences et Techniques, Fès, Marocco

Ghanmi Mohamed
Centre de Recherche Forestière, Rabat, Marocco

EL Ghadraoui Lahsen
Faculté des Sciences et Techniques, Fès, Marocco

Kerdouh Benaissa
Centre de Recherche Forestière, Rabat, Marocco

Bakkali Yakhlef Salah Eddine
Centre de Recherche Forestière, Rabat, Marocco

Ayman A. Diab
Molecular Markers and Genome Mapping Department, Agricultural Genetic Engineering Research Institute (AGERI), Agricultural Research Center (ARC), Giza, 12619-Egypt

Ahmed M. K. Nada
Plant molecular Biology department, Agricultural Genetic Engineering Research Institute (AGERI), Agricultural Research Center (ARC), Giza, 12619-Egypt

Ahmed Ashoub
Nucliec acid and protein structure department Center (ARC), Agricultural Genetic Engineering Research Institute (AGERI), Agricultural Research Center (ARC), Giza, 12619-Egypt

M.. Bhattacharjee
Biotechnology Division, CSIR-North-East Institute of Science and Technology, Jorhat-785006, Assam, India

B. GG. Uni
Biotechnology Division, CSIR-North-East Institute of Science and Technology, Jorhat-785006, Assam, India

S. Das
Biotechnology Division, CSIR-North-East Institute of Science and Technology, Jorhat-785006, Assam, India

M.Deka
Department of Biological Sciences, Institiute of Science and Technology, Gauhati University, Guwahati, 781014 Assam, India

P. G. Rao
Biotechnology Division, CSIR-North-East Institute of Science and Technology, Jorhat-785006, Assam, India

M. Govarthanan
Department of Biotechnology, Mahendra Arts and Science College, kalippatti, Namakkal, Tamilnadu-637501, India

A. Guruchandar
Department of Molecular Biology, School of Life sciences, University of Skovde, Sweden

S. Arunapriya
Department of Biotechnology, Vysya Arts and Science College, Salem, Tamilnadu-636 103, India

T. Selvankumar
Department of Biotechnology, Mahendra Arts and Science College, kalippatti, Namakkal, Tamilnadu-637501, India

K. Selvam
Department of Biotechnology, Mahendra Arts and Science College, kalippatti, Namakkal, Tamilnadu-637501, India

Sujata Dash
Microbiology laboratory, Regional Plant Resource Centre, Bhubaneswar -751015 Orissa, India

Nibha Gupta
Microbiology laboratory, Regional Plant Resource Centre, Bhubaneswar -751015 Orissa, India

Johnstone Neondo
Jomo Kenyatta University of Agriculture and Technology, P.O. Box 62000-00200, Nairobi, Kenya

Joseph Machua
Kenya Forestry Research Institute, P. O. Box 20412-00200, Nairobi, Kenya

Anne Muigai
Jomo Kenyatta University of Agriculture and Technology, P.O. Box 62000-00200, Nairobi, Kenya

Aggrey B. Nyende
Jomo Kenyatta University of Agriculture and Technology, P.O. Box 62000-00200, Nairobi, Kenya

Moses Munjuga
World Agroforestry Centre (ICRAF), P. O. Box 30677-00100, Nairobi, Kenya

Ramni Jamnadass
World Agroforestry Centre (ICRAF), P. O. Box 30677-00100, Nairobi, Kenya

Alice Muchugi
World Agroforestry Centre (ICRAF), P. O. Box 30677-00100, Nairobi, Kenya

Dagmawi Teshome Woldesenbet
Ethiopian Institute of Agricultural Research, Pawe Agricultural Research Center, Benishangul Gumuz, Ethiopia

Kassahun Tesfaye
Addis Ababa University, College of Natural Science, Microbial, Cellular and Molecular Biology Department, Addis Ababa, Ethiopia

Endashaw Bekele
Addis Ababa University, College of Natural Science, Microbial, Cellular and Molecular Biology Department, Addis Ababa, Ethiopia

N. Gasmelseed
National Cancer Institute, University of Gezira, Sudan

M. A. Aljak
National Cancer Institute, University of Gezira, Sudan

A. E. Elmadani
National Cancer Institute, University of Gezira, Sudan

M.E. Elgaili
Department of Pathology, Faculty of Medicine, University of Gezira, Sudan

O. K Saeed
Department of Internal Medicine, Faculty of Medicine, University of Gezira, Sudan

H. Imad
Department of Molecular Biology, Putra University, Kuala Lumpur City, Malaysia

Q. Cheah
Department of Molecular Biology, Putra University, Kuala Lumpur City, Malaysia

J. Mohammad
Department of Molecular Biology, Babylon University, Hilla City, Iraq

O. Aamera
Institute of Medico-legal in Baghdad, Ministry of Health of Iraq, Iraq

Smiullah
Department of Plant Breeding and Genetics, University of Agriculture, Faisalabad, Pakistan

Farooq Ahmed Khan
Department of Plant Breeding and Genetics, University of Agriculture, Faisalabad, Pakistan

Aqeel Afzal
Department of Plant Breeding and Genetics, University of Agriculture, Faisalabad, Pakistan

Abdullah
Department of Plant Breeding and Genetics, University of Agriculture, Faisalabad, Pakistan

Ambreen Ijaz
Department of Bioinformatics and Biotechnology, GC University, Faisalabad, Pakistan

Usman Ijaz
Department of Plant Breeding and Genetics, University of Agriculture, Faisalabad, Pakistan

Colombe Dadjo
Institute of Basic Sciences, Technology and Innovation, Pan African University, P. O. Box 62000-00200, Nairobi, Kenya

Jane Kahia
World Agroforestry Centre (ICRAF), Cote d' Ivoire Country Program Cocody Mermoz, Abidjan, Côte d'Ivoire

Catherine Muthuri,
ICRAF World Agroforestry Centre (ICRAF), P. O. Box 30677-00100, Nairobi, Kenya

Lucien Diby
World Agroforestry Centre (ICRAF), Cote d' Ivoire Country Program Cocody Mermoz, Abidjan, Côte d'Ivoire

Christophe Kouame
World Agroforestry Centre (ICRAF), Cote d' Ivoire Country Program Cocody Mermoz, Abidjan, Côte d'Ivoire

Peter Njenga
Jomo Kenyatta University of Agriculture and Technology, P. O. Box 62000-00200, Nairobi, Kenya

Modeste Kouassi
Centre National de Recherche Agronomique (CNRA), Laboratoire Central de Biotechnologies (LCB), 01 BP 1740 Abidjan 01, Côte d'Ivoire